Effects of Temperature on Ectothermic Organisms

Ecological Implications and Mechanisms of Compensation

Edited by

Wolfgang Wieser

With 126 Figures

Springer-Verlag Berlin · Heidelberg · New York 1973

Professor Dr. Wolfgang Wieser
Institut für Zoophysiologie der Universität Innsbruck
A-6020 Innsbruck

ISBN-13: 978-3-642-65705-4 e-ISBN-13: 978-3-642-65703-0
DOI: 10.1007/978-3-642-65703-0

Softcover reprint of the hardcover 1st edition 1973

Effects of Temperature on Ectothermic Organisms

Ecological Implications and Mechanisms of Compensation

Edited by

Wolfgang Wieser

With 126 Figures

Springer-Verlag New York · Heidelberg · Berlin 1973

Professor Dr. Wolfgang Wieser
Institut für Zoophysiologie der Universität Innsbruck
A-6020 Innsbruck

ISBN-13: 978-3-642-65705-4 e-ISBN-13: 978-3-642-65703-0
DOI: 10.1007/978-3-642-65703-0

Softcover reprint of the hardcover 1st edition 1973

Preface

The study of thermoregulation in endotherms has contributed much to the emergence of the concept of control theory in biology. By the same token, the study of temperature adjustment in ectotherms is likely to have a far-reaching influence on ideas on the regulation of metabolism in general. The reason for this is that ectotherms, in adapting to the vagaries of a thermally unstable environment, deploy a range of subtle molecular and organismic strategies. Thus the experimenter, using temperature changes as a tool, is well equipped to analyze some of these strategies.

This approach has enabled some important mechanisms of temperature-induced adaptation to be elucidated; the most striking of these are the effects on metabolism of changes in the conformation of enzymes and the transfer properties of membranes. Furthermore, there is a vague but persistent feeling among those working in this field that changes in the nervous system will ultimately prove to be the agency by which many of the molecular mechanisms of temperature adaptation are controlled. Should this indeed be the case, a new phase would soon begin in our understanding of the interactions between the systemic and the cellular levels of organization.

However, it is not only questions about the *causes* of temperature adaptation that can provide answers of potential importance to the general biologist; of equal significance are questions as to the *meaning* of temperature adaptation in a particular organism.

Not all the reactions observed *in vitro* allow conclusions to be drawn about their role *in vivo*; it will be necessary to make a careful analysis of the role that temperature plays in the ecology and biology of an organism before experimental findings at the molecular level can be seen in their proper perspective.

In discussions on the general problems relating to temperature adaptation one usually becomes painfully aware how little is known about, for example,

1. the temperature fluctuations in an ectotherm's micro-environment in the course of a day, let alone a year;
2. the effect of temperature on
   a) different forms of biological activity,
   b) different tissues, organs and physiological functions,
   c) different stages of development;
3. the effects of other ecological factors, e.g. photoperiod, on the expression of temperature-induced changes in metabolism or behaviour.

There is little doubt that answers to these questions would increase our understanding of countless biological problems; but, beyond this, they are indispensable for the formulation of further questions concerning the causes – molecular or systemic – of adaptation. In fact without the ecological meat, the molecular biologist would eat a lean meal.

It thus seems proper in all discussions on biological adaptation to recognize – and to stress – that there are two sides to the coin: *mechanisms* and *ecology*.

This is the concept on which we based the Symposium on "Effects of Temperature on Ectothermic Organisms", held at Obergurgl, Austria, from 4 to 8 September 1972.

I thank the many colleagues who helped to organize the meeting and those who by their efforts and contribution made it a successful venture. I should like to mention in particular the staff of the "Alpine Forschungsstelle Obergurgl", above all Dr. and Mrs. H. MOSER; the members of my Department at the University of Innsbruck; Professor D. NEUMANN of the Department of Zoology, University of Cologne, and Dr. R. NEWELL, Department of Zoology, Queen Mary College, University of London. We acknowledge with thanks financial support received from the Austrian "Bundesministerium für Wissenschaft und Forschung", the Tiroler Landesregierung, and the Volkswagenstiftung.

All the participants stayed at the Hotel Hochfirst and we are grateful to the proprietors, Mr. and Mrs. F. GSTREIN, for their efforts to create an atmosphere equally conducive to after-dinner discussions as to after-discussion dinners.

Innsbruck, October 1973 WOLFGANG WIESER

Contents

Mechanisms

Ecology

List of Participants

ALDRICH, J. C., Department of Zoology, Queen Mary College, University of London, Mile End Road, London E1 4 NS, England

BARNES, H., Dunstaffnage Marine Research Laboratory, Oban, Argyll, Scotland

BAYNE, B. L., Institute for Marine Environmental Research, 13/14 St. James Terrace, Citadel Road, Plymouth PL1 3AX, England

BRAUNE, H.-J., Zoologisches Institut der Universität Kiel, Lehrstuhl für Zoophysiologie, D-23 Kiel, Hegewischstraße 3

BUSCHINGER, ALFRED, Zoologisches Institut der Technischen Hochschule, D-61 Darmstadt, Schnittspahnstraße 3

COLLATZ, KLAUS-GÜNTER, Biologisches Institut I (Zoologie) der Albert-Ludwigs-Universität, D-78 Freiburg i. Br., Katharinenstraße 20

DUNCAN, NAN, Department of Zoology, Royal Holloway College, Englefield Green, Surrey, England

ENGELMANN, SUSANNE, Institut für Strahlenschutz, Reaktorenzentrum Seibersdorf, A-1082 Wien, Lenaugasse 10

GABBOTT, P. A., University College of North Wales, N.E.R.C. Unit, Marine Science Laboratories, Menai Bridge, Anglesey, Great Britain

GRAINGER, J. N. R., University of Dublin, Department of Zoology and Comparative Anatomy Trinity College, Dublin 2, Ireland

GRONOW, G., Zoologisches Institut der Universität Kiel, Lehrstuhl für Zoophysiologie, D-23 Kiel, Hegewischstraße 3

GRUBER, K., Institut für Zoophysiologie der Universität Innsbruck, A-6020 Innsbruck, Peter-Mayr-Straße 1 a

GUNKEL, W., Biologische Anstalt Helgoland, D-2192 Helgoland

HALBACH, U., Zoologisches Institut der Universität München, D-8 München 2, Luisenstraße 14

HARRI, MIKKO, N. R., Department of Zoology, University of Turku, SF-20500 Turku 50, Finland

HAZEL, J. R., School of Life Sciences, University of Nebraska, Oldfather Hall Lincoln, NB 68508, USA

HEITLINDEMANN, H., Zoologisches Institut der Universität Kiel, Lehrstuhl für Zoophysiologie, D-23 Kiel, Hegewischstraße 3

HOCHACHKA, P. W., Department of Zoology, University of British Columbia, Vancouver 8, Canada

HOFER, RUDOLF, Institut für Zoophysiologie der Universität Innsbruck, A-6020 Innsbruck, Peter-Mayr-Straße 1 a

HOFFMANN, KLAUS, II. Zoologisches Institut, D-852 Erlangen, Bismarckstraße 10
HORSTMANN, KLAUS, Institut für Angewandte Zoologie, D-87 Würzburg, Röntgenring 10
JANKOWSKY, D., Zoologisches Institut der Universität Kiel, D-23 Kiel, Hegewischstraße 3
JOHNSON, LELAND, G., Augustana College, Department of Biology, Sioux Falls, SD 57102, USA
KIRCHNER, WALTER, Pädagogische Hochschule Rheinland, Abteilung Bonn, Seminar für Biologie, D-53 Bonn, Römerstraße 164
KNAPP, WALTER, Institut für Zoophysiologie der Universität Innsbruck, A-6020 Innsbruck, Peter-Mayr-Straße 1a
KREBS, FALK, Zoologisches Institut der Universität Kiel, Lehrstuhl für Zoophysiologie, D-23 Kiel, Hegewischstraße 3
KURECK, ARMIN, Zoologisches Institut, D-5 Köln 41, Weyertal 119
LAGERSPETZ, KARI, Department of Zoology, University of Turku, SF-20500 Turku 50, Finland
LOCKER, ALFRED, Institut für Strahlenschutz, Reaktorenzentrum Seibersdorf, Biologisch-Medizinische Abteilung, A-1082 Wien, Lenaugasse 10
MALASEK, ALFRED, Institut für Strahlenschutz, Reaktorenzentrum Seibersdorf, A-1082 Wien, Lenaugasse 10
MANGUM, C. P., Department of Biology, College of William and Mary, Williamsburg, VA 23185, USA
MARSDEN, ISLAY, D., Department of Zoology, Queen Mary College, University of London, Mile End Road, London E1 4NS, England
MEINCKE, K.-F., II. Zoologisches Institut, D-852 Erlangen, Bismarckstraße 10
NEUMANN, D., Zoologisches Institut der Universität Köln, D-5 Köln-Lindenthal, Weyertal 119
NEWELL, R., Department of Zoology, Queen Mary College, University of London, Mile End Road, London E1 4NS, England
NOPP, H., II. Zoologisches Institut der Universität Wien, A-1010 Wien, Luegerring 1
PATTEE, E., Département de Biologie Animale et Zoologie, Université Claude Bernard, 43, Bd. du 11 Novembre 1918, F-69 Villeurbanne
PFLÜGER, W., Zoologisches Institut III der Universität Köln, D-5 Köln 41, Weyertal 119
PYE, VERONICA, Department of Biophysics and Physical Biochemistry, University of Pennsylvania, Philadelphia, PA 19104, USA
RATTE, HANS-TONI, D-5 Köln 1, Thürmchenswall 35
REICHEL, K., Institut für Zoophysiologie der Universität Innsbruck, A-6020 Innsbruck, Peter-Mayr-Straße 1a
REMMERT, H., II. Zoologisches Institut der Universität Erlangen-Nürnberg, D-8520 Erlangen, Bismarckstraße 10
ROUX, C., and ROUX, A. L., Département de Biologie Animale et Zoologie, Bat. 403, F-69 Villeurbanne, 43, Bd. du 11 Novembre 1918
RUTTNER-KOLISKO, A., Biologische Station Lunz am See, A-3293 Lunz am See
SANTARIUS, K. A., Botanisches Institut der Universität Düsseldorf, D-4 Düsseldorf, Ulenbergstraße 127
SATTLER, CHRISTA, Institut für Zoophysiologie der Universität Innsbruck, A-6020 Innsbruck, Peter-Mayr-Straße 1a

SCHAUB, GÜNTER, Institut für Angewandte Zoologie der Universität Bonn, D-53 Bonn, An der Immenburg 1

SPELLERBERG, IAN F., Department of Biology, The University, Southampton SO9 5NH, England

STÖVER, HELMA, Universität Regensburg, Fachbereich Biologie, D-84 Regensburg, Universitätsstraße 31

STREICHHAN, PETER, I. Zoologisches Institut der Universität Göttingen, D-34 Göttingen, Berliner Straße 28

TECKELMANN, U., II. Zoologisches Institut der Universität Erlangen, D-8520 Erlangen, Bismarckstraße 10

THEEDE, H., Institut für Meereskunde der Universität Kiel, D-23 Kiel, Niemannsweg 11

TICHY, WALTRAUD, Lehrkanzel für Meeresbiologie der Universität Wien, A-1090 Wien, Währingerstraße 17

WEGENER, UTE, Zoologisches Institut der Universität Kiel, Lehrstuhl für Zoophysiologie, D-23 Kiel, Hegewischstraße 3

WELBERS, P., Zoologisches Institut III, Universität Köln, D-5 Köln 41, Weyertal 119

WIESER, WOLFGANG, Institut für Zoophysiologie der Universität Innsbruck, A-6020 Innsbruck, Peter-Mayr-Straße 1 a

WINTER, CH., Fachbereich Biologie der Universität Frankfurt, D-6 Frankfurt a. M. Siesmayerstraße 70

WODTKE, E., Zoologisches Institut der Universität Kiel, Lehrstuhl für Zoophysiologie, D-23 Kiel, Hegewischstraße 3

Contributing Authors Not Present at Meeting

BEHRISCH, HANS W., Institute of Arctic Biology, University of Alaska, Fairbanks, AK 99701, USA

COYER, P. E., Department of Zoology, University of Massachusetts, Amherst, MA 01002, USA

DELOLME, R., Département de Biologie Animale et Zoologie, Université Claude Bernard Lyon I, F-69 Villeurbanne, 43, Bd. du 11 Novembre 1918

HEINRICH, B., College of Agricultural Sciences, Division of Entomology, University of California, Berkeley, CA 94720, USA

LASCOMBE, C., Département de Biologie Animale et Zoologie, Université Claude Bernard Lyon I, F-69 Villeurbanne, 43, Bd. du 11 Novembre 1918

THOMPSON, R. J., Departement of Biology, Dalhousie University, Halifax, Nova Scotia, Canada

WIDDOWS, J., Institute for Marine Environmental Research, St. James Terrace, Plymouth, England

Temperature Relations of Ectotherms: A Speculative Review

WOLFGANG WIESER

With 6 Figures

The interpretation of the temperature relationships of ectothermic animals has already run through two phases and we now appear to be experiencing the beginning of a third. In the first phase ectotherms were regarded as being at the mercy of the environment, an interpretation best expressed by Krogh's so-called "normal curve" (Krogh, 1914). In the second phase, dating from about 1936 (Fox, 1936; Spärck, 1936; Thorson, 1936; Schlieper, 1950; Precht *et al.*, 1955; Bullock, 1955; Prosser, 1958), the emphasis was placed upon the homeostatic character of metabolic processes. And now, in the third phase, the metabolism of such organisms is seen to be considerably more complicated: various temperature ranges are thought to be associated with different systemic states, involving, to some extent, specific ionic distributions as well as alternative enzyme and metabolic patterns.

The two latter interpretations revolve around the question of the control of physiological states under the influence of temperature. In the following, I want to attempt a survey of the problems brought up by this question and at the same time substantiate the third interpretation, i. e. the view that ectothermic animals (sometimes at least) are neither at the mercy of environmental temperatures nor are they simple homeostatic systems. They may, in fact, be considered as "multistable" systems (in the sense of Ashby, 1952), in which in answer to a change in environmental temperature that state of the system is called into action which is best suited to the prevailing or anticipated temperature regime.

Adaptation as the Maintenance of One Systemic State

It is possible to consider the metabolism of an ectothermic organism as a network in which the flow of material is controlled by transducing or switching elements, largely represented by enzymes and membranes. At a given temperature, other conditions remaining constant, a steady state is achieved which may be characterized by specific concentrations and turnover rates of body constituents.

Krogh's normal curve leads, or misleads, to the assumption that changes in environmental temperature bring about alterations in the rates of flow but not in the functional relationships of the metabolic network itself. Now, although harmonic accelerations or decelerations of flow rates in response to changes in temperature may occur frequently in ectotherms, a few limitations to the general validity of a model of this nature must be pointed out:

Fall in Temperature : Just as certain aspects of behaviour (such as flying; see Heinrich, this symposium) cannot be maintained below a critical frequency of oscillations, so are many physiological functions dependent upon the supply of a minimum number of molecules per unit of time. If the speed of reaction sinks below a critical value the entire function is forced to a standstill. This applies to synaptic transmission processes, particularly to the production of miniature endplate potentials for which the secretion of a critical number of transmitter molecules per unit of time appears to be essential (Katz, 1966); or to muscular contractions, for which a critical rate of ATP hydrolysis is required (Mutchmor and Richards, 1961). The same may hold true for the serial transcription and translation processes involved in DNA replication and protein synthesis, although detailed knowledge on this point is still lacking (but see Friedman, Lu and Rich, 1969).

Furthermore, upon cooling, the resistance of control elements may rise, leading to the impairment of specific transport processes. This is true of biological membranes, which can only fulfil their function when in a characteristic state of fluidity and which, upon cooling to below the critical temperature, assume a pseudocrystalline state with new transfer properties (Steim *et al.*, 1969; Träuble, 1971; Hazel, this symposium).

Rise in Temperature : The acceleration of metabolism is limited by substrates and by sources of energy. Apart from this a large increase in rate of turnover might lead to the accumulation of harmful metabolites. Long before protein denaturation takes place a rise in temperature can increase the permeability of membranes to such an extent that the unequal distribution of ions and small molecules on either side of the membrane can no longer be maintained.

These factors place limitations upon the extent to which the metabolism of ectotherms can change in proportion to environmental temperature. Exceeding the critical threshold necessarily leads to non-linear changes in the system. This is invariably the case near the limits of the biological temperature range, but if the danger arises that even within the biological temperature range such non-linear changes may occur then appropriate compensatory mechanisms must be of selective advantage. In the *cold* the speed of certain reactions would have to be raised, and this could be achieved by an acceleration of enzymatic processes as well as by a reduction of membrane resistances. On the other hand, a suppression of catabolic processes would prolong the residence time of key metabolites such as, for example, acetyl choline. In the *warm*, conversely, the speed of reactions would have to be reduced, membrane resistance raised and those metabolic processes accelerated which serve to remove waste products from the system (Hazel and Prosser, 1970). Such mechanisms would be of the nature of homeostatic devices maintaining a particular systemic state by compensating for the temperature-induced irregularities in the flow of material and energy. Under certain circumstances therefore, the apparently continuous course of Krogh's normal curve may be the expression of complex control processes by means of which the irregularities in an $R:T$ curve are smoothed out. Somero, Giese and Wohlschlag (1968), for example, showed that in an antarctic fish, although the dynamic state remained unaltered following warm acclimation, the sensitivity of the tissues to metabolic inhibitors was changed. From this it may be concluded that, despite an unbroken $R:T$ curve, changes had taken place in the metabolic organization of the fish.

There is a second group of phenomena of which it is tacitly assumed that they are not able to vary *ad libitum* in proportion to the temperature. This applies to the distribution and storage processes that take place at a certain environmental temperature whereas the stores are utilized at another temperature. Examples are provided by zooplankton that feed in the warm surface water but digest in the cold water of the deeps (McLaren; 1963, Halbach, this symposium), and by many terrestrial animals that feed by day but digest by night or in cooler habitats (Hofer, 1972). Here, too, should be listed the processes of development and maturation which commence in the cold season and have to lead to a new generation within an exactly prescribed interval of time.

To sum up, it can be said that for molecular as well as for ecological reasons it seems unlikely that the concept that ectothermic animals are at the mercy of environmental temperatures is of general applicability.

Independently of these deductive conclusions such a bulk of data had been collected by 1955 that the temperature independence of physiological processes in ectotherms was recognised as a distinct phenomenon and studied systematically. The first generalizations are due to Precht (1955, 1958) and Prosser (1958, 1961). Essentially, two kinds of response were distinguished, having in common a plateau in the curve obtained by plotting rate of reaction against temperature. In the one case, this plateau is only seen in acclimatized animals (abscissa: TA = TE), whereas in the other the plateau occurs in the course of "acute" experimental changes in temperature (abscissa: =TE). Prosser refers in the first case to a *translation* (of the $R:T$ curve) and in the latter to a *rotation* (see Fig. 1).

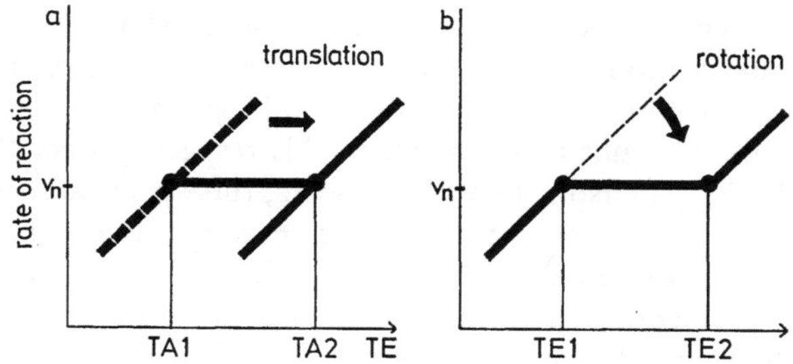

Fig. 1 a and b. The two main types of response to changes of temperature usually termed 'adaptive' in ectotherms. Basically, a reflects long term, b short term changes of metabolism. In a the animals were acclimated to each temperature (TA) at which they were subsequently measured; in b no acclimation to the experimental temperature (TE) took place

From the start, such relationships were regarded explicitly or implicitly as an expression of homeostatic mechanisms, the role of which was to keep constant the rates of metabolism in the face of fluctuations in temperature. Precht (1955) even referred to cases that did not fit into this scheme but reacted to warm acclimation with a *rise* instead of a *fall* in metabolism as "paradoxical" or "inverse".

Only fragmentary information is available as to the mode of action of homeostatic mechanisms in ectotherms. It is tempting to regard this kind of control as being analogous to the temperature regulation of endothermic animals and to explain it in terms of control theory. However, as late as 1968 Precht stated sceptically: "Man könnte daran denken, besonders die ideale Kompensation als Regelkreis darzustellen, bei der ein konstanter *Sollwert* durch einen sich über Tage erstreckenden Prozeß nach jeder Veränderung der Akklimatisationstemperatur wieder eingestellt wird ... doch führt eine solche Betrachtung kaum weiter, da wesentliche Glieder (Fühler, Art der Rückmeldung usw.) noch weitgehend unbekannt sind".

Nevertheless it is helpful to introduce some cybernetical models in order to test the extent to which they cover the observable phenomena. The logic of the model may lead to new hypotheses and new approaches to the problem.

Networks

In attempting to construct network models for the metabolic regulation of ectothermic animals the following facts have to be taken into consideration.

a) The temperature independence of an ectothermic animal can express itself either at the level of the whole organism or at the cellular level, or both, so that a further distinction has to be made between *systemic* and *cellular regulation.*

b) Interaction between cellular and systemic levels are known to exist. One of the most fascinating questions is that of how central commands are transmitted to the compensating tissue. It is assumed that certain cellular reactions require the continuous influence of the central mechanism, whereas others function even after its removal. In consequence, Precht (1964a) made a distinction between *direct effects (Direkteffekte)* and *after effects (Nacheffekte).*

All compensatory reactions of isolated tissues of acclimated animals may represent after-effects but need not do so since they might have arisen without the intervention of a central mechanism. A means of distinguishing unequivocally between these two possibilities is still lacking, but the use of cell and tissue cultures will no doubt be profitable in this respect (Joyce, 1967; Precht and Christoffersen, 1965; Anderson, 1970; Lagerspetz, this symposium).

c) In response to a change in temperature immediate compensatory reactions may be observed as well as others that set in after days or even weeks of pretreatment (acclimation). A distinction should therefore be made between *short-term* and *long-term*, or *rapid* and *slow* effects.

It can be assumed that the long-term reactions are the manifestation of complicated structural changes involving RNA and protein *syntheses*, or modifications in membrane structures. The rapid reactions can be explained as being modulations of enzyme *activities*, or compensatory reactions of the nervous system directly or indirectly governed by temperature.

d) Whether or not an organism or a particular tissue exhibits compensatory reactions, of whatever nature, is not a fixed property of the organism or tissue in question, but depends upon other factors, particularly season and stage of development. Thus the existence of *reference inputs* which determine the involvement, intensity and range of compensatory reactions has to be assumed.

Although no model has been described that represents all of the interactions of molecular and supramolecular structures in temperature compensation it can be predicted that connections of the following nature must be involved (Fig. 2):

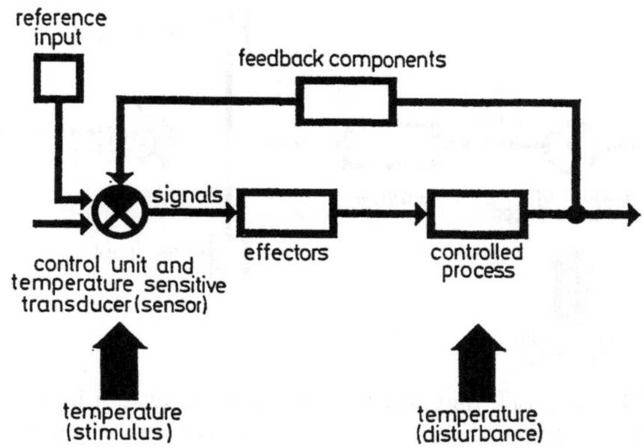

Fig. 2. Block diagram of a network representing some of the components which might participate in the control of ectotherm metabolism by temperature

Under the influence of the changing temperature of the milieu the transducer would have to alter its transfer properties in such a manner as to bring about a change in the concentration or the turnover rates of specific metabolites or ions. Some of these could again assume the role of signals influencing the effectors, and thus restore the output values that had been changed by the thermal disturbance. Temperature would thus represent a "stimulus" to the transducer, but a "disturbance" to the rest of the system. In Fig. 2a negative feedback component is included to indicate the possibility that after a disturbance the controlled process can itself set off changes in the system, thus leading to a restoration of the original values. A mechanism of this sort was speculated upon by Prosser (1967): he assumed that the temperature-dependent accumulation of a substrate (due, for example, to an overshoot) could lead directly to inhibition or enhancement of enzyme activity, or to repression or induction of enzyme syntheses and thus to an adjustment of the processes concerned. It is not possible to determine yet whether or not such feedback systems are, in fact, involved in the metabolism of ectothermic organisms. Oscillations which are usually considered characteristic of closed loop control systems with a negative feedback link were recently described by Wernick und Künnemann (1973). These authors found the substrate affinity of LDH of fish muscle to oscillate for about 18 days after the animals' transfer from 10 to 25 C.

However, it should be stressed that compensatory regulation are possible even in simple open loop control systems without negative feedback. There is overwhelming evidence, for example, that the activity of many enzymes involved in the metabolism of ectotherms can be modulated by pH and small cations such as K^+, Na^+, NH_4^+, Ca^{++}, Mg^{++} (Bygrave, 1967; Hubert *et al.*, 1970; Freed, 1971a; Somero, 1972; Behrisch, 1971, 1972). If the ionic milieu were to change with temperature, a specific modulation of key enzymes could compensate for the effect of temperature on metabolic variables.

This would be comparable to the control of metabolic processes by signals according to the following block diagram (Fig. 3):

The open-loop chain may even be shorter, as for example, when *lowering* the temperature *raises* the substrate affinity of an enzyme directly.

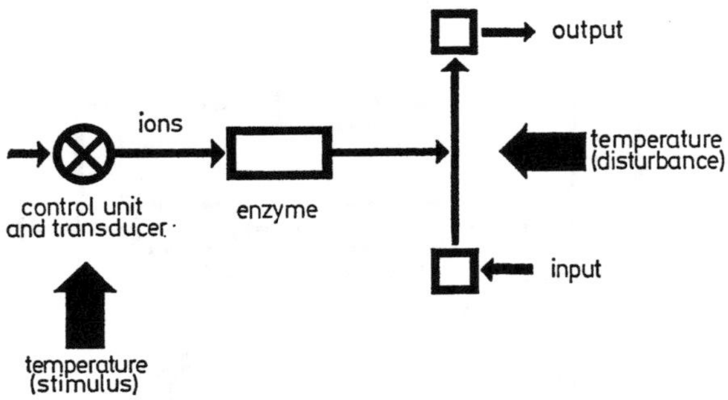

Fig. 3. Block diagram of an open loop system in which the rate of a reaction is controlled by temperature via conformational changes of an enzyme

Feedback does not play a part in any such cases.

All of the elements mentioned in this scheme have, in recent years, been shown to exist in ectothermic organisms, although it is not known into what type of network they can be fitted. Some examples of elements with potential roles in such networks are discussed below.

Temperature-sensitive Transducers. This function can be assumed by 1. any enzyme system which produces metabolites, cofactors or ions with a demonstrable modulating effect upon other enzymes; 2. enzymes whose substrate affinity undergoes a compensatory change under the direct influence of temperature; 3. membranes whose permeability to metabolites, cofactors or ions varies with temperature.

Since numerous ions and cofactors exert a modulating effect on enzymes (see Dixon and Webb, 1964, for review), there must be an even greater number of reaction systems with a potential transducer function.

A direct, compensating effect of temperature on the kinetic properties of enzymes has been described, e.g., by Massey *et al.* (1966), Somero (1969a, b) and Freed (1971b). Hochachka (this symposium), using *acetylcholinesterase* as an example, has speculated on one of the mechanisms involved. In these cases the enzyme plays the dual role of sensor and effector since the temperature effect can be translated directly into a homeostatic function (see Yielding, 1971).

Physiochemical investigations revealing a definite connection between temperature, lipid composition and permeability strongly indicate a possible role for the biomembrane as a temperature sensor (Steim *et al.*, 1969; Reinert and Stein, 1970; Träuble, 1971), and there are many examples of the temperature dependence of water and ionic fluxes across membranes of ectothermic organisms. As a particularly well analyzed case that of the eel, *Anguilla anguilla*, may be cited: the permeability of the gill epithelium to sodium relative to that of water increases strikingly in the cold (Motais and Isaia, 1972).

If the temperature stimulus acts directly on the membrane-dependent transport processes of nerve cells, thus affecting impulse frequency, the membrane involved could also be considered as a temperature-sensitive device. Compensatory reactions in the nervous system have been described by Burkhardt (1959), Kerkut and Ridge (1961, 1962), Miller (1967), Prosser, Precht and Jankowsky (1965), d'Ajello, Bettini and Grasso (1967), Dierolf and Brink (1973), Winter (this symposium) and others.

Signals. The concentrations and turnover rates of a large number of substances that can modulate enzyme activity or influence physiological processed vary with temperature. This is true of *pH* (Straub, 1957; Hochachka and Lewis, 1970; Howell *et al.*, 1970), *ions* (Straub, 1957; Hickman *et al.*, 1964; Rao, 1962, 1967; Houston and Madden, 1968; Houston, Madden and DeWilde, 1970; Prosser, McKay and Kato, 1970; Umminger, 1969; Motais and Isaia, 1972; Maetz, 1972), *metabolites* and *cofactors* (Freed, 1971a; Rao, 1967; Hazel and Prosser, 1970) and *neurotransmitters* (Lagerspetz and Tirri, 1968; Jensen, 1972; Lagerspetz, this symposium; Harri, this symposium). Both blood (Saroja, 1962; Rao, 1962, 1967; Jankowsky, 1964; Precht, 1964b; Joyce, 1967) and nervous system (see above references) have been implicated as conveyors of compensatory commands.

Effectors: At a *cellular* level enzymes as well as membranes can be considered as possible effectors, the properties of which may alter under the influence of specific signals so as to compensate for the effect of thermal energy on metabolism. As far as *enzyms* are concerned, the following changes have been considered: acceleration of synthesis (or deceleration of degradation) of a particular enzyme; synthesis of those isoenzymes which are the most efficient under the temperature conditions prevailing; mobilization of specific isoenzymes from an existing set; changes in substrate affinity by a change in the conformation of an enzyme. The two latter mechanisms would be responsible for *rapid*, the two former for *slow* compensatory effects. The entire problem has been reviewed several times in recent years (Hochachka, 1972; Hochachka and Somero, 1971, 1973; Somero, 1972; Somero and Hochachka, 1971; Behrisch, 1972).

The role of *biomembranes* as effectors is suggested by all investigations in which a temperature-dependent alteration in their composition has been demonstrated. Examples are offered by the mitochondrial membranes in fish gills (Caldwell and Vernberg, 1970), membrane lipids of the digestive tract (Smith and Kemp, 1971), or by the plasmalogen composition of the nervous system (Roots and Prosser, 1962; Roots and Johnston, 1968). In all of these cases the proportion of unsaturated fatty acids increases with cold acclimatization, this in turn leading to an alteration in the transfer properties of the membranes. The period of oscillation of the volume of the mitochondria (Packer, 1970) and the activity of membrane-bound enzymes (Caldwell, 1969; Raison, Lyons and Thomson, 1971; Hazel, this symposium), too, depend upon the proportion of unsaturated fatty acids in the membrane lipids. Mechanisms responsible for the reorganization of lipid membranes were discussed by Fulco (1972).

Muscles and glands can be considered as *systemic effectors* the activity of which may undergo a compensatory change under the influence of humoral or neural mechanisms. The heart of the terrestrial snail provides a particularly good example. Nopp (1965) was able to demonstrate an ideal compensation of the rate of heart beat: it showed an overshoot in response to rapid changes in temperature, but none in response to slower changes. A relationship between temperature and impulse frequency has been reported by Burkhardt (1959) and Kerkut and Ridge (1962), in exact agreement with the responses analysed by Nopp for the intact pulmonate heart. Further compensatory reactions of systemic effectors have been reported by Richards (1963), Smit (1967), Miller and Dehlinger (1969), Seymour (1972) and others.

Adaptation as an Alternation between Different Systemic States

During the past 20 years numerous *R*:*T* curves have been described which can be represented by the curve in Fig. 4a. It is irrelevant to our argument whether plateaus of this kind occur in the acclimatized or in non-acclimatized animals since such cases only differ with respect to the responsible mechanism and not in principle.

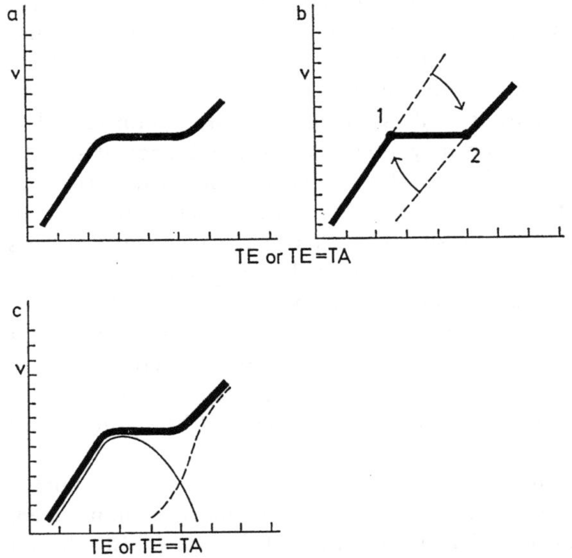

Fig. 4a–c. Diagram representing some of the concepts discussed in connection with R:T curves of the form shown in a; b effect of position of set point (1 or 2) on rotation of R:T curve; c the plateau may be caused by the overlap of two separate R:T curves. No distinction is made between long term and short term responses, the abscissa representing acute or acclimated temperature changes (TE, or TE = TA)

Precht (1968), interpreting data published by McLeese and Wilder (1958), claims that a curve of this type contained "ein durch eine Leistungsadaptation bedingtes Konstanzintervall" (of, in this case, the speed of locomotion of the American lobster). This formulation would seem to imply that the acclimatized animal maintains a constant locomotory activity within a certain temperature range by means of active regulation. Two points must be considered here as being of fundamental importance in the interpretation of all *R*:*T* curves.

a) The Problem of the "Set-point" in Temperature Compensation. Assuming that the plateau does indeed result from a kind of metabolic control, it would be essential to know whether the deviation from Krogh's normal curve is due to *acceleration* of metabolism in the lower part of the plateau or to *deceleration* of metabolism in the upper part. This would be analogous to asking whether an air-conditioning device cools or heats over a particular temperature range.

Obviously this question is answerable only if it can be determined whether the system is accelerating or decelerating at a particular moment, or if the set point of the system is known. If this set point is near the lower temperature end of the plateau ("1" in Fig. 4b) the metabolism must decelerate at higher temperatures, if it is near the high temperature end ("2" in Fig. 4b) then metabolism must accelerate at lower temperatures.

b) Homeostasis versus Ultrastability : The plateau in Fig. 4a could also be explained by the mobilization first of a cold-active and then of a warm-active reaction system in the course of acclimation. The plateau would thus be the result of the overlapping of the two $R:T$ curves, as is shown in Fig. 4c. The maximum speed of reaction of the two systems need not necessarily be equal.

The decisive adaptive mechanism here would be the temperature-dependent change from one systemic state to another which, in accordance with Ashby (1952), will be termed a mechanism of *ultrastability*. The plateau of the $R:T$ curve, that is to say the interval of constancy supposedly arising from a capacity adaptation, would in this case represent nothing more than a side effect of the reorganization of metabolism.

In the following, some attempts, partly speculative, will be made at further elucidating these two problems involved in temperature adaptation.

The Plateau as a Metabolic 'Ceiling'

The course of the $R:T$ curves shown in Figs. 4a and b represents a large number of measurements on ectothermic animals. It is typical that the curve rises steeply in the lower temperature range and then flattens out. In the lower temperature range the Q_{10} is usually well above 2, often between 6 and 10, and in the middle range it is between 1 and 2.

Variations of this type of curve have been reported for coelenterates (Sassaman and Mangum, 1970; Mangum *et al.*, 1972), gastropods (Davies, 1966; Wieser, Fritz and Reichel, 1970; Newell and Pye, 1971a), crustaceans (Vernberg, 1959a; Laserre and Renaud-Mornant, 1971; Roux, 1972), insects (Richards, 1963, 1964: Davison, 1971), fish (Meuwis and Heuts, 1957; Roberts, 1964), amphibians (Bishop and Gordon, 1967), reptiles (Aleksiuk, 1971) and in many other cases. The common occurrence of higher Q_{10} values in the lower temperature range has been emphasised by Scholander *et al.* (1953), Mutchmor and Richards (1961), Kenney and Richards (1955), Rao and Bullock (1954).

This type of curve shows a remarkable similarity to the respiration curve of mitochondria as they change from state 3 to state 4 (see Pye, this symposium). State 3 is characterized by plentiful supplies of ADP, which *in vivo* would be the result of high rates of ATP hydrolysis in energy-requiring processes. In state 4, on the other hand, ATP accumulates and the energy charge is high. In this case, if no further ATP-requiring reactions are mobilized, the energy consumption and thus the Q_{O_2} must reach a steady state that is more or less independent of temperature. This kind of pattern with respect to temperature is shown by the mitochondria of mung beans, potatoes and frogs (Pye, this symposium). Thus regarded, *high* Q_{10} values appear to represent conditions in which the energy consumption is able to respond with great sensitivity to temperature changes. Warming leads to a large acceleration of ATP-consuming processes, cooling to an equally large deceleration. It is perhaps significant that high Q_{10} values are so often to be encountered at the lower end of biological temperature ranges, where ectothermic organisms are obliged to speed up their metabolism as much as possible following hibernation or cold torpor, but have to be able to arrest it equally rapidly in the event of a renewed fall in temperature. A compensation in the sense of a "positive control" (set point 2 in Fig. 4b) would expend too much energy in this phase (Jungreis and Hooper, 1968). The possibility is not excluded, however, that in other cases where the animal has to fulfill a prescribed programme of behaviour, true compensatory control of metabolism at low temperatures may occur (see Vernberg, 1959b; Fitzpatrick, Bristol and Stokes, 1971; Mangum, 1972). This holds particularly true for animals that spend their entire life at low temperatures (Scholander *et al.*, 1953; Somero, 1972; Somero, Giese and Wohlschlag, 1968).

The middle phase of the typical $R:T$ curve shown in Fig. 4a is marked by *low* Q_{10}-values. Such temperature-insensitive phases frequently occur around the mean temperature to be expected in the environment at the time of collecting the experimental animal. If over this range the $R:T$ curve were to continue with the same temperature coefficient as is typical of the cold phase, the organism would have to expend a disproportionate amount of energy during the occasional peak temperatures. The view that plateaus in the $R:T$ curve of ectothermic animals represent adaptive *mechanisms for saving energy* (set point "1" in Fig. 4b) is supported by, among other observations, the fact that such phases are seen most distinctly in those animals that have to economize with their energy reserves. The best examples are provided by animals of the tidal regions, which are only able to renew their energy stores periodically (Newell and Northcroft, 1965, 1967; Davies, 1966; Newell and Pye, 1970a, b, 1971a, b, c; Widdows and Bayne, 1971). In many of these cases the standard but not the active respiration, especially in the upper temperature ranges, is to a large extent insensitive to temperature. Over the course of the season or following acclimatization, the $R:T$ plateau is displaced in such a manner as to come to lie over the whole of the prevailing temperature range or over its upper portion. Similar examples were presented by Nopp (1965) for land snails, in which the respiration of, above all, summer animals in dry habitats is almost completely insensitive to temperature between 20 and 40 C. In other animals, such as the polychaetes of the sublittoral living submerged in the sand, there is no difference between standard and active respiration, and plateaus are much less common (Croyer and Mangum, this symposium; Mangum, 1972). In other cases the routine respiration is temperature compensated whereas standard or active respiration is not (Bayne *et al.*, this symposium). Observations along the same lines are that the respiration of starving animals is less dependent upon temperature than that of fed animals (Marsden, this symposium), or that smaller individuals of a species are less dependent upon temperature than larger ones (Rao and Bullock, 1954, Meuvis and Heuts, 1957).

It seems feasible that an explanation for the various patterns of the $R:T$ curve can be found in the ecology and biology of the animals concerned. It is true, however, that generalizations may require a more detailed knowledge of organism environment relationships than is usually available. This is exemplified by a recent study by Mangum (1972) who showed that polychaetes from various marine habitats displayed $R:T$

patterns that could not be predicted from the temperature regimes under which the different species lived.

If ADP is regarded as a substrate for coupled respiration its exhaustion in state 4 of the mitochondria must lead to a more or less temperature-independent upper limit of the Q_{02} which can only be exceeded by the mobilization of new ATP-splitting processes. Similarly, limited supplies of other substrates may result in an upper limit of the Q_{02} and thus lead to a temperature-insensitive phase in respiration. Newell and Pye (1971a, b) have demonstrated this in mitochondria of *Littorina littorea*, the respiration of which remained unaltered with 0.01 mM pyruvate between 5 and 25 C, but was approximately doubled with 0.5 mM pyruvate (Newell, this symposium see also Wernick and Künnemann 1973). The observation that in cells many metabolites occur at concentrations far below that required to saturate key enzymes (Freed, 1971a) strengthens the hypotheses that some of these intermediates may exert a controlling influence on metabolism-temperature relationships.

Temperature-specific Mobilization of Metabolic Processes

If deviations of the $R{:}T$ curve of ectothermic organisms from the Arrhenius relationship be regarded as an expression of a homeostatic mechanism then the metabolic processes stimulated by cold appear to be logical and those stimulated by warmth seem at first sight paradoxical. This view was held by Precht when he established his well known scheme of adaptation types (Precht, 1955). However, since then it has been shown that temperature independence of metabolism requires the acceleration of energy-producing processes at low temperatures as well as the acceleration of degradative processes at high temperatures (Hazel and Prosser, 1970). Thus both cold- and warm-stimulation may be part of the homeostatic control of metabolism in ectotherms and there is no need to invoke the concept of a "paradoxical" adaptation.

If one looks beyond homeostasis one realizes that the stimulation of metabolism or of activity by high-temperature acclimation and acclimatization is a very general phenomenon in ectotherms. This type of effect is summarized in Figs. 5a–c.

If T_1 and T_2 represent two acclimation or experimental temperatures then at T_1 both systems are either independent of the acclimation temperature (5a) or exhibit cold stimulation (5b) or warm stimulation (5c). At T_2 warm stimulation occurs in every case. Such cases have been encountered recently in crustaceans (Jungreis and Hooper, 1968), insects (Buffington, 1969; Davison, 1971), gastropods (Wieser, Fritz and Reichel, 1970; Newell and Pye, 1970a, b, 1971a), amphibia (Dunlap, 1969, 1971; Miller and Mizell, 1972; Stier and Bock, 1966; Hofer, 1972) and fishes (Wodtke, this symposium).

Two aspects of these temperature relationships should be emphasised:

1. Their most important characteristic is not temperature independence, but the temperature-specific stimulation of metabolic processes. This supplements the scheme presented by Newell (this symposium), where importance is placed exclusively on the *constancy* of metabolism over critical temperature ranges. However, maintenance of a constant rate of metabolism is only one of several possibilities in the temperature relationships of ectotherms. It is certainly unjustifiable to share

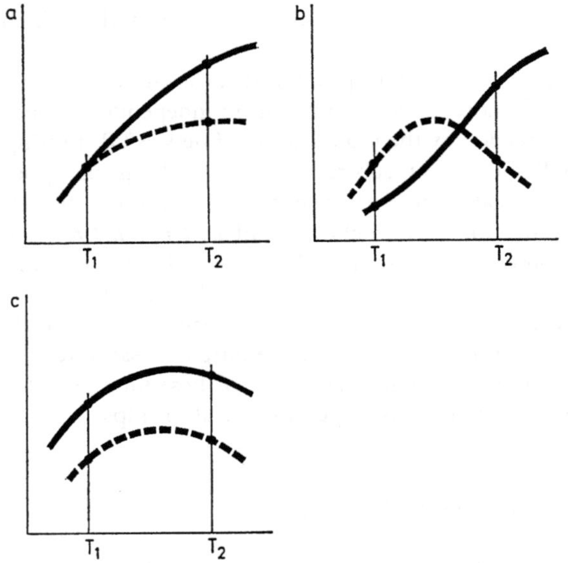

Fig. 5a–c. Three patterns of R:T curves in which warm-acclimated specimens (solid line) have higher rates of metabolism than cold-acclimated specimens (dashed line) at either one or two experimental temperatures (T_1, T_2)

Buffington's (1969) opinion that a rise in metabolic turnover rates due to warm acclimation is "non-adaptive".

2. The types of $R:T$ curves presented in Fig. 5 may be induced or modified by factors connected with seasonal changes or by different photoperiods (Vernberg, 1952; Hoar, 1956; Evans *et al.*, 1962; Roberts, 1964; Kasbohm, 1967; Wieser, Fritz and Reichel, 1970; Hofer, 1972). This indicates that the changes are governed by transducing elements that are not only temperature sensitive but probably also light sensitive.

The Nature of Temperature-specific Processes

The enzymatic mechanisms already mentioned (p. 7) can be regarded as models for temperature-specific metabolic processes. Various isozymes with distinct temperature relationships may operate within a single reaction series, so that under the influence of different temperature – and perhaps of other environmental factors – specific isozyme patterns are mobilized (Aleksiuk, 1971; Moon and Hochachka, 1971; Somero, 1972; Baldwin and Hochachka, 1970; Pandey, 1972; Newell, this symposium). It may be of significance that some enzyme systems show much greater variability in ectotherms than in endotherms, a striking example being provided by proteolytic enzymes (Pfleiderer and Zwilling, 1972).

A much more far-reaching effect of temperature on metabolism lies in the enhancement or suppression of specific metabolic sequences. If two enzymes compete for one and the same substrate and if the substrate affinity of the enzyme is raised by a change in temperature then the turnover number of one chain of reactions must be greater than that of the other. This possibility was pointed out recently by Somero (1972) for example, although the enhancement or inhibition of reactions that compete for the same co-factors or substrates is a regulatory mechanism that has long been under discussion (Holzer, 1963). Various metabolic systems with different substrate and co-factor patterns can thus become active in the organism according to temperature.

There are sufficient indications that in cold acclimation extramitochondrial metabolism, including glycolysis, gluconeogenesis, glycogen syntheses, lipogenesis and the pentosephosphate shunt increase at the expense of mitochondrial metabolism (Ekberg, 1958, 1961; Hochachka and Hayes, 1962; Hochachka, 1967, 1968; Kanungo and Prosser, 1959; Dean, 1969). Protein synthesis, too, is raised in certain animals and tissues following cold acclimation (Das and Prosser, 1967; Haschemeyer, 1969). Acute measurements of temperature dependence of melanin synthesis in the skin of a newt revealed that at low temperatures a dopa-oxidase, and at high temperatures a true tyrosinase play the main role (Wieser and Nopp-Pammer, 1968). At high temperatures the proportion of nitrogen compounds involved in energy production is larger than at low temperatures (Atherton and Aitken, 1970; Wieser, 1972). At 2 C a change in carbohydrate metabolism takes place in the wax moth (Lenartowicz and Niemierko, 1968), at 10 C an enzymatic change in the medusa *Chrysaora quinquecirrha* (Blanquet, 1972), and in plants the C_4-dicarboxylic acid pathway of photosynthesis is regarded as an adaptation to high temperatures (Björkman *et al.*, 1972).

A number of these transformations may be connected with homeostatic mechanisms, but it is much more likely that they are expressions of a fundamental reorganization of metabolism under the influence of environmental temperatures (Fry and Hochachka, 1970, p. 124). This reorganization may be connected with the rearrangement of lipid membranes (Fulco, 1972; Hazel, this symposium), or with the attainment of a specific energy charge. It has already been pointed out (p. 10) that many temperature-independent plateaus can be regarded as economy measures on the part of the organism, brought about by the control of substrate supplies or of ATP-consuming reactions. It is most likely that central mechanisms play an important role in this type of control, influencing functions of the intact organism and those of target tissues in the form of "Nacheffekte" (Roberts, 1966; Jankowsky, 1966; Lagerspetz and Tirri, 1968; Davison, 1971; Lagerspetz, this symposium). Thus a large part of the temperature-specific changes described at the molecular and cellular level may be connected with the whole organism adapting its energy consumption to prevailing or anticipated conditions in the natural habitat. This would mean in turn that the temperature-specific activation of isozyme patterns is a secondary effect, whereas the primary effect lies in the stimulation of central nervous and endocrine mechanism.

Temperature-specific Processes: A Device for Measuring Temperature?

The deviations of the $R:T$ curves from the Arrhenius relationship, as seen in many ectothermic animals, is perhaps explained best by assuming that different temperature ranges are associated with specific metabolic patterns. However, this explanation is not completely satisfactory. It cannot be overlooked, for example, that many animals exhibit a classical temperature relationship with a Q_{10} of about 2 over the entire biological temperature range (see Valen, 1958; Fuhrman and Fuhrman, 1959; Mangum and Sassaman, 1969; Tribe and Bowler, 1968; Cassuto, 1971; and many others). The temperature-specific switchover of metabolism therefore is neither a biochemical nor a physiological necessity. Further, investigations concerned not only with temperature relationships but taking into consideration the ecology and biology of the various species, have brought up the possibility that the apparently so characteristic pattern of the $R:T$ curve is not only induced or modified by the acclimation temperature but by seasonal factors or by photoperiod as well (Vernberg, 1952; Hoar, 1956; Evans *et al.*, 1962; Roberts, 1964; Stier and Bock, 1966; Buffington, 1969; Künnemann, Laudien and Precht, 1970; Miller and Mizell, 1972; Newell and Pye, 1971a; Harwood and Takata, 1965; Hofer, 1972). This would indicate that the fundamental reorganization of metabolism can not only be elicited by the influence of the actual temperature itself, but in anticipation of the temperature *to be expected.* A particularly good example of this has been described by Hofer (1972) in two species of frog in which the activity of digestive enzymes changes according to season, temperature and photoperiod. The rate of trypsin activation in *Rana ridibunda* is invariably raised by warm acclimation, whereas in *Rana temporaria* it is raised by warm (20 C) acclimation in spring following spawning, and in summer by cold (10 C) acclimation (Fig. 6).

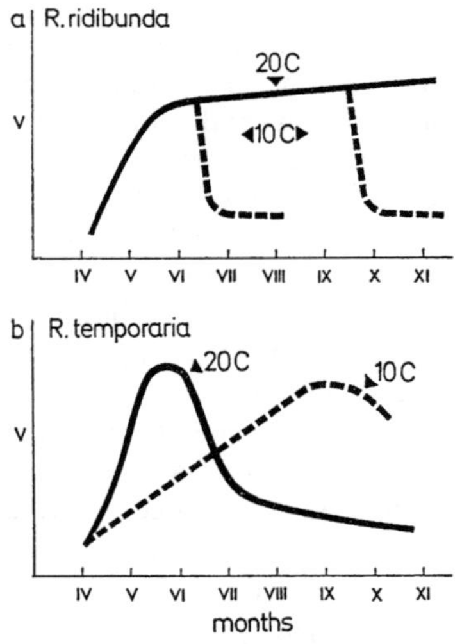

Fig. 6 a and b. Schematic representation of seasonal changes of R:T curves of warm (20 C) and cold (10 C) acclimated groups of two species of frogs. The rate measured is either enterokinase or trypsin activity. Experimental temperature remained constant (30 C). (Based on Hofer, 1972)

These differences can be correlated with the ecology of the two species. *R. ridibunda* is a warm-loving species whereas *R. temporaria* seeks warm feeding places immediately after spawning, but retires to a cooler woodland habitat in summer. The trypsin activation in *R. ridibunda* can, furthermore, be accelerated in summer by long day, whereas in *R. temporaria* it is completely independent of length of day.

This example demonstrates that in ectothermic animals there is not simply a "cold" and a "warm" metabolism, but that, according to endogenic and exogenic conditions, one and the same temperature can either stimulate or inhibit one and the same mechanism. Two conclusions are possible: either, the metabolism of some ectotherms is adjusted to the temperature regime anticipated, that type of metabolism being activated which "fits" the temperature to be expected, or, this adjustment in metabolism provides the animals with a chance *to select their own optimum environmental temperature.*

The latter hypothesis is based upon the fact that the temperature-specific processes already mentioned must lead to specific patterns of metabolites and ions and the ratio of specific metabolites or ions must be characteristic for a given body temperature. This brings to mind the observation that in homoiotherms the set point in the temperature control centre of the hypothalamus is supposed to be dependent upon a caracteristic Na^+/Ca^{++} ratio, each deviation from the programmed optimum value leading to a change in body temperature (Myers and Veale, 1970). Temperature-induced deviations from the programmed ionic ratios could similarly lead to changes in behaviour in an animal, inducing it to seek anew a certain temperature range. It has been shown clearly that the temperature of preference varies with season, for example in fishes (Zahn, 1963). The hypothesis is to some extent supported by the consideration that measuring the body temperature of soft-skinned terrestrial animals must be an exceedingly difficult task. What is indeed tne body temperature of a frog sitting on a cool grassy patch, its head exposed to the sun, and the peripheral temperature of which is influenced by the evaporation of water from the skin? This technical difficulty is perhaps the reason why the peripheral thermoreception of amphibia is so underdeveloped (Murray, 1962; Brattstrom, 1970). If it be assumed that one reaction mechanism is stimulated by low temperatures and another by high temperatures and that both produce their own specific metabolites the metabolic pattern in the blood would represent a compromise between the various temperature stimuli to which the whole organism is exposed. The metabolic pattern could be translated by a central transducer into a temperature signal to which the animal would respond. If, due to the influence of a reference input (which itself must be subject to the influence of seasonal or biological factors), the set point of the thermosensor is changed, then the animal would respond differently to one and the same temperature stimulus at various seasons. Applied to the example analysed by Hofer (1972) this would mean that the control centre of *R. temporaria* activates metabolism in spring in answer to biochemical warmth signals and that cold signals lead to inhibition. In summer the reaction pattern is reversed. The behaviour of the animals is influenced in such a manner that they seek that environmental situation which leads to an activated state of metabolism. This would be a warm habitat in spring subsequent to spawning, and a cool one in summer.

A Question of Time

One of the problems connected with this hypothetical mechanism is the *time factor*. On the one hand, like all thermometers, the biochemical temperature sensor has to respond rapidly to temperature changes, but on the other hand we know that many of the metabolic reorganizations in ectotherms take from several days to weeks for their completion. Therefore the existence of a series of temperature-sensitive control processes with various time constants must be postulated in one and the same organism (see also Smith and Kemp, 1971). The measurement of temperature must be completed within minutes or hours, and metabolic reactions that influence the behaviour of the animal not much later. A particularly striking example of adaptive changes that are completed in a matter of hours has been described by Furch (1972). Still, we are left with the enigma of the very long intervals required for the completion of other phenomena of metabolic adaptation, which, according to Widdows and Bayne (1971), can take up to 14 days in *Mytilus edulis*, whilst the induction of protein syntheses by hormones generally has a time lag of not much more than one day (see, for example, Clever, 1961).

As an explanation it may be considered that during the long period of acclimation a sort of selection process between different enzymes and reaction sequences is going on. It might be that the temperature stimulus at first initiates the production of a broad spectrum of isozymes from which, by a process of elimination, only the most useful are left finally, i.e. those that achieve the most nearly perfect transformation of the metabolism to its optimally adapted state. The time-determining process would therefore not be the induction of a new enzyme but rather *the half-life of an enzyme population*, i.e. the speed with which a "badly" adapted isozyme population could be replaced by one "better" adapted. It may be significant that recently the life history of an enzyme, *aldolase*, has been determined, revealing a half life of about eight days (Midelfort and Mehler, 1972).

The course of temperature acclimation can thus in some respects be compared with the course of immune reactions, one more indication of the similarity of the process of adaptation to that of memory, alluded to by Lagerspetz (this symposium).

Conclusion

Different investigators experimenting with different species at different times using different methods have reached different conclusions as to the nature of temperature relationships of ectotherms. Some have claimed proportionality of metabolism with temperature, some have stressed homeostasis, and some have described even more complex relations of the type termed "multistable" (in the sense of Ashby, 1952) in this paper. The best way of reconciling these diverging findings and interpretations is by assuming that they all have their place and time in the ectothermic realm of nature. Too little is known yet of the detailed ecology and of the requirements and demands of the life cycles of even a few ectotherms to permit construction of an adequate model of the subtle interrelationships existing in this class of organisms between temperature on the one hand, and metabolism, development and behaviour on the other.

It should be realized that it is the ecologists, developmental biologists and ethologists who have to frame the questions to which the answers discovered by biochemists and physiologists can be fitted. Since the purely descriptive phases in the former sciences have largely passed it is the application of strictly analytical methods that will lead to the formulation of the questions that are to occupy biochemists and physiologists of the future. Examples are provided by the recent interest in the effects of fluctuating temperatures on different life stages of ectotherms, or in the interplay of season, photoperiod and temperature in controlling metabolism and development.

Furthermore, the study of the interaction of systemic and cellular levels of organization may very well profit from the use of temperature and temperature changes as analytical tools, and will most certainly require increased collaboration between neurophysiologists, pharmacologists and molecular biologists. Some of the important biological problems of the next decade have become visible through the fabric of the research outlined in this review.

References

Aleksiuk, M.: Temperature-dependent shifts in the metabolism of a cool temperate reptile, *Thamnophis sirtalis parietalis*. Comp. Biochem. Physiol. **39 A**, 495–503 (1971).

Anderson, T. R.: Temperature adaptation and the phospholipids of membranes in goldfish *(Carassius auratus)*. Comp. Biochem. Physiol. **3**, 663–687 (1970).

Ashby, W. R.: Design for a Brain. New York: John Wiley & Sons 1952.

Atherton, W. D., Aitken, A.: Growth, nitrogen metabolism and fat metabolism in *Salmo gairdneri*, Rich. Comp. Biochem. Physiol. **36**, 719–747 (1970).

Baldwin, J., Hochachka, P. W.: Functional significance of isoenzymes in thermal acclimatization: acetylcholinesterase from trout brain. Biochem. J. **116**, 883–887 (1970).

Behrisch, H. W.: Temperature and the regulation of enzyme activity in poikilotherms. Regulatory properties of fructose diphosphatase from muscle of the Alaskan king crab. Biochem. J. **121**, 399–409 (1971).

Behrisch, H. W.: Molecular mechanism of adaptation to low temperature in marine poikilotherms. Some regulatory properties of dehydrogenases from two arctic species. Mar. Biol. **13**, 267–275 (1972).

Bishop, L. G., Gordon, M. S.: Thermal Adaptation of Metabolism in Anuran Amphibians. In: Prosser, C. L. (Ed.): Publ. no. 84, 263–280. Am. Assoc. Adv. Sci., Washington, D. C. 1967.

Björkman, O., Pearcy, R., Harrison, A. T., Mooney, H.: Photosynthetic Adaptation to High Temperatures: A Field Study in Death Valley, California. Science **175**, 786–789 (1972).

Blanquet, R. S.: Temperature acclimation in the medusa, *Chrysaora quinquecirrha*. Comp. Biochem. Physiol. **43 B**, 717–723 (1972).

Brattstrom, B. H.: Amphibia. In: Whittow, G. C. (Ed.): Comparative Physiology of Thermoregulation, Vol. I, p. 135–166. New York and London: Academic Press 1970.

Buffington, J. D.: Temperature acclimation of respiration in *Culex pipiens pipiens* (Diptera: Culicidae) and the influence of seasonal selection. Comp. Biochem. Physiol. **30**, 865–878 (1969).

Bullock, T. H.: Compensation for temperature in the metabolism and activity of poikilotherms. Biol. Rev. **30**, 311–342 (1955).

Burkhardt, D.: Die Erregungsvorgänge sensibler Ganglienzellen in Abhängigkeit von der Temperatur. Biol. Zentr. **78**, 22–62 (1959).

Bygrave, F. L.: The ionic environment and metabolic control. Nature (London) **214**, 667–671 (1976).

Caldwell, R. S.: Thermal compensation of respiratory enzymes in tissues of the goldfish *(Carassius auratus L.)*. Comp. Biochem. Physiol. **31**, 79–93 (1969).

Caldwell, R. S., Vernberg, F. J.: The influence of acclimation temperature on the lipid composition of fish gill mitochondria. Comp. Biochem. Physiol. **34**, 179–191 (1970).

Cassuto, Y.: Oxidative activities of liver mitochondria from mammals, birds, reptiles and amphibia as a function of temperature. Comp. Biochem. Physiol. **39 B**, 919–923 (1971).

Clever, U.: Genaktivitäten in den Riesenchromosomen von *Chironomus tentans* und ihre Beziehungen zur Entwicklung. I. Genaktivierung durch Ecdyson. Chromosoma (Berlin) **12**, 607–675 (1961).

D'Ajello, V., Bettini, S., Grasso, A.: Effect of temperature on the endogeneous activity of *Periplaneta americana* L. nerve cord. Riv. Parassit **28**, 71–78 (1967).

Das, A. B., Prosser, C. L.: Biochemical changes in tissues of goldfish acclimated to high and low temperatures – I. Protein synthesis. Comp. Biochem. Physiol. **21**, 449–467 (1967).

Davies, P. S.: Physiological ecology of *Patella*. I. The effect of body size and temperature on metabolic rate. J. Marine Biol. Assoc. U. K. **46**, 647–658 (1966).

Davison, T. F.: The effect of temperature on oxidative phosphorylation in isolated flight muscle sarcosomes. Comp. Biochem. Physiol. **38 B**, 21–34 (1971).

Dean, J. M.: The metabolism of tissues of thermally acclimated trout *(Salmo gairdneri)*. Comp. Biochem. Physiol. **29**, 185–196 (1969).

Dierolf, B. M., Brink, P. R.: Effects of thermal acclimation on cable constants of the earthworm median giant axon. Comp. Biochem. Physiol. **44 A**, 401–406 (1973).

Dixon, M., Webb, E. C.: Enzymes, 2nd ed. London: Longmans, Green and Co. 1964.

Dunlap, D. G.: Influence of temperature and duration of acclimation, time of day, sex and body weight on metabolic rates in the hylid frog, *Acris crepitans*. Comp. Biochem. Physiol. **31**, 555–570 (1969).

Dunlap, D. G.: Acutely measured metabolic rate-temperature curves in the cricket frog, *Acris crepitans*. Comp. Biochem. Physiol. **38 A**, 1–16 (1971).

Ekberg, D. R.: Respiration in tissues of goldfish adapted to high and low temperatures. Biol. Bull. **114**, 308–316 (1958).

Ekberg, D. R.: Aerobe und anaerobe Stoffwechselprozesse bei der Temperaturadaptation von Karauschen. Zool. Anz. **24**, 78–80 (1961).

Evans, R. M., Purdie, F. C., Hickman Jr., C. P.: The effect of temperature and photoperiod on the respiratory metabolism of rainbow trout *(Salmo gairdneri)*. Can. J. Zool. **40**, 107–118 (1962).

Fitzpatrick, L. C., Bristol, J. R., Stokes, R. M.: Thermal acclimation and metabolism in the Allegheny mountain salamander *Desmognathus ochrophaeus*. Comp. Biochem. Physiol. **40 A**, 681–688 (1971).

Fox, H. M.: The activity and metabolism of poikilothermal animals in different latitudes. I, p. 945–955. Proc. Zool. Soc., London 1936.

Freed, J. M.: Properties of muscle phosphofructokinase of cold- and warm-acclimated *Carassius auratus*. Comp. Biochem. Physiol. **39 B**, 747–764 (1971 a).

Freed, J. M.: Temperature effects on muscle phosphofructokinase of the Alaskan King Crab, *Paralithodes camtschatica*. Comp. Biochem. Physiol. **39 B**, 765–774 (1971 b).

Friedman, H., Lu, P., Rich, A.: Ribosomal Subunits produced by Cold Sensitive Initiation of Protein Synthesis. Nature **223**, 909–913 (1969).

Fry, F. E. J., Hochachka, P. W.: Fish. In: Whittow, G. C. (Ed.): Comparative Physiology of Thermoregulation, Vol. I, p. 79–134. New York: Academic Press 1970.

Fuhrman, J., Fuhrman, A.: Oxygen consumption of animals and tissues as a function of temperature. J. Gen. Physiol. **42**, 715–722 (1959).

Fulco, A. J.: The biosynthesis of unsaturated fatty acids by bacilli. IV. Temperature-mediated control mechanisms. J. Biol. Chem. **247**, 3511–3519 (1972).

Furch, K.: Der Einfluß einer Vorbehandlung mit konstanten und wechselnden Temperaturen auf die Hitzeresistenz von *Gammarus salinus* und *Idotea baltica*. Mar. Biol. **15**, 12–34 (1972).

Harwood, R. F., Takata, N.: Effect of photoperiod and temperature on fatty acid composition of the mosquito *Culex tarsalis*. J. Insect Physiol. **11**, 711–716 (1965).

Haschemeyer, A. E. V.: Studies on the control of protein synthesis in low temperature acclimation. Comp. Biochem. Physiol. **28**, 535–552 (1969).

Hazel, J., Prosser, C. L.: Interpretation of inverse Acclimation to Temperature. Z. Vergleich. Physiol. **67**, 217–228 (1970).

Hickman, C. P., McNabb, R. A., Nelson, J. S., van Breeman, E. D., Comfort, D.: Effect of cold acclimation on electrolyte distribution in rainbow trout *(Salmo gairdneri)*. Can. J. Zool. **42**, 577–597 (1964).

Hoar, W. S.: Photoperiodism and thermal resistance of goldfish. Nature **178**, 364–365 (1956).

Hochachka, P. W.: Organization of metabolism during temperature compensation. In: Prosser, C. L. (Ed.): Molecular Mechanisms of Temperature Adaptation, p. 177–203. Am. Assoc. Adv. Sci., Washington 1967.

Hochachka, P. W.: Action of temperature on branch points in glucose and acetate metabolism. Comp. Biochem. Physiol. **25**, 107–118 (1968).

Hochachka, P. W.: Enzymatic Adaptations to Deep-Sea Life. Oregon State Press (in press).

Hochachka, P. W., Hayes, F. R.: The effect of temperature acclimation on pathways of glucose metabolism in the trout. Can. J. Zool. **40**, 261–270 (1962).

Hochachka, P. W., Lewis, J. K.: Enzyme variants in thermal acclimation: trout liver citrate synthetases. J. Biol. Chem. **245**, 6567–6573 (1970).

Hochachka, P. W., Somero, G. N.: Biochemical adaptation to the environment. In: Hoar, W. S., Randall, D. J. (Eds.): Fish Physiology, Vol. VI, p. 99–156. New York: Academic Press 1971.

Hochachka, P. W., Somero, G. N.: Biochemical Strategies in Environmental Adaptation. Philadelphia: Saunders Publ. 1973.

Hofer, R.: Einfluß von Temperatur, Photoperiode und Jahreszeiten auf Verdauung und Atmung zweier Froscharten: *Rana ridibunda* (bzw. *R. esculenta*) und *Rana temporaria*. Zool. Jb. Physiol. **76**, 507–530 (1972).

Holzer, H.: Intrazelluläre Regulation des Stoffwechsels. Naturwiss. **50**, 260–270 (1963).

Houston, A. H., Madden, J. A.: Environmental temperature and plasma electrolyte regulation in the carp, *Cyprinus carpio*. Nature **217**, 969–970 (1968).

Houston, A. H., Madden, J. A., DeWilde, M. A.: Environmental temperature and the body fluid system of the freshwater teleost-IV. Water-Electrolyte regulation in thermally acclimated carp, *Cyprinus carpo*. Comp. Biochem. Physiol. **34**, 805–818 (1970).

Howell, B. J., Baumgardner, F. W., Bondi, K., Rahn, H.: Acid-base balance in cold-blooded vertebrates as a function of body temperature. Am. J. Physiol. **218**, 600–606 (1970).

Hubert, E. J., Villanueva, A., Gonzales, A., Marcus, F.: Univalent cation activation of fructose 1,6-diphosphatase. Arch. Biochem. Biophys. **138**, 590–597 (1970).

Jankowsky, H.-D.: Der Einfluß des Blutes auf den Sauerstoffverbrauch des isolierten Muskelgewebes von Schleien *(Tinca tinca* L.*)*. Zool. Anz. **172**, 233–239 (1964).

Jankowsky, H.-D.: The effect of temperature on the metabolic level of the eel *Anguilla vulgaris* L. Helgoländer Wiss. Meeresuntersuch. **13**, 402–407 (1966).

Jensen, D. W.: The effect of temperature on transmission at the neuromuscular junction of the sartorius muscle of *Rana pipiens*. Comp. Biochem. Physiol. **41** A, 685–695 (1972).

Joyce, C. M.: The effect of environmental temperature on succinic dehydrogenase activity in the ear skin of the rabbit. Proc. Roy. Irish Acad., Sect. B, no. 21, **65**, 425–435 (1967).

Jungreis, A. M., Hooper, A. B.: Physiological properties of cold resistance adaptation in the fresh-water crayfish *Orconectes virilis*. Comp. Biochem. Physiol. **26**, 91–100 (1968).

Kanungo, M. S., Prosser, C. L.: Physiological and biochemical adaptation of goldfish to cold and warm temperatures. I. Standard and active oxygen comsumptions of cold- and warm-acclimated goldfish at various temperatures. J. Cellular Comp. Physiol. **54**, 259–263 (1959).

Kasbohm, P.: Der Einfluß des Lichtes auf die Temperaturadaptation bei *Rana temporaria*. Helgoländer Wiss. Meeresuntersuch. **16**, 157–178 (1967).

Katz, B.: Nerve, Muscle and Synapse. New York: McGraw-Hill 1966.

Kenney, J. W., Richards, A. G.: Differences between leg and flight muscle of the giant water bug, *Lethocerus americanus*. Entomol. News **66**, 29–36 (1955).

Kerkut, G. A., Ridge, R. M. A. P.: The effect of temperature changes on the resting potential of crab, insect and frog muscle. Comp. Biochem. Physiol. **3**, 64–70 (1961).

Kerkut, G. A., Ridge, R. M. A. P.: The effect of temperature changes on the activity of the neurons of the snail, *Helix aspersa*. Comp. Biochem. Physiol. **5**, 283–295 (1962).

Krogh, A.: The quantitative relation between temperature and standard metabolism in animals. Intern. Z. Physik.-chem. Biol. **1**, 491–508 (1914).
Künnemann, H., Laudien, H., Precht, H.: Der Einfluß von Temperaturänderungen auf Enzyme der Fischmuskulatur. Versuche mit Goldorfen *Idus idus*. Mar. Biol. **7**, 71–81 (1970).
Lagerspetz, K. Y., Tirri, R.: Transmitter substances and temperature acclimation in *Anodonta* (Pelecypoda). Ann. Zool. Fenn. **5**, 396–400 (1968).
Laserre, P., Renaud-Mornant, J.: Interprétation écophysiologique des effets de température et de salinité sur l'intensité respiratoire de *Derocheilocaris remanei biscayensis* Delamare 1953 (Crustacea, Mystacorarida) C. R. Acad. Sc. Paris **272**, 1159–1162 (1971).
Lenartowicz, E., Niemierko, S.: The effect of low temperature and starvation on carbohydrate metabolism in larvae of *Galleria mellonella* L. J. Insect Physiol. **14**, 451–462 (1968).
Maetz, J.: Branchial sodium exchange and ammonia excretion in the goldfish *Carassius auratus*. Effects of ammonia-loading and temperature changes. J. Exptl. Biol. **56**, 601–620 (1972).
Mangum, C. F.: Temperature sensitivity of metabolism in offshore and intertidal onuphid polychaetes. Mar. Biol. **17**, 108–114 (1972).
Mangum, C. F., Oakes, M. J., Shick, J. M.: Rate-Temperature responses in scyphozoan medusae and polyps. Mar. Biol. **15**, 298–303 (1972).
Mangum, C. F., Sassaman, C.: Temperature sensitivity of active and resting metabolism in a polychaetous annelid. Comp. Biochem. Physiol. **30**, 111–116 (1969).
Massey, V., Curti, B., Ganther, H.: A temperature dependent conformational change in D-amino acid oxidase and its effect on catalysis. J. Biol. Chem. **241**, 2347–2357 (1966).
McLaren, I. A.: Effects of temperature on growth of zooplankton, and the adaptive value of vertical migration. J. Fisheries Res. Board Can. **20**, 685–727 (1963).
McLeese, D. W., Wilder, D. G.: The activity and catchability of the lobster (*Homarus americanus*) in relation to temperature. J. Fisheries Res. Board Can. **15**, 1345–1354 (1958).
Meuwis, A., Heuts, M. J.: Temperature dependence of breathing rate in carp. Biol. Bull. (Woods Hole) **112**, 97–107 (1957).
Midelfort, C. F., Mehler, A. F.: Deamidation in vivo of an asparagine residue of rabbit muscle aldolase. Proc. U.S. Natl. Acad. Sci. **69**, 1816–1819 (1972).
Miller, L. K.: Caudal nerve function as related to temperature in some Alaskan mammals. Comp. Biochem. Physiol. **21**, 679–686 (1967).
Miller, L. K., Dehlinger, P. J.: Neuromuscular function at low temperatures in frogs from cold and warm climates. Comp. Biochem. Physiol. **28**, 915–921 (1969).
Miller, L. C., Mizell, S.: Seasonal variation in heart rate response to core temperature changes. Comp. Biochem. Physiol. **42A**, 773–779 (1972).
Moon, T. W., Hochachka, P. W.: Effect of thermal acclimation on multiple forms of the liver soluble $NADP^+$-linked isocitrate dehydrogenase in the family Salmonidae. Comp. Biochem. Physiol. **40B**, 207–213 (1971).
Motais, R., Isaia, J.: Temperature-dependence of permeability to water and to sodium of the gill epithelium of the eel *Anguilla anguilla*. J. Exptl. Biol. **56**, 587–600 (1972).
Murray, R. W.: Temperature Receptors. In: Lowenstein, O. (Ed.): Adv. Comp. Physiol. Biochem, Vol. 1, p. 117–175. New York and London: Academic Press 1962.
Mutchmor, J. A., Richards, A. G.: Low temperature tolerance in insects in relation to the influence of temperature on muscle apyrase activity. J. Insect Physiol. **7**, 141–158 (1961).
Myers, R. D., Veale, W. L.: Body temperature: possible ionic mechanism in the hypothalamus controlling set point. Science **170**, 95–97 (1970).
Newell, R. C., Northcroft, H. R.: The relationship between cirral activity and oxygen uptake in *Balanus balanoides*. J. Marine Biol. Assoc. U.K. **45**, 387–403 (1965).
Newell, R. C., Northcroft, H. R.: A re-interpretation of the effect of temperature on the metabolism of certain marine invertebrates. J. Zool. **151**, 277–298 (1967).
Newell, R. C., Pye, V. I.: Seasonal changes in the effect of temperature on the oxygen consumption of the winkle *Littorina littorea* (L.) and the mussel *Mytilus edulis* (L.). Comp. Biochem. Physiol. **34**, 367–383 (1970a).
Newell, R. C., Pye, V. I.: The influence of thermal acclimation on the relation between oxygen consumption and temperature in *Littorina littorea* (L.) and *Mytilus edulis* (L.). Comp. Biochem. Physiol. **34**, 385–397 (1970b).

Newell, R. C., Pye, V. I.: Quantitative aspects of the relationship between metabolism and temperature in the winkle *Littorina littorea* (L.). Comp. Biochem. Physiol. **38 B**, 635–650 (1971 a).

Newell, R. C., Pye, V. I.: Temperature-induced variations in the respiration of mitochondria from the winkle, *Littorina littorea* (L.). Comp. Biochem. Physiol. **40 B**, 249–261 (1971 b).

Newell, R. C., Pye, V. I.: Variations in the relationship between oxygen consumption, body size and summated tissue metabolism in the winkle *Littorina littorea*. J. Marine Biol. Assoc. U.K. **51**, 315–338 (1971 c).

Nopp, H.: Temperaturbezogene Regulationen des Sauerstoffverbrauches und der Herzschlagrate bei einigen Landpulmonaten. Z. Vergleich. Physiol. **50**, 641–656 (1965).

Packer, L.: Relation of structure to energy coupling in rat liver mitochondria. Federation Proc. **29**, 1533–1540 (1970).

Pandey, K. K.: Isozyme Specificity to Temperature. Nature New Biol. **239**, 27–28 (1972).

Pfleiderer, G., Zwilling, R.: Die molekulare Evolution proteolytischer Enzyme. Naturwiss. **59**, 396–405 (1972).

Precht, H.: Wechselwarme Tiere und Pflanzen. In: Precht, H., Christophersen, J., Hensel, H. (Eds.): Temperatur und Leben, S. 1–177. Berlin–Göttingen–Heidelberg: Springer 1955.

Precht, H.: Concepts of the temperature adaptation of unchanging reaction systems of cold-blooded animals. In: Prosser, C. L. (Ed.): Physiological Adaptation, p. 50–78. Am. Physiol. Soc., Washington, D. C. 1958.

Precht, H.: Anpassungen wechselwarmer Tiere im normalen Temperaturbereich und ihre Ursachen. Naturw. Rundschau **11**, 438–442 (1964).

Precht, H.: Über die Bedeutung des Blutes für die Temperaturadaptation von Fischen. Zool. Jb. Physiol. **71**, 313–327 (1964).

Precht, H.: Der Einfluß „normaler" Temperaturen auf Lebensprozesse bei wechselwarmen Tieren unter Ausschluß der Wachstums- und Entwicklungsprozesse. Helgoländer Wiss. Meeresuntersuch. **18**, 487–548 (1968).

Precht, H., Christophersen, J.: Temperaturadaptation des Cilienepithels isolierter Kiemen und Fühlerspitzen von Mollusken. Z. Wiss. Zool. **171**, 197–209 (1955).

Prosser, C. L.: General Summary: The Nature of Physiological Adaptation. In: Prosser, C. L. (Ed.): Physiological Adaptation, p. 167–180. Am. Physiol. Soc., Washington, D.C. 1958.

Prosser, C. L.: In: Prosser, C. L., Brown, Jr., F. A. (Eds.): Comparative Animal Physiology, 2nd ed. Philadelphia, Pa.: Saunders 1961.

Prosser, C. L.: Molecular Mechanisms of Temperature Adaptation in Relation to Speciation. Am. Ass. Adv. Sci., Washington, D. C. **84**, 351–376 (1967).

Prosser, C. L., Mackay, W., Kato, K.: Osmotic and ionic concentrations in Alaskan fish and goldfish from different temperatures. Physiol. Zool. **43**, 81–89 (1970).

Prosser, C. L., Precht, H., Jankowsky, H. D.: Nervous control of metabolism during temperature acclimation of fish. Naturwiss. **52**, 168–169 (1965).

Raison, J. K., Lyons, J. M., Thomson, W. W.: The influence of membranes on the temperature-induced changes in the kinetics of some respiratory enzymes of mitochondria. Arch. Biochem. Biophys. **142**, 83–90 (1971).

Rao, K. P.: Physiology of acclimation to low temperature in poikilotherms. Science **137**, 682–683 (1962).

Rao, K. P.: Biochemical correlates of Temperature Acclimation. In: Prosser, C. L. (Ed.): Molecular Mechanisms of Temperature Adaptation, p. 227–244 (1967).

Rao, K. P., Bullock, T. H.: Q_{10} as a function of size and habitat temperature in poikilotherms. Am. Naturalist **88**, 33–44 (1954).

Reinert, J. C., Steim, J. M.: Calorimetric detection of a membrane-lipid phase transition in living cells. Science **168**, 1580–1582 (1970).

Richards, A. G.: The effect of temperature on heart beat frequency in the cockroach, *Periplaneta americana*. J. Insect. Physiol. **9**, 597–606 (1963).

Richards, A. G.: The generality of temperature effects on developmental rate and on oxygen consumption in insect eggs. Physiol. Zool. **37**, 199–211 (1964).
Roberts, J. L.: Metabolic responses of fresh-water sunfish to seasonal photoperiods and temperatures. Helgoländer Wiss. Meeresuntersuch. **9**, 459–473 (1964).
Roberts, J. L.: Systemic versus cellular acclimation to temperature by poikilotherms. Helgoländer Wiss. Meeresuntersuch. **14**, 451–465 (1966).
Roots, B. I., Johnston, P. V.: Plasmalogens of the nervous system and environmental temperature. Comp. Biochem. Physiol. **26**, 553–560 (1968).
Roots, B. I., Prosser, C. L.: Temperature acclimation and the nervous system in fish. J. Exptl. Biol. **39**, 617–630 (1962).
Roux, C.: Les variations de la courbe métabolisme température de *Gammarus lacustris* G.O. Sas (Crustacé, Amphipode) sous l'influence de divers facteurs ecologiques. Crustaceana, Suppl. **3**, 287–296 (1972).
Saroja, K.: Studies on the oxygen consumption in earthworms. Ph. D. Thesis, Sri Venkateswara Univ., India (1962).
Sassaman, C., Mangum, C. P.: Patterns of temperature adaptation in North American Atlantic coastal actinians. Mar. Biol. **7**, 123–130 (1970).
Schlieper, C.: Temperaturbezogene Regulationen des Grundumsatzes bei wechselwarmen Tieren. Biol. Zentr. **69**, 216–226 (1950).
Scholander, P. F., Flagg, W., Walters, V., Irving, L.: Climatic adaptation in arctic and tropical poikilotherms. Physiol. Zool. **26**, 67–92 (1953).
Seymour, M. K.: Effects of temperature change on irrigation rate in *Arenicola marina* (L.). Comp. Biochem. Physiol. **43A**, 553–564 (1972).
Smit, H.: Influence of temperature on the rate of secretion in the brown bullhead *Ictalurus nebulosus*. Comp. Biochem. Physiol. **21**, 125–132 (1967).
Smith, M. W., Kemp, P.: Parallel temperature-induced changes in membrane fatty acids and in the transport of amino acids by the intestine of goldfish (*Carassius auratus* L.). Comp. Biochem. Physiol. **39B**, 357–365 (1971).
Somero, G. N.: Pyruvate kinase variants of the Alaskan king crab: evidence for a temperature-dependent interconversion between two forms having distinct and adaptive kinetic properties. Biochem. J. **114**, 237–241 (1969).
Somero, G. N.: Enzymic mechanisms of temperature compensation: immediate and evolutionary effects of temperature on enzymes of aquatic poikilotherms. Am. Naturalist **103**, 517–530 (1969).
Somero, G. N.: Molecular mechanisms of temperature compensation in aquatic poikilotherms. In: South, F. E. *et al.* (Eds.): Hibernation and Hypothermia, Perspectives and Challenges, p. 55–80. Amsterdam: Elsevier Publ. Comp. 1972.
Somero, G. N., Giese, A. C., Wohlschlag, D. E.: Cold adaptation of the antarctic fish *Trematomus bernacchii*. Comp. Biochem. Physiol. **26**, 223–233 (1968).
Somero, G. N., Hochachka, P. W.: Biochemical adaptation to the environment. Amer. Zool. **11**, 159–167 (1971).
Spärck, R.: On the relation between metabolism and temperature in some marine lamellibranchs and its ecological and zoogeographical importance. Kgl. Danske Videnskab. Selskab, Biol. Medd. **13**, 1–27 (1936).
Steim, J. M., Tourtelotte, M. E., Reinert, J. C., McElhaney, R. N., Rader, R. L.: Calorimetric evidence for the liquid-crystalline state of lipids in a biomembrane. Proc. Natl. Acad. Sci. U.S. **63**, 104–109 (1969).
Stier, T. J. B., Bock, H. C.: Seasonal changes of heart rate-temperature relationships in toads. Proc. Soc. Exptl. Biol. Med. **123**, 149–151 (1966).
Straub, M.: Weitere Untersuchungen zur Temperaturadaptation der Sauerstoffbindung des Blutes von *Rana esculenta* L. Z. Vergleich. Physiol. **39**, 507–523 (1957).
Thorson, G.: On larval development, growth and metabolism of arctic marine bottom invertebrates compared with those of other seas. Medd. Grønland **100**, 1–155 (1936),
Träuble, H.: Phasenumwandlungen in Lipiden. Mögliche Schaltprozesse in biologischen Membranen. Naturwiss. 58. Jg. H. **6**, 277–284 (1971).
Tribe, M. A., Bowler, K.: Temperature dependence of "standard metabolic rate" in a poikilotherm. Comp. Biochem. Physiol. **25**, 427–436 (1968).

Umminger, B. L.: Physiological studies on supercooled killifish *(Fundulus heteroclitus)*. I. Serum inorganic constituents in relation to osmotic and ionic regulation at subzero temperatures. J. Exptl. Zool. **172**, 283–302 (1969).

Valen, E.: Oxygen consumption in relation to temperature in some poikilotherms. Acta Physiol. Scand. **42**, 358–362 (1958).

Vernberg, F. J.: The oxygen consumption of two species of salamanders at different seasons of the year. Physiol. Zool. **25**, 243–249 (1952).

Vernberg, F. J.: Studies on the physiological variation between tropical and temperate zone fiddler crabs of the genus *Uca*. II. Oxygen consumption of whole organisms. Biol. Bull. **117**, 163–184 (1959).

Vernberg, F. J.: Studies on the physiological variation between tropical and temperate zone fiddler crabs of the genus *Uca*. III. The influence of temperature acclimation on oxygen consumption of whole organisms. Biol. Bull. **117**, 582–593 (1959).

Wernick, A., Künnemann, H.: Der Einfluß der Temperatur auf die Substrat-Affinität der Laktat-Dehydrogenase aus Fischen. Mar. Biol. **18**, 32–36 (1973).

Widdows, J., Bayne, B. L.: Temperature acclimation of *Mytilus edulis* with reference to its energy budget. J. Marine Assoc. U.K. **51**, 827–843 (1971).

Wieser, W.: O/N ratios of terrestrial isopods at two temperatures. Comp. Biochem. Physiol. **43A**, 859–868 (1972).

Wieser, W., Fritz, H., Reichel, K.: Jahreszeitliche Steuerung der Atmung von *Arianta arbustorum* (Gastropoda). Z. Vergleich. Physiol. **70**, 62–79 (1970).

Wieser, W., Nopp-Pammer, E.: Effects of temperature and moulting cycle on melanin synthesis in the newt, *Triturus cristatus*. Comp. Biochem. Physiol. **24**, 1015–1025 (1968).

Yielding, K. L.: Regulation of protein activity and turnover through specific modifications in structure. In: Rechcigl, M. (Ed.): Enzyme Synthesis and Degradation in Mammalian Systems, p. 141–164. Basel: Karger 1971.

Zahn, M.: Jahreszeitliche Veränderungen der Vorzugstemperaturen von Scholle (*Pleuronectes platessa* Linne) und Bitterling (*Rhodeus sericeus* Pallas), S. 652–580. Verhandl. Deut. Zool. Ges., München. Leipzig: Akad. Verlag 1963.

Temperature Acclimation and Adaptive Changes in the Nervous System

KARI Y. H. LAGERSPETZ

With 6 Figures

I wish to begin by quoting Sir Charles Sherrington who, in his book "The Integrative Action of the Nervous System", says that "the nervous system is in a certain sense the highest expression of that which the French physiologists term the *milieu interne*" (Sherrington, 1906, p. 3–4), a reference to the concept of Claude Bernard. The nervous system, through its sensory input, is the primary organ of adaptation to the environment. In addition, pharmacological research has shown that the functional contacts between two nerve cells, the synapses, and the functional contacts between nerve cells and effectors, are in many cases the most sensitive and adaptive parts of the nervous system and of the whole animal.

It is reasonable to assume that an adaptive change in a function of a living animal is often based on an adaptive change in the system controlling that function. The control systems of multicellular animals include the endocrines and the nervous system, the endocrines again being for the most part controlled by the nervous system. The adaptiveness of the nervous system thus deserves special attention when phenomena of adaptation to environmental changes are considered. I will limit myself to certain examples of temperature adaptation in ectothermic animals.

The first example pertains to a primitive motile mechanism controlled at tissue level by a tissue hormone or perhaps by neuroid transmission. Precht and Christophersen (1965) showed that cold resistance of the ciliary activity of the ciliated epithelium on the antennae of the snail *Planorbis corneus* was increased by keeping the isolated tips of the antennae at low temperature for three days. Working with the isolated gills of the fresh water mussel, *Anodonta cygnea*, we found that one day at 21 C was sufficient to increase the heat resistance of the ciliary activity significantly above that found in the other isolated gill of the same animal, kept at 5 C for one day (Lagerspetz and Dubitscher, 1966).

In the marine mussel, *Mytilus edulis*, the ciliary activity of the gills is known to be controlled by acetylcholine (ACh) (Bülbring, Burn and Shelley, 1953; Milton, 1969) and by serotonin (5-HT) at tissue level (Aiello, 1957; Gosselin, 1961). We next studied the control of the ciliary activity in the gills of *Anodonta*, and found it also to be controlled by ACh at tissue level, although an external nervous control by nerves releasing 5-HT was not excluded (Lagerspetz, Länsimies, Impivaara and Senius, 1970). In support of our hypothesis that a particular function in an organism may adapt to temperature changes by adaptation of the control mechanism controlling that function we subsequently found that ACh greatly increased the heat resistance of the ciliary activity in isolated gills (Lagerspetz, Impivaara and Senius, 1970). In addition, if isolated gills were kept in atropine solution during the acclimation period or during the experiments, no acclimation effect was seen (Fig. 1). The activity of acetylcholinesterase in the gills was increased by warm acclimation, and the acetylcholine content of the warm-acclimated gills decreased more rapidly at 39 C than that of the cold-acclimated gills, showing that ACh was both liberated and hydrolyzed more rapidly in the warm-acclimated gills. It was suggested that the frequency of ACh-receptor combinations was important in determining heat resistance.

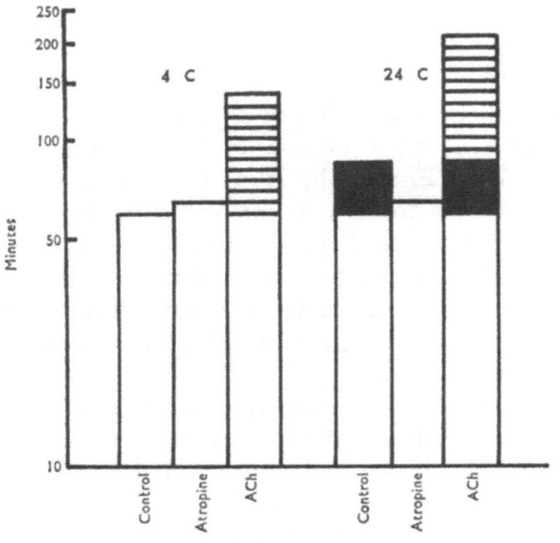

Fig. 1. The average thermal resistance time of the ciliary activity in isolated *Anodonta* gills kept for 20–24 hours at 4 C and at 24 C. The measurement of the thermal resistance is made at 39 C, in tap water (Control), or in the presence of 10^{-4} g per ml atropine sulphate (Atropine), or 10^{-3} g per ml acetylcholine chloride (ACh). The black part of the columns indicates the acclimation effect and the ruled part the ACh effect. (From Lagerspetz, Impivaara and Senius, 1970)

Our interests were thus focussed on the possibilities offered by the adaptiveness of neurotransmitter systems and neurotransmitter-receptor interactions in explaining the phenomena of temperature adaptation. Earlier evidence pointing to the importance of neurotransmitters in temperature adaptation in the central nervous system of rats (Corrodi, Fuxe and Hökfelt, 1967) and mice (Harri and Tirri, 1968), also stimulated this interest. The heat resistance of the heart beat of *Anodonta* was chosen for study. The strong stimulatory action of 5-HT on the *Anodonta* heart was already known (Erspamer and Ghiretti, 1951), as was the 5-HT-ergic and cholinergic control of the heart (S.-Rózsa and Pécsi, 1967). In addition, 5-HT was known to enhance the heat resistance of the heart beat in this species (Sakharov and Pécsi, 1965).

Following transfer from 4 to 21 C the heart beat of winter animals showed a significant increase in heat resistance after only 2 days. After 9 to 10 days the heat resistance had increased further (Lagerspetz and Tirri, 1968). Simultaneously, the 5-HT concentration of the heart increased by 38% (Table 1). In summer animals the heart beat was highly heat resistant and the 5-HT concentration of such hearts

Table 1. 5-HT content of the heart and the upper temperature limit of the heart beat in cold-acclimated (4 C) and warm-acclimated (21 C) winter and summer specimens of *Anodonta*. (From Lagerspetz and Tirri, 1968)

	5-HT ng/g	N	P	limit C	P
winter cold-accl.	477 ± 22	22	<0.001	29.1	< 0.001
winter warm-accl.	659 ± 45	22		43.7	
summer cold-accl.	687 ± 72	12	n.s.	39.5	n.s.
summer warm-accl.	757 ± 84	12	n.s.	40.8	n.s.

was elevated. In this case neither variable was affected by acclimation to different temperatures. It thus appeared that high heat resistance of the heart beat, whether induced by acclimation to warmth or by seasonal acclimatization, is associated with an elevated 5-HT concentration in the heart. Together with the known thermo-protective action of exogenous 5-HT these results support the view that neuro-transmitter substances are involved in the mechanism of temperature acclimation. Further supporting data concerning the role of neurotransmitter substances in temperature acclimation in the frog are given by Harri (this symposium). It is also of interest that, in a recent study, Jensen (1972) has shown that the neuromuscular junction of the frog is blocked more readily by cold than either the nerve or the muscle, and that cold acclimation of the animal lowers the temperature at which neuromuscular block occurs. Changes in the limiting step, the thermal resistance of either synaptic or neuromuscular transmission, may thus well be one of the mechanisms involved in the adaptation of thermal resistance.

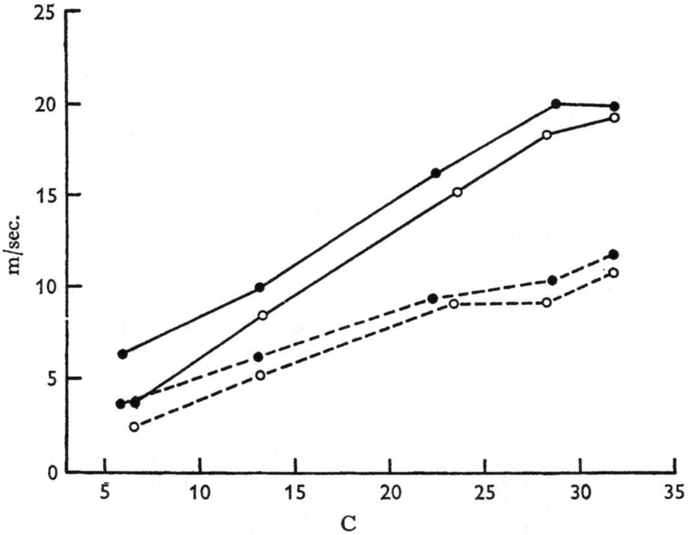

Fig. 2. Conduction velocity in the giant nerve fibres of the earthworm at different experimental temperatures. Solid circles, animals acclimated to 13 C; open circles, animals acclimated to 23 C. Solid lines, median fibre; broken lines, lateral fibres. The acclimation effects are significant (P < 0.001) only at the lowest experimental temperature. (From Lagerspetz and Talo, 1967)

Another approach to the problem of the role of the nervous system in temperature adaptation of ectotherms is the electrophysiological one. Using isolated nerve cords from the earthworm *Lumbricus terrestris*, we were able to show that single giant nerve fibres exhibit temperature compensation in several of their functional parameters. Thus the conduction velocity was higher at low temperatures in giant nerve fibres from earthworms acclimated to 13 C than in those from earthworms acclimated to 23 C (Fig. 2). Similarly, the duration of the rising and falling phases of the action potential showed significant temperature compensation when these parameters were measured at 6 or 13 C (Fig. 3; Lagerspetz and Talo, 1967). The same was

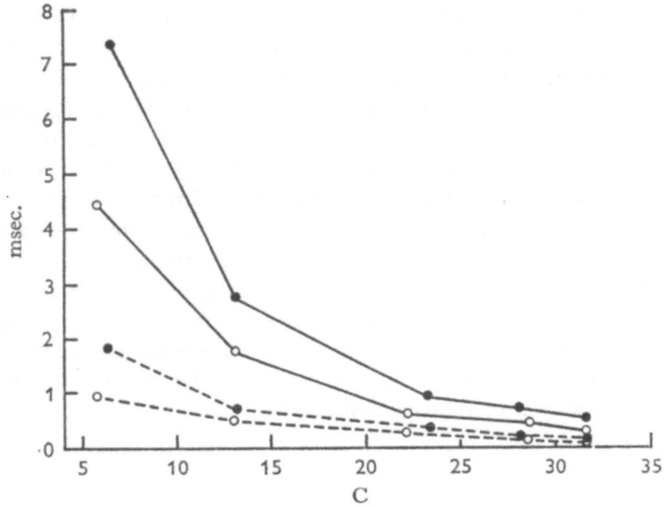

Fig. 3. Duration of the rising and falling phases of action potential in the median giant nerve fibre of the earthworm at different experimental temperatures. Solid circles, animals acclimated to 23 C; open circles, animals acclimated to 13 C. Solid lines, falling phase; broken lines, rising phase. The acclimation effects are statistically significant only at the two lowest experimental temperatures. (From Lagerspetz and Talo, 1967)

Table 2. Effect of temperature acclimation on the duration of the absolute refractory period in the nerve of the earthworm at different temperatures. (From Talo and Lagerspetz, 1967)

(AT, Acclimation temperature; ET, experimental temperature; *n*, number of experiments.)

	ET (C.)		Absolute refractory period (msec.) Median giant fibre			Laterial giant fibres		
AT (C.)	Variation	Mean	Variation	Mean	*n*	Variation	Mean	*n*
13	4.6–6.8	6.0	4.3–8.1	6.50[a]	5	5.3–8.2	6.70	6
23	5.5–7.2	6.2	6.2–19.6	12.04[a]	8	5.65–14.0	9.28	10
13	11.7–13.5	12.8	2.5–3.6	3.09[b]	8	1.5–3.9	2.96	8
23	12.2–13.8	13.1	2.7–4.3	3.72[b]	12	2.2–5.3	3.50	11
13	22.1–24.2	22.7	1.0–1.7	1,24	6	1.10–2.15	1.53	6
23	22.4–23.3	23.1	1.15–2.0	1.47	12	1.00–2.80	1.50	11

[a] Difference between the mean values significant at the level $P < 0.05$.
[b] Difference between the mean values significant at the level $P < 0.01$.

true of the absolute refractory period, and of the maximal response frequency of the giant fibres (Tables 2 and 3; Talo and Lagerspetz, 1967). In our work we used an extracellular recording technique. Later, Dierolf and McDonald (1969) were able to corroborate and extend our findings for earthworm giant fibres using an intracellular recording technique.

The observed adaptive changes point to the occurrence of changes in the time relations of ionic conductance in the giant nerve fibres. It was therefore of interest to study whether the cation transport mechanisms of nerve-cell membranes show similar compensatory temperature relationships. There is much evidence for the view that the Na^+—K^+-stimulated adenosinetriphosphatase (ATPase) activity found in homogenates and microsomal preparations of nervous tissue represents the active cation transport mechanism necessary for the maintenance of the resting potential. We therefore studied the effect of temperature acclimation on the temperature dependence of the Mg^{2+}- and Na^+—K^+-activated ATPases using nerve-cord homogenates from earthworms. Whereas no acclimation effect was seen in the temperature dependence of the Mg^{2+}-ATPase activity, the Na^+—K^+-ATPase activity was much less sensitive to low and intermediate temperatures (below 35 C) in animals acclimated to 14 C than in animals acclimated to 25 C (Fig. 4; Lagerspetz, Kohonen

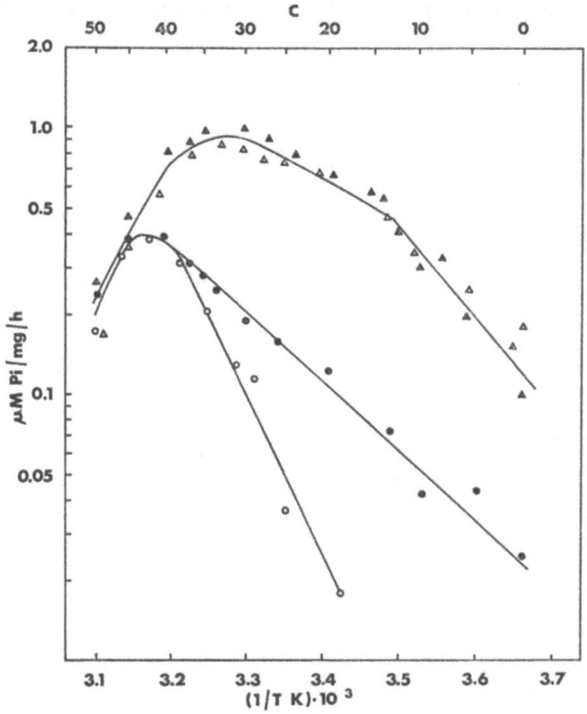

Fig. 4. Effect of temperature on the activities of Mg^{2+} ATPases (the upper curve) and Na^+—K^+ ATPases (the two lower curves) in nerve cord homogenates from earthworms acclimated to 14 C (closed triangels and circles) and for three weeks to 25 C (open triangles and circles). The points are mean values from 3 to 8 experiments. (From Lagerspetz, Kohonen and Tirri, 1973)

Table 3. Effect of temperature acclimation on the maximum response frequency of the giant nerve fibres of the earthworm. (From Talo and Lagerspetz, 1967)

AT (C.)	ET mean (C.)	Mean maximum response frequency (1/sec.)	*n*	*P* <
13	5.3	68.5	7	0.01
23	5.7	38.3	6	

and Tirri, 1973). This might be due to a conformational change in the enzyme during acclimation, but experiments with other animals rendered another explanation more plausible.

When the temperature dependence of Na^+—K^+-ATPase activities were studied in brain homogenates from a cold-water fish, the roach *(Leuciscus rutilus)* and from a warm-water fish, the kissing gourami *(Helostoma temmincki)*, the Arrhenius plots of the enzyme activities showed a clear double-peaked temperature dependence in the cold-water fish, the higher activity maximum under our experimental conditions being at 41 C and the lower at 20 C (Fig. 5; Kohonen, Tirri and Lagerspetz, 1973). A double-peaked temperature dependence was found also in homogenates as well as in microsomal fractions from the brain of another cold-water fish, the rudd

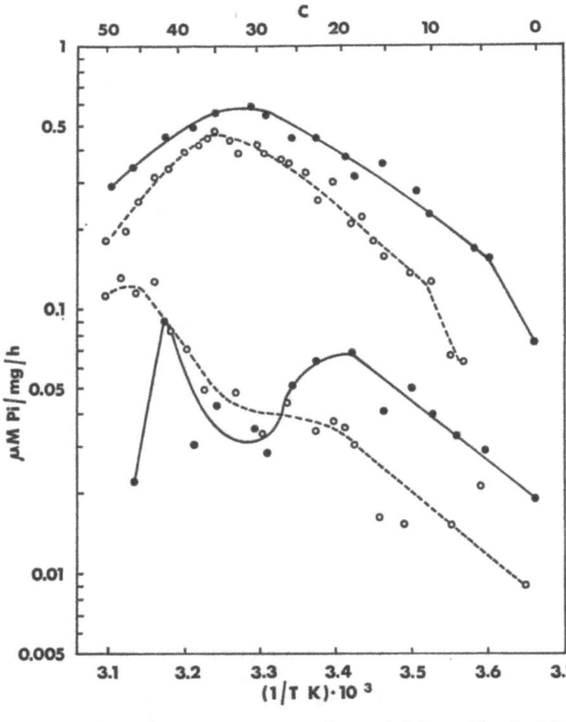

Fig. 5. Effect of temperature on the activities of brain Mg^{2+} ATPases (the two upper curves) and Na^+—K^+ ATPases (lower curves) of the roach (closed circles, solid lines) and of the gourami (open circles, broken lines). Data for the roach are mean values from 5, for the gourami from 6 separate experiments. (From Kohonen, Tirri and Lagerspetz, 1973)

(Scardinius erythrophthalmus) (unpublished observations). This type of temperature dependence is difficult to explain, unless the existence of two or more Na^+—K^+-ATPase isoenzymes with different temperature characteristics is assumed. Further evidence for the validity of this explanation was gained by studying the temperature dependence of the Na^+—K^+-ATPase activities in brain homogenates from young rats which had not yet developed an efficient regulation of the body temperature. In rats aged from 6 to 11 days, the Na^+—K^+-ATPase activity showed, in addition to the usual high activity maximum at about 45–50 C, progressive development of a second activity maximum at 20–24 C (Fig. 6; Tirri, Lagerspetz and Kohonen, 1973).

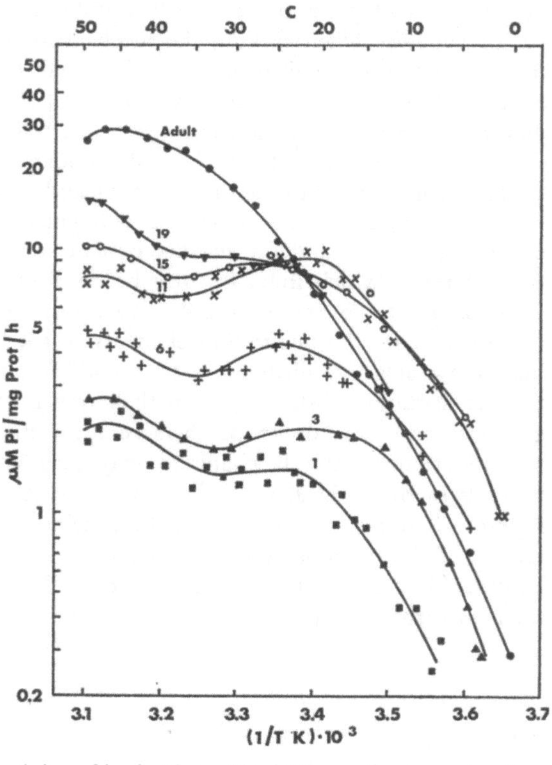

Fig. 6. Effect of temperature on the activity of brain Na^+—K^+ ATPase of rats aged 1, 3, 6, 11, 15, and 19 days and of adult rats . The age of the animals is indicated on each curve. Data from two or three experiments are represented by each curve. (From Tirri, Lagerspetz and Kohonen, 1973)

When homeothermy had gradually been attained by the young rats after the age of 11 days, this second maximum also disappeared. These results seemed to indicate the existence of at least two Na^+—K^+-ATPase isoenzymes, one of which was induced by the low temperatures occurring, at times, in the bodies of young rats before the attainment of homeothermy.

If the explanation advanced above is valid, i.e. if there exist in the nerve-cell membranes inducible Na^+—K^+-ATPase isoenzymes with different temperature characteristics, the mechanisms of temperature adaptation of the nervous system is more easily understandable. Selective synthesis of proteins, especially enzymes, has often been evoked as one of the possible explanations of acclimation phenomena, e.g. by Prosser (1969). However, the significance of specific isozymic patterns in the adaptation of animals with different acclimation histories has been established in only a few cases, one of the most notable being the multiple acetylcholinesterases of the trout brain, studied by Baldwin and Hochachka (1970). It may well be that the most likely place in which to look for adaptive changes in enzymes is the nervous system, which is often the controlling and limiting factor of physiological functions. In the nervous system, the cation transport mechanism may be more sensitive, limiting, and adaptive than most other systems in cellular metabolism.

To sum up, when seeking the mechanisms of temperature adaptation in ectotherms, or the mechanisms of any other acclimation phenomenon, the appropriate control system, in most cases a nervous one, should be studied. Adaptive changes may occur in several parameters of the nervous control system. The conducting elements, i.e. the nerve fibres, may show adaptive changes in the time relations of the ionic conductance of their membranes, and these may be connected with inducible cation transport enzymes possessing different temperature characteristics. Adaptive changes in the conduction velocity or in the refractory period, i.e. in the highest frequencies at which impulses may be conducted, may affect the rate and size of many peripheral responses, and thus be a possible basis for capacity adaptations. On the other hand, the integrating elements of the control mechanisms, the synapses, may show adaptive changes in the synthesis, uptake, release and breakdown of their transmitter substances, or in their transmitter-receptor interactions, or in the synthesis and breakdown of their receptor molecules. Adaptive changes in the synaptic transmission mechanisms may play an important part in the adaptive changes in the resistance limits of many peripheral functions, and thus be at the root of all resistance adaptations. In conclusion, the extreme plasticity so characteristic of the nervous system with regard to learning and memory, must also be considered as an explanatory principle in the primitive analogy of learning and memory furnished by acclimation phenomena.

Summary

1. The ciliary activity of the epithelium in isolated surviving gills of the freshwater mussel, *Anodonta cygnea* shows heat resistance adaptation. Acetylcholine (ACh), which controls ciliary activity in the tissues, also greatly increases the heat resistance. The activity of acetylcholinesterase is increased by warm acclimation, and atropine inhibits this effect.

2. High heat resistance of the heart beat in *Anodonta*, whether induced by warm acclimation or by seasonal acclimatication, is associated with a high 5-hydroxytryptamine (5-HT) concentration in the heart. 5-HT also increases the heat resistance of the heart beat.

3. These data, together with those concerning the role of neurotransmitter substances in the temperature adaptation of the frog (Harri, this symposium) support

the view that neurotransmitter systems are involved in the mechanism of temperature adaptation.

4. Electrophysiological evidence shows that acclimation to different temperatures leads to compensatory changes in the time relations of ionic conductance in the giant nerve fibres of the earthworm *Lumbricus terrestris*, especially at low (6–13 C) temperatures. Furthermore, the Na^+—K^+-stimulated ATPase activity of nerve-cord homogenates is much less sensitive to low and intermediate temperatures in animals acclimated to 14 than in those acclimated to 25 C.

5. The $R:T$ curve of the Na^+—K^+-ATPase activity in brain homogenates from a cold-water fish, the roach (*Leuciscus rutilus*), has two peaks, whereas that from a warm-water fish, the kissing gourami (*Helostoma temmincki*) has only one. A double-peaked temperature dependence of the Na^+—K^+-ATPase activity is also found both in brain homogenates and in brain microsomal fractions of another cold-water fish, the rudd (*Scardinius erythrophthalmus*).

6. In young rats with undeveloped thermoregulation, the temperature dependence of the brain Na^+—K^+-ATPase activity is double-peaked in homogenates.

7. The existence of inducible Na^+—K^+-ATPase isoenzymes with different temperature characteristics in the nervous system of ectotherms is suggested.

8. Both the integrating and conducting elements of the nervous control systems of animals show adaptive changes which may be important mechanisms of adaptations in thermal resistance and capacity found in ectotherms.

References

Aiello, E. L.: The influence of the branchial nerve and of 5-hydroxytryptamine on the ciliary activity of *Mytilus* gill. Biol. Bull. (Woods Hole) **113**, 325 (1957).

Baldwin, J., Hochachka, P. W.: Functional significance of isoenzymes in thermal acclimatization. Biochem. **116**, 883–887 (1970).

Bülbring, E., Burn, J. H., Shelley, H.: Acetylcholine and ciliary movement in the gill plates of *Mytilus edulis*. Proc. Roy. Soc., B **141**, 445–466 (1953).

Corrodi, H., Fuxe, K., Hökfelt, T.: A possible role played by central monoamine neurones in thermoregulation. Acta Physiol. Scand. **71**, 224–232 (1967).

Dierolf, B. M., McDonald, H. S.: Effects of temperature acclimation on electrical properties of earthworm giant axons. Z. Vergleich. Physiol. **62**, 284–290 (1969).

Erspamer, V., Ghiretti, F.: The action of enteramine on the heart of molluscs. J. Physiol. **115**, 470–481 (1951).

Gosselin, R. E.: The cilioexcitatory activity of serotonin. J. Cellular Comp. Physiol. **58**, 17–25 (1961).

Harri, M., Tirri, R.: Brain monoamines in the temperature acclimation of mice. Acta Physiol. Scand. **75**, 631–635 (1969).

Jensen, D. W.: The effect of temperature on transmission at the neuromuscular junction of the sartorius muscle of *Rana pipiens*. Comp. Biochem. Physiol. **41 A**, 685–695 (1972).

Kohonen, J., Tirri, R., Lagerspetz, K. Y. H.: Temperature dependence of the ATP-ase activities in brain homogenates from a cold-water fish and a warm-water fish. Comp. Biochem. Physiol. **44 B**, 819–821 (1973).

Lagerspetz, K. Y. H., Dubitscher, I.: Temperature acclimation of the ciliary activity in the gills of *Anodonta*. Comp. Biochem. Physiol. **17**, 665–671 (1966).

Lagerspetz, K. Y. H., Impivaara, H., Senius, K.: Acetylcholine in the thermal resistance acclimation of the ciliary activity in the gills of *Anodonta*. Comp. Gen. Pharmacol. **1**, 236–240 (1970).

Lagerspetz, K. Y. H., Kohonen, J., Tirri, R.: Temperature acclimation of the ATP-ase activites in the nerve cord of the earthworm, *Lumbricus terrestris* L. Comp. Biochem. Physiol. **44 B**, 823–827 (1973).

Lagerspetz, K. Y. H., Länsimies, H., Impivaara, H., Senius, K.: Control of the ciliary activity in the gills of *Anodonta* by acetylcholine. Comp. Gen. Pharmacol. **1**, 152–154 (1970).

Lagerspetz, K. Y. H., Talo, A.: Temperature acclimation of the functional parameters of the giant nerve fibres in *Lumbricus terrestris* L. I. Conduction velocity and the duration of the rising and falling phase of action potential. J. Exptl. Biol. **47**, 471–480 (1967).

Lagerspetz, K. Y. H., Tirri, R.: Transmitter substances and temperature acclimation in *Anodonta* (Pelecypoda). Ann. Zool. Fenniae **5**, 396–400 (1968).

Milton, A. S.: Choline acetylase in the gill plates of *Mytilus edulis*. Proc. Roy. Soc. B **150**, 240–244 (1959).

Precht, H., Christophersen, J.: Temperaturadaptation des Cilienepithels isolierter Kiemen und Fühlerspitzen von Mollusken. Z. Wiss. Zool. **171**, 197–209 (1965).

Prosser, C. L.: Principles and general concepts of adaptation. Environ. Res. **2**, 404–416 (1969).

Sakharov, D. A., Pécsi, T.: Protecting and reactivating effect of serotonin on the heat-inactivated heart of the fresh water mussel. Ann. Biol. Tihany **32**, 117–121 (1965).

Sherrington, C. S.: The integrative action of the nervous system. Constable, London p. 413 (1906). Reprinted 1961.

S.-Rózsa, K., Pécsi, T.: Comparative studies of the effect produced by biologically active agents on the isolated hearts of *Helix pomatia* L. and *Anodonta cygnea* L. Ann. Biol. Tihany. **34**, 59–72 (1965).

Talo, A., Lagerspetz, K. Y. H.: Temperature acclimation of the functional parameters of the giant nerve fibres in *Lumbricus terrestris* L. II. The refractory period. J. Exptl. Biol. **47**, 481–484 (1967).

Tirri, R., Lagerspetz, K. Y. H., Kohonen, J.: Temperature dependence of the ATP-ase activities in brain homogenates during the postnatal development of the rat. Comp. Biochem. Physiol. **44 B**, 473–480 (1973).

Neural Control of Temperature Adaptation in *Rana temporaria*

Mikko N. E. Harri

With 5 Figures

Introduction

Despite the contradictory results of several investigations it seems probable that the thyroid gland plays an important role in temperature acclimation not only in mammals but also in amphibians (c. f. Jankowsky, 1960, 1964; Gorbman, 1964). Temperature acclimation in mammals is known to be controlled both by endocrine glands and the nervous system (e.g., Hsieh and Carlsson, 1957; Depocas, 1960; Héroux, 1962; LeBlanc, 1969), whereas in poikilothermic vertebrates the role of the nervous system in temperature regulation is still uncertain. Differences exist even within homeotherms since the calorigenic response to noradrenaline, well documented in mammals, seem sto be lacking in birds (Hart, 1962; Hissa and Palokangas, 1970; Palokangas and Hissa, 1971).

Effect of Temperature Acclimation on Neurotransmitters

Changes in the ambient temperature are reported to cause activation of certain neurons in several mammalian species. This effect can be detected by changes in the level of the transmitter produced by these neurons (e.g., Toh, 1960; Moore *et al.*, 1961; Leduc, 1961; Musacchia *et al.*, 1962; Harri and Tirri, 1969; Spafford and Pengelley, 1971). Therefore in order to investigate the neural control of temperature acclimation in the frog the first step would be to ascertain whether the tissue levels of known transmitters, like 5-*hydroxytryptamine* (5-HT, *serotonin*) and *adrenaline* (Brodie and Bogdanski, 1964; Kärki and Lahovaara, 1965; Azuma *et al.*, 1965; Angelakos *et al.*, 1965; Segura *et al.*, 1967), change in the course of acclimation.

It can be shown (Harri, 1972a, b) that after transferring 25 C-acclimated winter frogs to 5 C the level of brain 5-HT decreases and that of brain and adrenal adrenaline increases. Transfer in the other direction leads to opposite changes. All changes are temporary and are not seen in summer animals. On the other hand, the concentrations of intestinal 5-HT or of heart adrenaline do not change during temperature acclimation. The levels of brain and of intestinal 5-HT are at their highest during the summer and lowest during the winter, the opposite being the case for the levels of brain, heart and adrenal adrenaline. Seasonal changes of these transmitters are not simple effects of temperature since changes caused by the latter are temporary and smaller than the former and do not occur in all tissues studied.

These results suggest that both season and temperature acclimation might induce alterations in the activity of the neurons producing 5-HT or adrenaline as transmitters. However, no conclusions can be drawn as to the rate of utilization of these substances. Changes in neural activity may (Harri and Tirri, 1969; Spafford and Pengelley, 1971), or may not (Draskoczy and Lyman, 1967), be accompanied by changes in the level of the neuromodulatory substance. The nature of this relationship can be analyzed by inhibiting the synthesis of a transmitter by means of specific enzyme inhibitors. This causes a depletion in the level of the transmitter, and the rate of this depletion has been shown to reflect the activity of the neurons using the transmitter in question (Andén *et al.*, 1969).

When 5-HT biosynthesis was inhibited with p-chlorophenylalanine (PCPA) at 5 C its level decreased by 41 % in the brain of cold-acclimated winter frogs but only by 21 % in the summer frogs (Harri, 1972a). After transferring winter frogs to 25 C the rate of decrease of brain 5-HT fell, as is indicated by the increased content of this amine relative to the cold-acclimated animals (Fig. 1, left). In summer frogs,

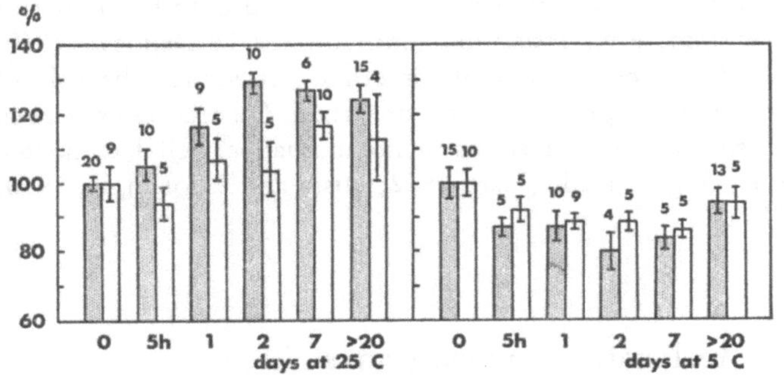

Fig. 1. Changes in the rate of depletion of brain 5-HT in winter (shaded columns) and summer frogs (open columns) after transferring the 5 C-acclimated frogs to 25 C (left) and the 25 C-acclimated animals to 5 C (right), when the 5-HT synthesis was inhibited by p-chlorophenylalanine (200 mg/kg). Depletion of brain 5-HT before transferring the animals to a new temperature is taken as 100% (= the first columns), and was 41% in the brains of 5 C-acclimated winter frogs and 21% in the brains of 5 C-acclimated summer frogs (left), and 5% and 9% in the brains of 25 C-acclimated winter and summer frogs (right), respectively. When the animals were transferred from 5 C to 25 C (left) they were kept for 2 hours after the drug injection at 5 C, and when transferred from 25 C to 5 C (right) they were kept at 25 C for 4 hours after the injection. The number of experiments is given at the top of each column. Vertical bars indicate ± S.E. (After Harri, 1972a)

under similar experimental conditions, utilization of brain 5-HT decreased only slightly. The results suggest that brain 5-HT neurons were less active in cold-acclimated summer frogs than in cold-acclimated winter frogs. On the other hand, in the 25 C-acclimated winter and summer frogs only 5 and 9 % depletion, respectively, was observed in brain 5-HT after inhibition of its synthesis 4 hours before killing the animals. After transferring the 25 C-acclimated frogs to 5 C the rate of utilization of brain 5-HT increased and the amine content of the neurons fell correspondingly (Fig. 1, right). Utilization of intestinal 5-HT did not change significantly after exposure of animals to a new temperature, and no difference was seen between winter and summer frogs.

When disulfiram was used to inhibit catecholamine synthesis the interval between drug administration and the killing of animals was of the same order or longer (i.e., 2 hours when measured at 25 C and 4 hours when measured at 5 C) than that generally used in mammals (e.g. Svensson and Waldeck, 1969; Julku *et al.*, 1970; Tirri, 1970). The results show that 4 hours after the injection of disulfiram adrenaline depleted by 17% in the brain of cold-acclimated winter frogs and by 35% in the brain of coldacclimated summer frogs when measured at 5 C, indicating that the activity of brain adrenergic neurons was greater in summer than in winter frogs (Harri, 1972b). However, owing to the slow turnover of catecholamines in the frog, these experiments give only little information about the changes in the utilization of adrenaline in the course of temperature acclimation.

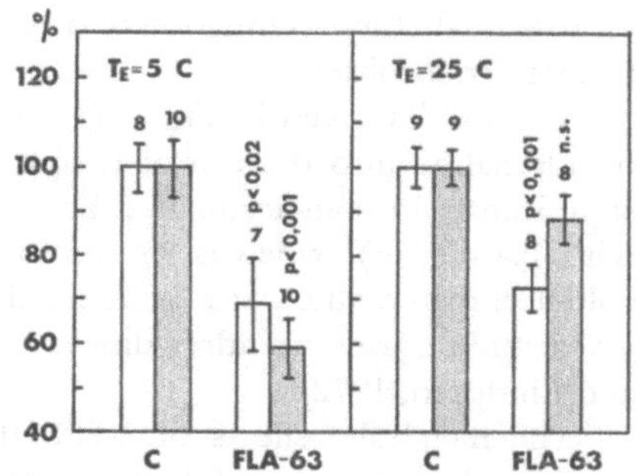

Fig. 2. Brain adrenaline levels in the 5 C- (open columns) and 25 C-acclimated (shaded columns) winter frogs after inhibition of the catecholamine synthesis with FLA-63 (50 mg/kg) at 5 C (left) and at 25 C (right). C = control animals injected with the solvent only and kept at the experimental temperature (T_E) after the injection. In assays at 5 C the interval between drug administration and killing was 20 hours, and in those at 25 C it was 8 hours. The number of experiments is given at the top of each column. Vertical bars indicate ± S.E. Statistical significances of the differences between the control and treated animals are indicated at the top of the columns of treated animals. (After Harri, 1972b)

A new disulfiram-like dopamine-β-hydroxylase inhibitor, FLA-63 (Svensson and Waldeck, 1969), was also used to inhibit catecholamine synthesis (Fig. 2; Harri, 1972b). In these experiments the interval between drug administration and killing the animals was 8 hours when measured at 25 C and 20 hours when measured at 5 C. The results indicate that depletion of adrenaline in the brains of cold-acclimated winter frogs was significant both when measured at 5 C (Fig. 2, left) and at 25 C (Fig. 2, right). On the other hand, under similar experimental conditions, the depletion of adrenaline in the brains of warm-acclimated animals was significant only when measured at 5 C, indicating that in this experimental group cold exposure had a much larger effect on adrenaline utilization than warm exposure.

In mammals changes in utilization of 5-HT and/or noradrenaline in the hypothalamus after alteration of the ambient temperature have been proposed as the primary reactions which initiate and control temperature acclimation (e.g., Feldberg and Myers, 1963, 1964; Corrodi *et al.*, 1967; Harri and Tirri, 1969; Tirri, 1970, 1971). The observed changes in the utilization of 5-HT and adrenaline in the frog brain very soon after transferring the animals to a new temperature may indicate that they are important for the control and perhaps for the initiation of temperature acclimation in amphibians as well.

Effect of Temperature Acclimation on the Metabolic Response to Neurotransmitters

It has been shown in mammals that the increased release of noradrenaline from the sympathetic nerve endings in cold is accompanied by an increased metabolic response of cold-acclimated animals to this amine (LeBlanc and Pouliot, 1964). Therefore, because the results presented above indicate increased activity of the 5-HT and the adrenaline neurons in frogs after exposure to cold, it seemed of interest to test whether acclimation of frogs to cold also increases their metabolic sensitivity to these transmitters.

The results given in Fig. 3 (Harri and Hedenstam, 1972) indicate that injection of adrenaline into the dorsal lymph sac elevated the rate of oxygen consumption significantly in cold-acclimated but not in warm-acclimated winter frogs ($p < 0.02$ with 0.1 mg/kg), whereas in summer frogs it was elevated significantly in both cold-acclimated and warm-acclimated animals ($p < 0.05$ with 0.1 mg/kg). A hyperglycaemic response to adrenaline was observed only in cold-acclimated frogs (Harri and Lindgren, 1972).

The metabolic effects of 5-HT (Fig. 4; Harri *et al.*, 1972) show that in cold-acclimated winter frogs 0.5 mg/kg of 5-HT elevated the oxygen consumption 21 % over the values found after the injection of saline ($p < 0.02$). In warm-acclimated winter frogs, however, only doses above 1.0 mg/kg ($p < 0.02$) stimulated oxygen consumption. Furthermore, summer frogs were less sensitive to the metabolic action of 5-HT since the doses needed to produce a significant increase in oxygen consumption were greater than in winter frogs. In addition, the doses needed to produce a significant increase in oxygen consumption were smaller in cold-acclimated ($p < 0.02$ with 2.0 mg/kg) than in warm-acclimated animals ($p < 0.05$ with 8.0 mg/kg), indicating that the metabolic sensitivity to 5-HT increases markedly during cold-acclimation in summer frogs, too.

As stated above, the utilization of adrenaline increased after exposing warm-acclimated frogs to the cold. Correspondingly, the metabolic response of winter frogs to adrenaline was seen only in cold-acclimated animals. Further, in summer frogs, in which adrenaline utilization tended to be faster than in winter frogs, the metabolic response to this amine was observed in warm-acclimated animals as well. As with adrenaline, the metabolic sensitivity of frogs to 5-HT increased only in those groups in which its rate of utilization had also increased. The effects of adrenaline in the frog agree well with those seen in mammals, whereas 5-HT is generally shown to be ineffective in stimulating the oxygen consumption in mammals (Garattini and Valzelli, 1965) and in birds (Freeman, 1970; Palokangas and Hissa, 1971).

Metabolic Effects of Repeated Injections of Neurotransmitters

In order to test whether an increased utilization of 5-HT or adrenaline is necessary for producing cold acclimation in the frog, utilization of these amines was

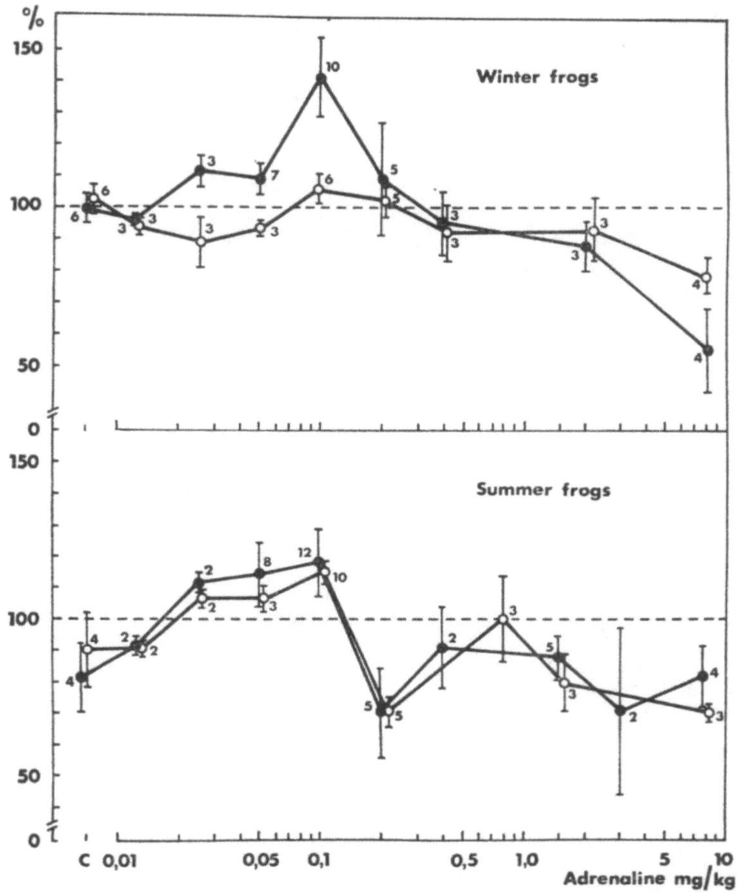

Fig. 3. The maximal effects of different doses of adrenaline (in mg/kg) on the oxygen consumption of cold- (8–10 C) (solid circles) and warm-acclimated (25 C) frogs (open circles) measured during the winter and summer at 25 C. All animals were injected with physiological saline and the oxygen consumption measured for 30 mins. Then the animals were injected with either saline again (controls, C) or with adrenaline, and the oxygen consumption was recorded for a further 20–30 mins. 100% = oxygen consumption after the first (saline) injection. Oxygen consumption after the second (saline or adrenaline) injection was expressed as a percentage of the first (saline) injection. The absolute values of the oxygen consumption (in ml/kg per hr) were 104.6 ± 7.1 and 81.0 ± 6.3 for the cold- and warm-acclimated winter frogs and 122.9 ± 1.6 and 117.0 ± 4.5 for the summer frogs, respectively. The number of experimental animals in each group is indicated in the figure. Vertical bars indicate ± S.E. (After Harri and Hedenstam, 1972)

artificially increased by injecting them daily for 30 days into warm-acclimated animals (Harri, unpublished). After the adrenaline injections the rate of oxygen consumption of the treated animals increased by 20% whilst 5-HT treatment did not elevate the metabolic rate. Furthermore, chronic 5-HT treatment, and to lesser degree, chronic adrenaline treatment, decreased the thermal maximum of oxygen consumption in the warm-acclimated winter frogs. However, decrease in the thermal maximum of

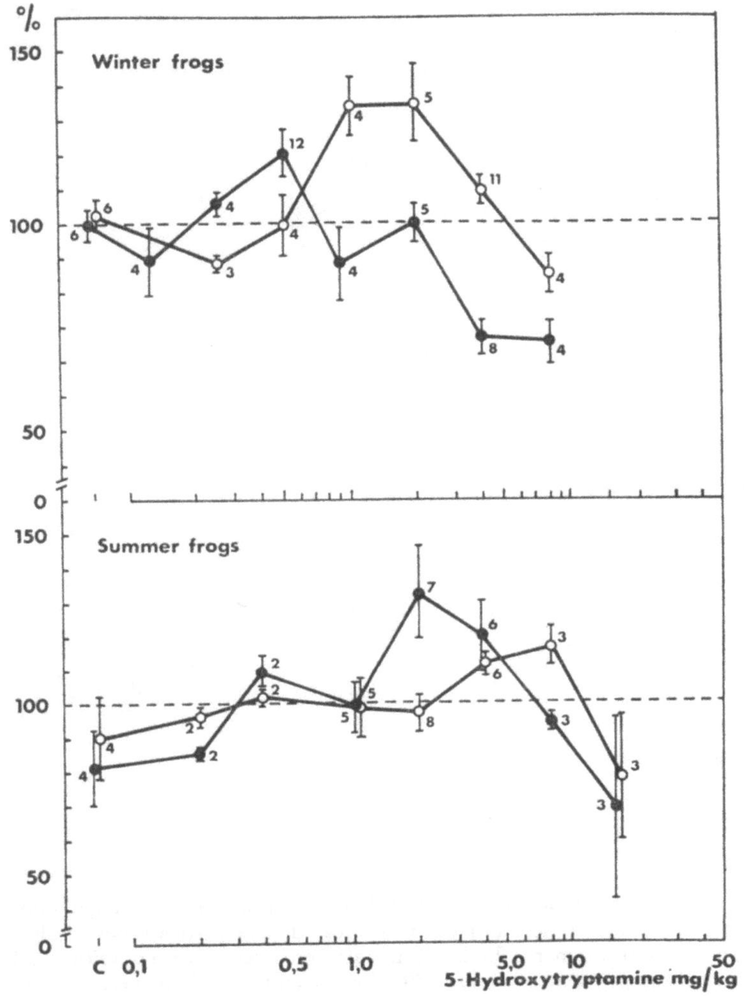

Fig. 4. The maximal effects of different doses of 5-HT (in mg/kg) on the oxygen consumption of cold- (5 C) (solid circles) and warm-acclimated (25 C) (open circles) frogs measured during the winter and summer at 25 C. Other explanations as in Fig. 4. (After Harri *et al.*, 1972)

oxygen consumption is not necessarily associated with an increase in cold resistance but may be a consequence of toxic effects caused by chronic drug administration. Measurement of cold resistance was, however, so difficult as to be practically unsuccessful (Holzapfel, 1937; Brattstrom, 1968). Neither chronic adrenaline treatment nor chronic 5-HT treatment changed the metabolic sensitivity of frogs to adrenaline, whereas both types of treatment increased the metabolic sensitivity to 5-HT.

It has been shown recently that catecholamines and 5-HT induce secretion of thyroid hormone by direct action on the thyroid follicle cells in mice (Melander and Sundler, 1972; Melander *et al.*, 1972). Therefore it was of interest to study the effects of chronic adrenaline and 5-HT treatments on the thyroid gland of the frog. The results indicate that chronic adrenaline treatment, but not chronic 5-HT treatment, caused a significant decrease in the thyroid colloid content, an indication of increased thyroid hormone secretion. The colloid decrease caused by adrenaline treatment was, however, significantly less than that caused by exposure of the animals to cold.

Thus, although not all the criteria used were changed by chronic adrenaline or 5-HT injections to the degree caused by exposure of the animals to cold, the results indicate that repeated injections of these transmitters into warm-acclimated frogs can produce some of the effects of cold acclimation. It is concluded that the increased utilization of adrenaline and 5-HT in the cold may initiate and control the peripheral adaptive changes associated with cold acclimation in amphibians. To sum up, Fig. 5 shows schematically the main reactions which, in agreement with the above results, may control cold acclimation in the frog. It is proposed that the course of events in warm acclimation is similar but reversed.

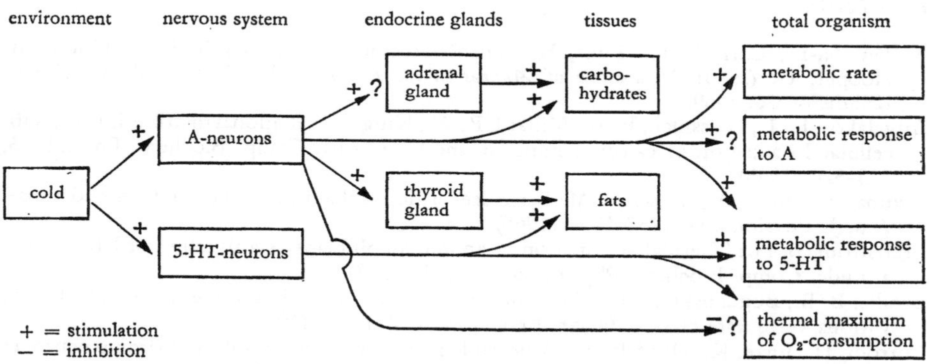

Fig. 5. Diagram illustrating the observed and some assumed reactions which may control cold acclimation in the frog

Summary

Working on the assumption that neural activation can change the levels of 5-hydroxytryptamine (5-HT) or adrenaline, both important transmitters in amphibians, the levels of these amines were investigated as a function of season and of temperature acclimation in *Rana temporaria*. After transfer of the 25 C-acclimated winter frogs to 5 C the level of brain 5-HT decreased and that of brain and adrenal adrenaline increased. Transfer in the other direction led to opposite changes. All changes were temporary and were not seen in the summer frogs. The contents of intestinal 5-HT or heart adrenaline did not change during temperature acclimation. The levels of brain and intestinal 5-HT were at their highest during the summer and lowest during the winter, the opposite being the case for the levels of brain, heart and adrenal adrenaline. Seasonal changes of these transmitters are not simple effects of temperature, since changes caused by the latter are temporary and smaller than the former, and do not occur in all tissues studied. After inhibition of amine biosynthesis, depletion of brain 5-HT (activity of the 5-HT neurons) was greater in the winter frogs than in the summer frogs. The depletion was accelerated during cold acclimation and retarded during warm acclimation. Utilization of intestinal 5-HT did not change by warm acclimation. Activity of the adrenaline neurons was increased after exposing warm-acclimated frogs to the cold. The results also indicate that the metabolic sensitivity of frogs to both adrenaline and 5-HT was increased by cold acclimation. Furthermore, repeated injections of warm-acclimated frogs with 5-HT or adrenaline induced changes in the metabolic rate, in the thermal maximum of oxygen consumption, in the metabolic sensitivity to 5-HT and in the thyroid gland, similar to those seen on exposure of animals to cold. It can be thus concluded that the increased utilization of these amines in the cold may initiate and control the peripheral adaptive changes associated with cold acclimation. It is assumed that warm acclimation leads to the reverse reactions.

References

Andén, N.-E., Corrodi, H., Fuxe, K.: Turnover studies using synthesis inhibition. In: Hooper, G. (Ed.): *Metabolism of Amines in the Brain*, p. 38–47. St. Albans: Fisher, Knight & Co. 1969.

Angelakos, E. T., Glassman, P. M., Millard, R. W., King, M.: Regional distribution and subcellular localization of catecholamines in the frog heart. Comp. Biochem. Physiol. **15**, 313–324 (1965).

Azuma, T., Binia, A., Visscher, M. B.: Adrenergic mechanism in the bullfrog and turtle. Am. J. Physiol. **209**, 1287–1294 (1965).

Brattstrom, B. H.: Thermal acclimation in anuran amphibians as a function of latitude and altitude. Comp. Biochem. Physiol. **24**, 93–111 (1968).

Brodie, B. B., Bogdanski, D. F.: Biogenic amines and drug action in the nervous system of various vertebrate classes. Progr. Brain Res. **9**, 234–242 (1964).

Corrodi, H., Fuxe, K., Hökfelt, T.: A possible role played by central monoamine neurones in thermo-regulation. Acta Physiol. Scand. **71**, 224–232 (1967).

Depocas, F.: The calorigenic response of cold-acclimated white rats to infused noradrenaline. Can. J. Biochem. Physiol. **38**, 107–114 (1960).

Draskóczy, P. R., Lyman C. P.: Turnover of catecholamines in active and hibernating ground squirrels. J. Pharmacol. Exptl. Therap. **155**, 101–111 (1967).

Feldberg, W., Myers, R. D.: A new concept of temperature regulation by amines in the hypothalamus. Nature (London) **200**, 1325 (1963).

Feldberg, W., Myers, R. D.: Effects on temperature of amines injected into the cerebral ventricles. A new concept of temperature regulation. J. Physiol. (London) **173**, 226–237 (1964).

Freeman, B. M.: Thermoregulatory mechanisms of the neonate fowl. Comp. Biochem. Physiol. **33**, 219–230 (1970).

Garattini, S., Valzelli, L.: *Serotonin*, p. 392 Amsterdam: Elsevier 1965.

Gorbman, A.: Endocrinology of the amphibia. In: Moore, J. A. (Ed.): *Physiology of the Amphibia*,, p. 371–425. New York and London: Academic Press 1964.

Harri, M. N. E.: Effect of season and temperature acclimation on the 5-hydroxytryptamine level and utilization in the brain and intestine of the frog, *Rana temporaria*. Comp. Gen. Pharmacol. **3**, 11–18 (1972a).

Harri, M. N. E.: Effect of Season and temperature acclimation on the tissue catecholamine level and utilization in the frog, *Rana temporaria*. Comp. Gen. Pharmacol. **3**, 101–112 (1972b).

Harri, M. N. E., Hedenstam, R.: Calorigenic effect of adrenaline and noradrenaline in the frog, *Rana temporaria*. Comp. Biochem. Physiol. **41 A**, 409–419 (1972).

Harri, M. N. E., Hedenstam, R., Lindgren, E., Puuska, M.: Calorigenic effect of 5-hydroxytryptamine in the frog, *Rana temporaria*. Comp. Biochem. Physiol. **43 A**, 545–552 (1972).

Harri, M. N. E., Lindgren, E.: Adrenergic control of carbohydrate metabolism in the frog, *Rana temporaria*. Comp. Gen. Pharmacol. **3**, 226–234 (1972).

Harri, M. N. E., Tirri, R.: Brain monoamines in the temperature acclimation of mice. Acta Physiol. Scand. **75**, 631–635 (1969).

Hart, J. S.: Seasonal acclimatization in four species of wild birds. Physiol. Zool. **35**, 224–236 (1962).

Héroux, O.: Seasonal adjustments in captured wild Norway rats – II. Survival time, pelt insulation, shivering and metabolic and pressor responses to noradrenalin. Can. J. Biochem. Physiol. **40**, 405–414 (1962).

Hissa, R., Palokangas, R.: Thermoregulation in the titmouse (*Parus major* L.). Comp. Biochem. Physiol. **33**, 941–953 (1970).

Holzapfel, R. A.: The cyclic character of hibernation in frogs. Quart. Rev. Biol. **12**, 65–84 (1937).

Hsieh, A. C., Carlsson, D. L.: Role of adrenaline and noradrenaline in chemical regulation of heat production. Am J. Physiol. **190**, 243–245 (1957).

Jankowsky, H. D.: Über die hormonale Beeinflussung der Temperaturadaptation beim Grasfrosch (*Rana temporaria* L.). Z. Vergleich. Physiol. **43**, 392–410 (1960).

Jankowsky, H. D.: Die Bedeutung der Hormone für die Temperaturanpassung im normalen Temperaturbereich. Helgoländer Wiss. Meeresuntersuch. **9**, 412–419 (1964).

Julku, H., Lagerspetz, K. Y. H., Tirri, R.: The diurnal variation of thermoregulatory efficiency in mice. Ann. Zool. Fennici **7**, 375–377 (1970).

Kärki, N. T., Lahovaara, S.: Identification and assay of 5-hydroxytryptamine in brain and intestine of lower vertebrates. Ann. Med. Exptl. Fenniae **43**, 419–423 (1965).

Le Blanc, J.: Stress and interstress adaptation. Federation Proc. **28**, 996–1000 (1969).

Le Blanc, J., Pouliot, M.: Importance of noradrenaline in cold adaptation. Am. J. Physiol. **207**, 853–856 (1964).

Leduc, J.: Catecholamine production and release in exposure and acclimation to cold. Acta Physiol. Scand. **53** (Suppl. 183) (1961).

Melander, A., Nilsson, E., Sundler, F.: Sympathetic activation of thyroid hormone secretion in mice. Endocrinology **90**, 194–199 (1972).

Melander, A., Sundler, F.: Interactions between catecholamines, 5-hydroxytryptamine and TSH on the secretion of thyroid hormone. Endocrinology **90**, 188–193 (1972).

Moore, K. E., Calvert, D. N., Brody, T. M.: Tissue catecholamine content of cold-acclimated rats. Proc. Soc. Exptl. Biol. Med. **106**, 816–818 (1961).

Musacchia, X. J., Jellinek, M., Cooper, T.: Effect of hibernation and cold torpor on tissue catecholamine content. Proc. Soc. Exptl. Biol. Med. **110**, 856–857 (1962).

Palokangas, R., Hissa, R.: Thermoregulation in young blackheaded gull (*Larus ridibundus* L.). Comp. Biochem. Physiol. **38A**, 743–750 (1971).

Segura, E. T., Biscardi, A. M., Apelbaum, J.: Seasonal variations of brain epinephrine, norepinephrine and 5-hydroxytryptamine associated with changes in the EEG of the toad, *Bufo arenarum* Hensel. Comp. Biochem. Physiol. **22**, 843–850 (1967).

Spafford, D. C., Pengelley, E. T.: The influence of the neurohumor serotonin on hibernation in the golden-manteled ground squirrel, *Citellus lateralis*. Comp. Biochem. Physiol. **38**, 239–250 (1971).

Svensson, T. H. S., Waldeck, B.: On the significance of central noradrenaline for motor activity: experiments with a new dopamine-hydroxylase inhibitor. Eur. J. Pharmacol. **7**, 278–282 (1969).

Tirri, R.: Brain monoamines in the control of body temperature in mice. Ann. Zool. Fennici **7**, 323–328 (1970).

Tirri, R.: Central effects of 5-hydroxytryptamine and noradrenaline on body temperature and oxygen consumption in infant rats. Experientia **27**, 274–276 (1971).

Toh, C. C.: Effects of temperature on the 5-hydroxytryptamine (serotonin) content of tissues. J. Physiol. (London) **151**, 410–415 (1960).

The Influence of Temperature on Membrane Processes

CHRISTIAN WINTER

With 5 Figures

The question as to how sensory and nerve cells are influenced by changes of temperature is one which has interested neurophysiologists since the days of Bernstein (1912), touching as it does upon the basic problems of excitatory mechanisms.

Although the membrane potential of nerve-, sensory- and muscle-cells usually changes but little with temperature (Hodgkin and Katz, 1949; Guttmann, 1971), this slight effect nevertheless has to be explained.

According to the Hodgkin-Huxley hypothesis the membrane potential can be considered as a function of the absolute temperature and of the permeability and concentration of certain ions that are spatially separated from one another by the membrane.

$$E = f(T, C_i, P_i)$$

E = membrane potential
T = absolute temperature
C_i = ion concentration
P_i = ion permeability

This relationship is described in detail by the Goldman equation:

$$E = -\frac{\mathrm{R} \cdot \mathrm{T}}{\mathrm{F}} \cdot \ln \frac{\mathrm{P}_K [\mathrm{K}^+]_i + \mathrm{P}_{Na} [\mathrm{Na}^+]_i + \mathrm{P}_{Cl} [\mathrm{Cl}^-]_o}{\mathrm{P}_K [\mathrm{K}^+]_o + \mathrm{P}_{Na} [\mathrm{Na}^+]_o + \mathrm{P}_{Cl} [\mathrm{Cl}^-]_i}$$

R = general gas constant
F = Faraday constant

If it be assumed that the ion concentration and permeability themselves are independent of temperature then the membrane potential should be proportional to the absolute temperature.

What changes in membrane potential can then be expected? If the membrane potential at, for example, 15 C is —70 mV, it ought to have dropped to —67 mV at 5 C and to have risen to —73 mV at 25 C. The resting potential of the membrane of giant fibres of the crayfish lies approximately within these limits (Machne and Oroczo, 1968). Upon warming from 8 C to 20 C it rises by 6–7 mV. The low Q_{10} value of 1.1 for these measurements favours the hypothesis that the temperature effect depends mainly upon passive diffusion processes.

However, such results permit of no generalizations. The resting potential of giant fibres of the closely related lobster exhibits a much larger increase in potential on warming than is predicted by the Goldman equation (Dalton and Hendrix, 1962). Membranes whose potential arises not only from diffusion processes but is also influenced by temperature-dependent metabolic processes (electrogenic ion pump) behave in this manner (Senft, 1967). The influence of such ion pumps will be dealt with in more detail below.

Changes in temperature have a much more striking effect on the generation of propagated action potentials than on the membrane potential. Commencing at moderate temperatures a marked slowing down of the potential changes is produced by cooling, the total duration of the nerve impulse increases greatly (Fig. 1) and at the same time the rates of rise and fall of these potential changes decrease conspicuously (Table 1).

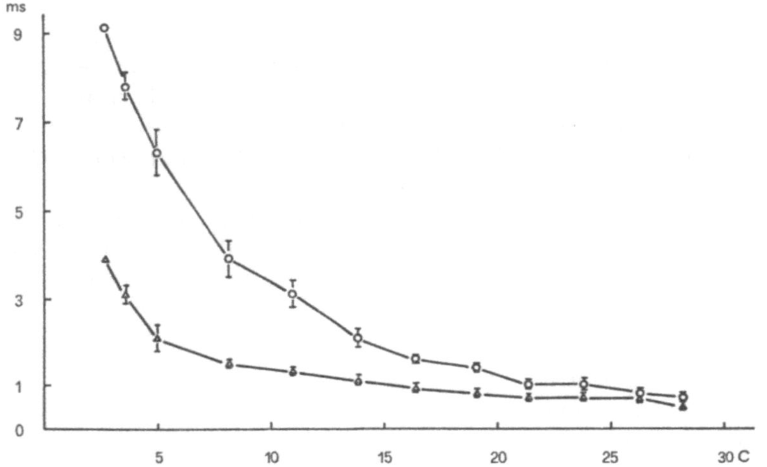

Fig. 1. Influence of temperature on the mean duration of rise and fall of the action potentials. Stretch receptor neurone, *Astacus fluviatilis.* Abscissa: Experimental temperature in C. Ordinate: Time in msec. Error shown: Mean error of the mean value. (After Winter, 1972)

Table 1. Q_{10}-values of the maximum rates of rise and fall measured from action potentials of various nerve preparations. (1) Hodgkin and Katz, 1949; (2) Winter, 1972; (3) Machne and Orozco, 1968; (4) Inman and Peruzzi, 1961

	rising phase			falling phase		
Temperature (C)	5–10	10–20	20–30	5–10	10–20	20–30
Loligo giant fibre (1)	2,25	1,92	1,39	4,65	3,17	1,93
Astacus stretch receptor neurone (2)	2,9	2,0	1,3	4,3	3,6	1,6
Procambarus giant fibre (3)	—	1,65	—	—	1,9	—
Cat Pacinian corpuscle (4)	—	—	2,59	—	—	4,19

Table 1 reveals that, within the same temperature range, the Q_{10}-values on warming are larger for the rate of fall than for the rate of rise of a nervous impulse. This means that the repolarization phase is more strongly accelerated on warming than the rising or depolarization phase and, as a result, the amplitude must decrease.

Conversely, the amplitude of the action potential ought to increase on warming if the *rise* is more rapid than the *fall*. Such a discrepancy in the effect of temperature on the rising and falling phases can be influenced experimentally by altering the Ca^{++} content of the bathing solution (Machne and Orozco, 1968: *Procambarus* giant fibres). With low Ca^{++} (3.37 mM = 1/4 of the normal Ca^{++} content) the rising phase is accelerated more strongly and, as a result, the recorded action potentials are larger.

Despite the considerable influence of temperature on the individual action potentials the efficiency of the nervous system of an animal remains unaffected as long as all-or-none-potentials can be initiated, which requires that, depending upon species, a given upper and lower temperature limit be not exceeded (stretch receptor neurone: cold block at <2–3 C, heat block at >33 C; see Burkhardt, 1959).

The speed at which an action potential can be passed along the information channel is decisive in determining the efficiency of a nervous system: temperature plays an important role at this point. Given the same fibre diameter and degree of myelinization, warm-blooded animals invariably exhibit higher speeds of conduction than cold-blooded ones (v. Muralt, 1959).

Since temperature changes in the animal's environment primarily affect sensory endings in the periphery of its nervous and receptor apparatus the important question arises as to how the afferent discharge pattern, which reflects the specific peripheral stimulus, can be altered by temperature. Does a peripheral compensation exist?

The discharge frequency of the muscle spindle *(extensor digiti)* of the frog, for example, increases linearly between 3 and 30 C under a constant mechanical test stimulus, falls off sharply above 30 C and is blocked at 40 C (Ottoson, 1965). If the pre-stretched muscle spindle is exposed to sudden temperature stimuli, discharge plateaux occur, depending upon the experimental temperature used (Winter, 1967): warming raises the discharge frequency, whereas cooling lowers it (Fig. 2). This behaviour is surprising on two accounts:

a) the "overshoot" excitation seen on cooling, as is described below for other structures, is missing.

b) the frequency of discharge cannot be explained by the effect of temperature on the corresponding generator potential. It falls on cooling, although following experimental stretching the generator potential is uninfluenced by temperature over a wide range (Ottoson, 1965).

A number of sensory and nerve cells offers a contrast to the muscle spindle of the frog in that their afferent discharge is characterized by a completely different type of behaviour. On sudden cooling the discharge frequency, following a phase of overshoot excitation, assumes a level of frequency that is low compared with the initial frequency. On sudden warming the impulse frequency is temporarily reduced but later rises to a higher level. Such a reaction, already familiar in various receptors, especially mechanoreceptors, was termed "paradoxical" by Sand (1938), because an increase of activity would be expected on increasing temperature, as in the muscle spindle of the frog.

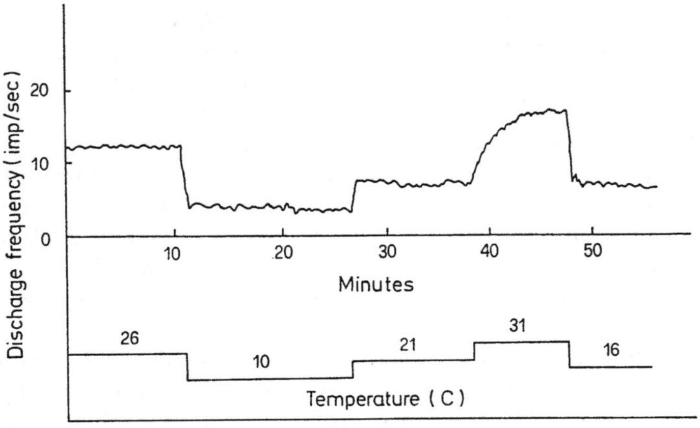

Fig. 2. Pattern of discharge frequency after sudden changes in temperature. Muscle spindles of *extensor digiti*, *Rana temporaria*. Abscissa: Time in min. after beginning of experiment. Ordinate: Discharge frequency in impulses/sec. Temperature is given underneath in C. Following a rise of temperature from 21 C to 31 C a delayed rise in discharge frequency was repeatedly recorded. (Winter, 1967)

Well-known examples of paradoxical behaviour are presented by the ampullae of Lorenzini, muscle receptors in selachians, lateral organs of *Xenopus*, neurones of abdominal ganglia of the ventral cord in *Astacus* and *Periplaneta*, and motoneurones in the spinal cord of the cat (Murray, 1962).

In Fig. 3 it is seen that in all preparations, taken exclusively from ectotherms, a sudden lowering of temperature is accompanied by excitation peaks and a sudden rise in temperature by inhibition peaks. A constant discharge frequency ensues, which is be ow the initial value if cooling occurs and above it in the case of heating. Such a response is of interest since it is comparable to the behaviour of the cold fibres of endotherms. In consequence, it could be asked whether such an impulse pattern following temperature stimulation is perhaps connected with thermoreception.

The causes of this peculiar response to temperature are not known in all cases, although certain clues as to the underlying processes are provided by the author's own work on the stretch receptors of the crayfish (Winter, 1972). This sensory organ, consisting of receptor muscle, sensory cells and afferent nerves can be investigated in an isolated state. The analysis necessarily includes the influence of temperature on the receptor muscle since it is impossible to investigate the neurone without it. Temperature-induced changes in tension in this muscle must have an effect on the state of activity of the neurone, and changes in tension of this nature do in fact exist, but are too small by a factor of 10 to account for the constant discharge frequencies observed. Moreover, they provide no clue as to the causes of the excitation peaks at the onset of cooling (Winter and Fleissner, 1971). The rest of the investigation was therefore devoted to the sensory cells. Since intracellular recording is usually only successful in stretched receptor muscle, the cell is depolarized beyond the firing threshold of —68 mV and produces action potentials.

In order to obtain uninterrupted measurements of the membrane potential it is

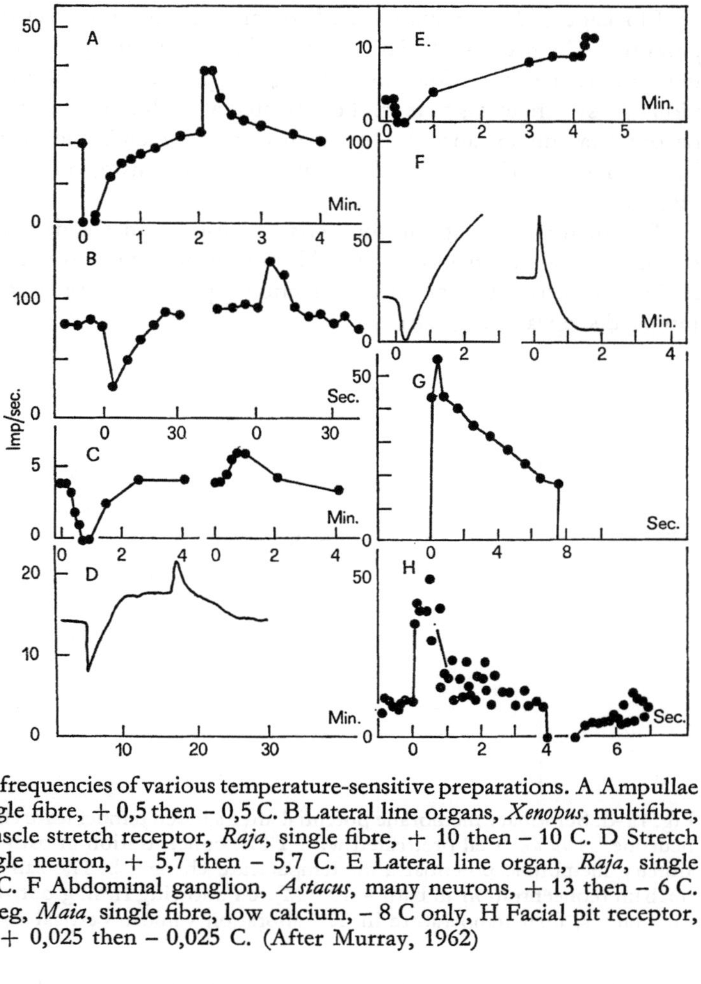

Fig. 3 A–H. Discharge frequencies of various temperature-sensitive preparations. A Ampullae of Lorenzini, *Raja*, single fibre, + 0,5 then – 0,5 C. B Lateral line organs, *Xenopus*, multifibre, + 6 then – 6 C. C Muscle stretch receptor, *Raja*, single fibre, + 10 then – 10 C. D Stretch receptor, *Astacus*, single neuron, + 5,7 then – 5,7 C. E Lateral line organ, *Raja*, single fibre, + 10 then – 10 C. F Abdominal ganglion, *Astacus*, many neurons, + 13 then – 6 C. G Motor nerve from leg, *Maia*, single fibre, low calcium, – 8 C only, H Facial pit receptor, *Crotalus*, single fibre, + 0,025 then – 0,025 C. (After Murray, 1962)

necessary to prevent the generation of impulses by blocking the influx of Na^+ with Tetrodotoxin, TTX ($2–9 \cdot 10^{-9}$ g/ml). Under such experimental conditions rapid temperature jumps produce a drop in membrane potential (= less negative) upon cooling and a rise upon warming (Fig. 4a). The membrane potential goes through a maximum during the temperature jump and falls subsequently to a lower plateau value which is more negative or more positive according to the direction of the temperature change (cold generator potential).

As already mentioned, the influence of temperature on the membrane potential can be estimated if, as a first approximation, it is assumed that ion concentration and ion permeability do not change with temperature. Accordingly, the membrane potential should be proportional to the absolute temperature. Fig. 5 shows that in the case investigated the values recorded (o– –o initial value, x—x plateau value) do not coincide with the theoretical values calculated from the Goldman equation (–.–).

Blockage of Na^+ influx by TTX or procaine has no effect on the cold generator potential. The greatest deviations from the Goldman equation are seen upon warming, and at the same time the drop to the plateau is at its most pronounced. Obviously a system exists that can be activated at high temperatures but which probably becomes rapidly exhausted. Disturbances of the membrane metabolism, for example by a reduction of the Na^+—K^+ transport mechanism, ought then to yield further information.

Following blocking of the active transport mechanism (Na^+—K^+ pump) by ouabain (= g-strophantin $4 \cdot 10^{-5}$ M) the displacement of the membrane potential in both positive and negative directions, previously seen following temperature jumps, disappears (Fig. 4b).

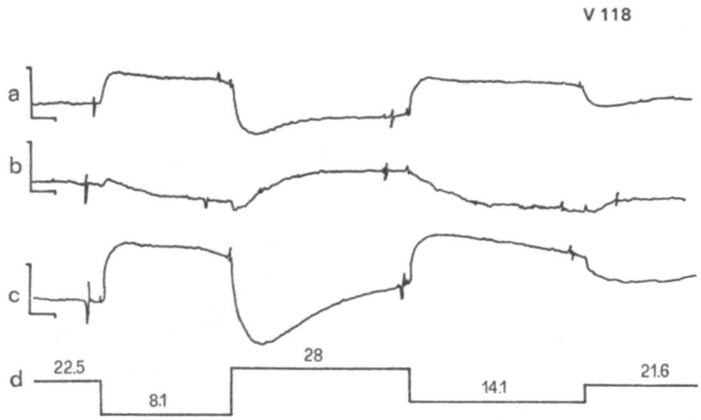

Fig. 4a–d. Stationary membrane potential after sudden temperature changes. Stretch receptor neurone, *Astacus*. (Cell pretreated with TTX, concentration in bath $5–7 \cdot 10^{-9}$ g/ml). a Stationary membrane potential after temperature change. b The same cell after addition of ouabain (concentration in bath $4 \cdot 10^{-5}$ M). c Following rinsing out of ouabain. d Temperature jumps, final temperature in C. Calibration values: Vertical 10 mV, horizontal 1 min

Similar results were obtained following inhibition with strophantidin and 2,4-dinitrophenol. If a connection exists between generation of the cold generator potential and Na^+—K^+-transport, the former ought to break down whenever transport is interrupted experimentally; this is achieved a) by replacing Na^+ ions inside the cell by Li^+ so that no more Na^+ is transported to the outside of the cell, b) by reducing K^+ on the outside of the cell.

How can a temperature-induced disturbance of the active transport mechanism lead to an alteration of membrane potential and thus control the discharge frequency of the cell?

The maintenance of concentration differences for K^+ and Na^+ in an excitable membrane requires the consumption of energy (ion pump). It is assumed that Na^+ transport from the inside to the outside of the cell is associated with the uptake of K^+. If the pump functions as a cation exchanger, Na^+ efflux and K^+ influx are coupled in the ratio of 1:1, so that the membrane potential is not affected by the transport process. In this case temperature changes should not precipitate any membrane effect (electroneutral Na^+—K^+ pump). However, if in this exchange process more Na^+ ions are brought to the outside than K^+ ions to the inside, then the activity of the pump must have an effect on the membrane potential (= electrogenic Na^+—K^+ pump, Baker, 1966). Inhibition of the pump (by cooling or by ouabain) should bring about a lowering of the membrane potential. This assumption is to a large extent confirmed by the experimental results.

After blocking the active transport system a new temperature dependence emerges which cannot be explained by proportionality with the absolute temperature (Fig. 5): on cooling the membrane potential increases slowly whereas it decreases on warming. The present author has not carried out experiments to elucidate this problem. Recent observations on neurones from the abdominal ganglia of *Aplysia* permit certain conclusions to be drawn: they, too, possess an ouabain sensitive portion of the membrane potential. Marchiafava (1970) was able to demonstrate that the temperature-induced changes in membrane potential following ouabain inhibition are due to the different temperature coefficients of permeability of Na^+ and K^+ ions.

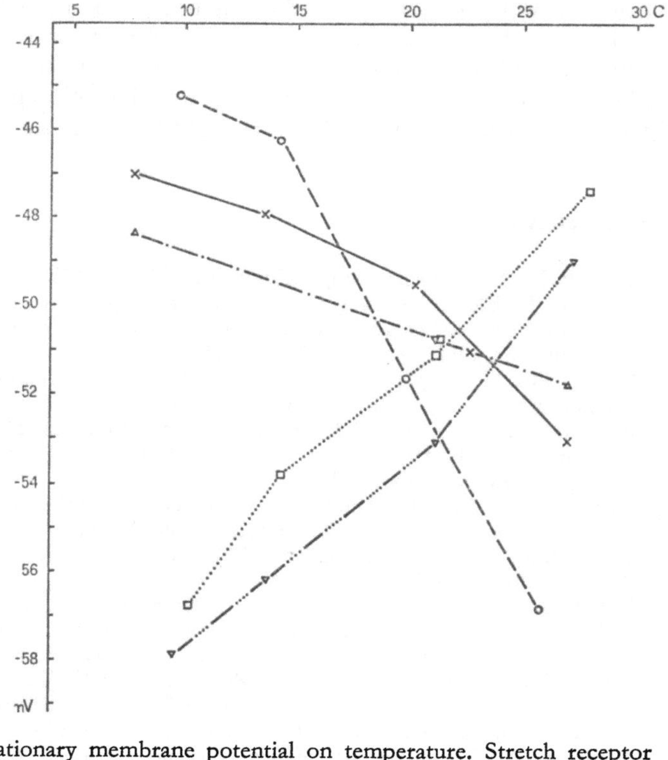

Fig. 5. Dependence of stationary membrane potential on temperature. Stretch receptor neurone, *Astacus fluviatilis*. Abscissa: Temperature in C. Ordinate: stationary membrane potential in mV. △–.–△ Calculated membrane potential proportional to absolute temperature; ○---○ Maximum initial displacement of membrane potential (MP) as function of temperature reached after 1 min; ×--× stationary MP as function of end temperature; □...□ Stationary MP following ouabain (concentration in bath $4 \cdot 10^{-5}$ M) as function of end temperature; △–...–△ Stationary MP as function of end temperature following replacement of Na in the bath solution by Li^+

The neurone membrane becomes more permeable to Na^+ ions in comparison to K^+ ions if temperature rises, and the inside of the cell becomes less negative.

In explaining the paradoxical course of the discharge frequency of stretch receptor neurones at least three mechanisms can be assumed to influence the membrane potential: 1. dependence of the membrane potential on the absolute temperature (Goldman equation). 2. Inhibition and activation of the electrogenic Na^+ pump by temperature changes. 3. Differing membrane permeabilities to Na^+ and K^+ following cooling and warming.

As already mentioned, the effect of temperature on any kind of receptor could lead to false information about peripheral stimuli. Unfortunately, the question as to compensatory reactions in the CNS is largely unanswered. The following kind of compensatory reaction would be conceivable in the abdominal stretch receptors of the crayfish: the neurone could be influenced by inhibitory accessory nerve fibres arising from the ventral cord, and if the activity of these nerve fibres exhibited the same temperature dependence as was found by Kerkut and Taylor (1956) for spontaneous ganglion activity (Fig. 3F) the temperature effect on the receptor cells could be abolished. However, this has not yet been confirmed experimentally. If ventral ganglia with an intact connection with the receptor organ are cooled, no significant difference from the isolated preparation can be found. So far, electrophysiologists have largely neglected the question of long-term adaptation to acclimation temperatures, a point which might be of special interest here.

References

Baker, P. F.: Die Natriumpumpe. Endeavour **25**, 166–173 (1966).

Bernstein, J.: Elektrobiologie, S. 212. Braunschweig: Vieweg und Sohn 1912.

Burkhardt, D.: Die Erregungsvorgänge sensibler Ganglienzellen in Abhängigkeit von der Temperatur. Biol. Zentr. **78**, 22–62 (1959).

Dalton, J. C., Hendrix, D. E.: Effects of Temperature on Membrane Potentials of Lobster Giant Axon. Am. J. Physiol. **202**, 491–494 (1962).

Guttman, R.: The Effect of Temperature on the Function of Excitable Membranes. In: Adelman, W. J. (Hrsg.): Biophysics and physiology of excitable membranes. Van Nostrand Reinhold Comp. 1971.

Hodgkin, A. L., Katz, B.: The Effect of Temperature on the Electrical Activity of the Giant Axon of the Squid. J. Physiol. **109**, 240–249 (1949).

Inman, D. R., Peruzzi, P.: The Effects of Temperature on the Responses of Pacinian Corpuscles. J. Physiol. **155**, 280–301 (1961).

Kerkut, G. A., Taylor, B. J. R.: Effect of Temperature on the Spontaneous Activity from the Isolated Ganglia of the Slug, Cockroach and Crayfish. Nature **178**, 426 (1956).

Machne, X., Orozco, R.: Effects of Temperature and Calcium Variation on Membrane Potentials of Crayfish Axons. Am. J. Physiol. **214**, 672–676 (1968).

Marchiafava, P. I.: The Effect of Temperature Change on Membrane Potential and Conductance in *Aplysia* Giant Nerve Cell. Comp. Biochem. Physiol. **34**, 847–852 (1970).

Muralt, v. A.: Neue Ergebnisse der Nervenphysiologie. Berlin-Göttingen-Heidelberg: Springer 1958.

Murray, R. W.: Temperature Receptors in Animals. Symp. Soc. Exptl. Biol. **16**, 245–266 (1962).

Ottoson, D.: The Effects of Temperature on the Isolated Muscle Spindle. J. Physiol. **180**, 636–648 (1965).

Sand, A.: The Function of the Ampullae of Lorenzini, with some Observations on the Effect of Temperature on Sensory Rhythms. Proc. Roy. Soc. (London), Ser. B **125**, 524–553 (1938).

Senft, J. P.: Effects of Some Inhibitors on the Temperature-Dependent Component of Resting Potential in Lobster Axon. J.Gen. Physiol. **50**, 1834–1848 (1967).

Winter, Ch.: Untersuchungen über die Temperaturabhängigkeit der Erregung von Mechanorezeptoren. Verhandl. Deut. Zool. Ges., Göttingen **30**, 234–239 (1967).

Winter, Ch.: Untersuchungen über den Einfluß der Temperatur auf die Erregungsvorgänge eines Mechanoreceptors. Habilitationsschrift, Regensburg 1972.

Winter, Ch., Fleissner, G.: Thermoelastic Tension Changes in the Muscles of the Abdominal Stretch Receptors of the Crayfish. Comp. Biochem. Physiol. **40 A**, 523–533 (1971).

The Regulation of Cellular Function by Temperature-induced Alterations in Membrane Composition

Jeffrey R. Hazel

With 4 Figures

A most pervasive and widespread effect of environmental temperature upon the physiology and biochemistry of ectothermic organisms is reflected in the lipid composition of their membrane systems. It is the intention of this review (1) to briefly summarize the evidence for temperature-induced alterations in membrane lipid composition, (2) to discuss the implications of such alterations in terms of the regulation of membrane function, both barrier and catalytic, and (3) to discuss the regulatory aspects of the lipid biosynthetic pathway which may account for the temperature sensitivity of lipid and membrane composition.

Temperature-induced Alterations in Membrane Composition

A large body of evidence suggests that in the membrane systems of ectothermic organisms the degree of fatty acid unsaturation is in some manner controlled by the environmental temperature. For crustacean plankton the iodine number varied inversely with water temperature, such that the melting point of membrane lipid was maintained just below the water temperature; colder temperatures resulted in the accumulation of tetra-, penta-, and hexaenoic acids of the C_{20} and C_{22} fatty acids (Farkas and Herodek, 1964). *Chlorella sorokiniana* grown at seven temperatures between 14 C and 38 C exhibited maximal unsaturation of membrane lipids at 22 C and a linear decrease in unsaturation with either an increase or decrease in culture temperature; the predominant acids at 14 C were diunsaturated (47.2%), at 22 C were triunsaturated (30.2%) and at 38 C were totally saturated (Patterson, 1970). Similarly, the proportion of unsaturated fatty acids in *Pseudomonas fluorescens* did not vary with the growth cycle, but was markedly increased by a reduction in culture temperature (Cullen *et al.*, 1971).

Liver fatty acids from cold acclimated frogs exhibited a greater degree of unsaturation due to increased proportions of polyenoic acids at low temperatures; this was referable to decreased levels of palmitate (16:0), stearate (18:0), and oleate (18:1) and relatively higher levels of linoleic (18:2), arachidonic (20:4) and docosatetraenoic (22:4) acids in the cold (Baranska and Wlodawer, 1969). Mosquitofish acclimated to 15 C were characterized by a greater degree of unsaturation in total body lipids due to decreased levels of palmitic (16:0) and stearic (18:0) acids and increased levels of docosahexaenoic (22:6), oleic (18:1) and docosapentaenoic (22:5) acids when compared with animals acclimated to 27 C (Knipprath and Mead, 1966).

The most thoroughly studied species in this regard has been the common goldfish, *Carassius auratus*. The iodine value for total body fat increased by 0.51 per degree centigrade drop in acclimation temperature and the cholesterol:phospholipid ratio was found to vary inversely with acclimation temperature (Hoar and Cottle, 1951). Muscle fatty acids exhibited an increased degree of unsaturation following a period of cold acclimation due primarily to increased levels of unsaturation in the C_{16} and C_{18} fatty acids (Knipprath and Mead, 1967). The fatty acids of goldfish brain also demonstrated an overall tendency for a greater degree of unsaturation at lower acclimation temperatures due to increased levels of docosapenta- and hexaenoic (22:5,6) and arachidonic (20:4) acids (Johnston and Roots, 1964). Analysis of phospholipids from goldfish brain indicated no significant changes in the relative phosphatide composition following cold acclimation, however the major phosphatides, phosphatidyl ethanolamine and lecithin, did become relatively more unsaturated at lower temperatures due to decreased levels of palmitic acid and increased levels of palmitoleic (16:1) acid (Roots, 1968). In addition, the plasmalogen content of brain tissue from goldfish acclimated to 30 C was significantly greater than that from fish acclimated to 5 C; lower acclimation temperatures mediated a greater degree of unsaturation in phosphatidyl ethanolamine plasmalogens due to an increase in oleyl aldehyde relative to stearaldehyde (Roots and Johnston, 1969). Similar patterns in fatty acid composition have been reported for goldfish intestinal lipids; the quantities of $C_{20:4}$ and $C_{22:6}$ present primarily in the phospholipid fraction were halved by a 20 C rise in acclimation temperature, whereas the quantity of $C_{18:0}$ increased 3-fold over the same temperature range (Kemp and Smith, 1970).

Studies of this type have been extended to the organelle level by comparing the lipid composition of mitochondria isolated from the gill tissue of goldfish acclimated to 10 C and 30 C; fatty acid unsaturation increased in the cold due to elevated levels of monounsaturated fatty acids and arachidonic, eicosapentaenoic (20:5) and docosapentaenoic (22:5) acids (Caldwell and Vernberg, 1970).

Such correlations between ambient temperature and membrane composition are not limited to acclimation studies. Brain fatty acids of the Antarctic fish *Trematomus* contain only trace amounts of C_{12} and C_{14} acids, whereas the C_{24}-polyunsaturated fatty acids comprise nearly 7% of the total (Morris and Schneider, 1969). In comparing the fatty acid composition of meso- and thermophilic fungi, oleic acid (18:1) was the most abundant fatty acid in thermophiles as opposed to linoleic acid in mesophiles; thermophiles contained between 0.56–1.01 double bonds per mole of fatty acid, whereas mesophiles contained between 0.96–1.60 (Mumma *et al.*, 1970). A comparison of the fatty acid composition of rat liver mitochondria with that of catfish liver mitochondria revealed a 15% increase in the degree of unsaturation in the mitochondrial fatty acids from fish; this was reflected in the linolenate (18:2)/linolenate (18:3) ratios which were 0.26:1 and 10:1 respectively for fish and rat liver mitochondria (Richardson and Tappel, 1962).

In summary, either cold acclimation or habitation of a chronically cold environment is invariably associated with an increased degree of unsaturation of the fatty acid moieties of phospholipids from a wide variety of species at all levels of biological organization; although the fatty acids involved demonstrate considerable specificity with respect to species and tissue, the overall trend is unambiguous. This environmentally induced alteration in fatty acid unsaturation is often accomplished by the conversion of saturated fatty acids characteristic of warmer temperatures to the corresponding mono- and dienoic acids, and frequently by elongation and desaturation resulting in an increased level of long chain polyunsaturated fatty acids in the cold. In contrast, the effects of temperature upon the relative composition of the various phosphatides are much less pronounced (Cullen *et al.*, 1971; Caldwell and Vernberg, 1970; Roots, 1968).

The Effects of Temperature-induced Alterations in Membrane Composition upon Membrane Function

Permeability Properties

The assumption has been continually made that temperature-induced alterations in membrane composition result in stabilized (temperature-independent) permeability and barrier properties of membranes largely by maintaining a specific liquid-crystalline phase in lipid-water or lipid-protein systems (Johnston and Roots, 1964; Caldwell and Vernberg, 1970), or in other words, by eliminating critical phase transitions at low environmental temperatures in cold acclimated organisms. This implies that for a given set of membrane constituents there exists an optimal temperature range for biological function within which suitable molecular interactions necessary for membrane stability and function may occur. Since in most cases variations in temperature do not significantly alter the distribution of phospholipid classes but instead alter only the hydrocarbon composition, it is the physical properties of these hydrocarbon chains which are particularly relevant. The selective advantage of such compensatory changes in lipid composition is thus believed referable to the fact that lipids laid down in the cold are characterized by lower melting points, greater areas per molecule in a monomolecular film, and greater mobility at a given temperature than their saturated analogues; incorporation of such lipids into membrane systems during cold acclimation preserves the expanded character of monolayer films and thus offsets the increased molecular cohesion resulting from reduced kinetic energy at low temperatures, thereby presumably maintaining membrane permeability and fluidity at least partially independent of environmental temperature fluctuations (Van Deenen, 1969). Most of the evidence supporting this hypothesis is circumstantial and has been derived from the permeability properties of liposomes of well-defined and relatively simple chemical composition.

That the melting point of a particular fatty acid of specified chain length is inversely proportional to the degree of unsaturation is a firmly established fact. The introduction of a *cis*-double bond into stearic acid (18:0) resulting in the formation of oleic acid (18:1) drastically decreases the melting point (from 69.6 to —9 C) and almost halves the chain interaction energies; this effect is greatest when the unsaturation is near the middle of the fatty acid chain (Cullen *et al.*, 1971). In mixtures of palmitic and linoleic or linolenic acids, an increase of only 5% in the degree of unsaturation reduces the melting point of the mixture by as much as 15 C (Raison *et al.*, 1971a). Therefore, small increases in the degree of unsaturation in membrane lipids of cold acclimated animals could abolish temperature-induced phase changes (solidification) at reduced environmental temperatures.

That thermotropic phase changes do actually occur in biological membranes has been clearly demonstrated by differential scanning calorimetry measurements of intact *Mycoplasma laidlawii* membranes of closely controlled lipid composition; in addition, the critical temperature at which the transition from a liquid crystalline to a crystalline state occurred was observed to decrease with increasing membrane unsaturation (Steim *et al.*, 1969). Such temperature-induced phase changes are due in part to the direct effect of temperature upon hydrocarbon chain mobility and hence membrane fluidity. Electron spin resonance studies of *Mycoplasma* membranes from oleate and stearate enriched cells have shown that spin label mobility and membrane fluidity increase markedly with increasing temperature; at all temperatures the membranes of oleate enriched cells were more fluid and characterized by a greater spin label mobility than those of stearate enriched cells (Tourtellote *et al.*, 1970). In addition, stearate enriched cells exhibited a thermotropic phase transition at 40 C which was accompanied by a reduction in membrane fluidity, whereas such a transition was not observed in oleate enriched cells at temperatures above 20 C. Thus incorporation of unsaturated fatty acids into membranes at low environmental temp ratures extends the range over which membrane fluidity (the liquid crystalline state) may be maintained to lower ambient temperatures due to the greater molecular mobility at any given temperature of unsaturated fatty acids compared with their saturated counterparts.

Force-area curves for the extractable lipid of *E. coli* grown at 20 C, 30 C and 40 C as well as for phosphatidyl ethanolamine extracted from *P. fluorescens* grown at 22 C and 5 C illustrated that at all film pressures the mean surface area per phospholipid molecule was lower at high growth temperatures (Haest *et al.*, 1969; Cullen *et al.*, 1971), indicating that thermal acclimation does indeed preserve the expanded character of monolayer films, thus tending to counteract enhanced molecular cohesion at low temperatures. Perhaps of greater importance, X-ray diffraction studies of phosphatidyl ethanolamine from *P. fluorescens* indicated that an organism grown at 22 C produced lipids that exhibited a specific phase behavior at 22 C, and when the organism was grown at 5 C it produced lipids that exhibited identical behavior at 5 C; an acute rise in temperature resulted in a gradual conversion from the lamellar to the hexagonal phase in each case. Although it is questionable whether such a mixed-phase system is present in biological membranes, there is apparently some control in the organism which dictates that the fatty acids produced at 22 C or 5 C when incorporated into phospholipids have characteristic and identical physical properties at the ambient temperatures at which they normally function (Cullen *et al.*, 1971).

Whatever the biochemical rationale for the increase in lipid unsaturation with decreased environmental temperature, such changes in the physiochemical properties of the membrane are reflected in membrane functions. The most obvious effect is upon membrane permeability. Liposomes prepared from phospholipids extracted from *E. coli* grown at 20 C were more sensitive to osmotic swelling (more permeable) at any given temperature than those prepared from cultures grown at higher temperatures; in both cases permeability was markedly temperature sensitive (Haest *et al.*, 1969). In addition, the increased osmotic fragility of the erythrocytes of heat-exposed hamsters was correlated with a lower proportion of linolenate in membrane phospholipids (Kuiper *et al.*, 1971). When the conductance of yeast cell *(S. cerevisiae)* suspensions was measured in 0.2 *M* mannitol, initially it increased linearly with temperature, but at a certain critical temperature there was a dramatic increase in conductance, presumably due to a drastic alteration in membrane structure permitting the passage of massive intracellular electrolytes; this transition occurred at 50 C for cells grown at 10 C, and at 60 C for cells grown at 40 C (Chang and Matson, 1972). Thus it is clear that temperature-induced alterations in membrane composition do influence the permeability and barrier properties of a cell membrane in such a way as to maintain these parameters at least partially independent of environmental temperature fluctuations. Based on the observation that the active uptake of alanine into intestinal mucosal cells of goldfish did not depend on the acclimation temperature, whereas the net transport of alanine across the cell did, it was postulated that the passive efflux of amino acids from the mucosal cells may be selectively regulated by the temperature-induced changes in membrane permeability (Smith and Kemp, 1971); this concept is supported by the fact that changes in the docosahexaenoic/stearic acid ratio exhibit a similar time course during acclimation from 15 C to 30 C as does amino acid transport.

Catalytic Functions

In addition to determining the barrier and permeability properties of a cell, membrane systems also subserve a catalytic function by providing an appropriate matrix for many membrane-bound enzyme systems; for example it has been demonstrated that the enzymes of the electron transport system require the lipid component of the membrane for activity (Fleischer *et al.*, 1962). In addition many such enzymes exhibit a partial or obligate requirement for unsaturated fatty acids (Rothfield and Finkelstein, 1968; Endo and Rothfield, 1969; Sekuzu *et al.*, 1963; Cunningham and Hager, 1971). The effect of membrane fatty acid composition upon mitochondrial function has been clearly demonstrated in unsaturated fatty acid auxotrophs of yeast in which the degree of unsaturation of mitochondrial lipids can be controlled by regulating the diet: the *P*/0 ratios of mitochondria containing high levels of unsaturated fatty acids were in the normal range for yeast mitochondria and respiratory control ratios were high, whereas in mitochondria with low levels of unsaturated fatty acids, the *P*/0 ratios approached zero and respiratory control was absent (Proudlock *et al.*, 1968). In liver mitochondria from essential fatty acid deficient rats, the frequency of oscillation in mitochondrial volumes associated with the energized transport of ions across the inner membrane was decreased, presumably related to the loss of fatty acids of the linoleate series (Williams *et al.*, 1972). The effect of membrane composition upon membrane-bound enzyme activity has been eloquently illustrated for the mechanistically and genetically distinct transport systems for glucose and galactose in fatty acid auxotrophs of *E. coli* (Wilson and Fox, 1971). When such mutants were grown on oleate medium, both transport systems exhibited a distinct break in their Arrhenius plots at 13–14 C, whereas cells grown on linoleate medium were characterized by a break at 7–8 C; the activation energies below the breaks were greater than those above. The temperatures at which these discontinuities occurred corresponded approximately to the temperatures of the phase transitions of the respective fatty acid. Since the same transport proteins were present regardless of the fatty acid supplement employed, and since the transition temperatures were correlated with the melting point of the fatty acid supplement, the breaks in catalytic activity were attributed to temperature-induced phase changes in membrane lipid.

Similar discontinuities in Arrhenius plots have been reported for succinate oxidation in endotherms and chilling sensitive plants (Raison *et al.*; Lyons and Raison, 1970b); however, discontinuities have not been detected from a variety of ectotherms (Raison and Lyons, 1971a; Lyons and Raison, 1970a) presumably due to the greater degree of unsaturation of mitochondrial lipids characteristic of ectotherms (Richardson and Tappel, 1962). In active (aroused) ground squirrels, a distinct break in the Arrhenius plot for succinate oxidation was also observed whereas no break could be discovered in hibernating animals (Raison and Lyons, 1971b). Incorporation of spin labels into these mitochondrial preparations indicated that the label undergoes a marked change in mobility at a temperature corresponding to the transition temperature for enzymatic activity (Raison *et al.*, 1971a) strongly implying that such discontinuities in the Arrhenius plots are due to thermotropic phase changes in membrane lipid. Further evidence to support this contention has been obtained by freeze fracture electron microscopy of *A. laidlawii* membranes above and below the transition temperature (Verkleij *et al.*, 1972); above the transition temperature the fracture faces appeared to be smooth, whereas below this temperature they exhibited distinct banding patterns characteristic of purified phospholipids in the gel phase.

Temperature-induced alterations in membrane composition during cold acclimation of ectothermic organisms may be aimed at maintaining a specific membrane fluidity at the temperature at which the animal must function, thus preventing the transition to the gel state. Freeze fracture electron microscopy studies have indicated that in such a transition to the gel state, membrane protein may actually be extruded from the lipid bilayer; this would explain the drastic decline in catalytic activity of membrane-bound enzymes in endotherms at low temperatures (Verkleij *et al.*, 1972). The above data clearly indicate that the nature of the lipid milieu in the immediate microenvironment of a membrane-bound enzyme may markedly influence its catalytic activity. It is tempting to speculate that environmentally-induced alterations in membrane structure not only maintain an appropriate membrane fluidity, but may in addition also modulate the activity of membrane-bound enzymes during thermal acclimation. Many enzymes do exhibit compensatory changes in specific activity following thermal acclimation (see Hazel and Prosser, 1970 for a review) and it is possible that in the case of membrane-bound enzymes this may be due to lipid modulation of activity (efficiency) rather than, or in addition to, quantitative changes in the amount of a given enzyme. This hypothesis is substantiated by the finding that the decreased Na/K-ATPase activity from the intestinal mucosa of warm acclimated goldfish is not accompanied by a decreased number of pump sites as measured by ouabain binding (Smith and Ellory, 1971) and the fact that enhanced respiratory activity in gill mitochondria of cold acclimated goldfish is not accompanied by changes in the cytochrome or cytochrome oxidase concentration

(Caldwell, 1969). This evidence strongly implies that the activity rather than the quantity of these enzymes is being regulated during the thermal acclimation process.

The validity of this hypothesis has recently been strengthened by studies of succinic dehydrogenase (SDH) in the epaxial muscle of goldfish (Hazel, 1972b). As indicated in Fig. 1, when SDH was extracted from the inner mitochondrial membrane of goldfish muscle mitochondria and prepared in a soluble, lipid-free form, the specific activity could be increased by recombination with soluble lipid preparations. Re-

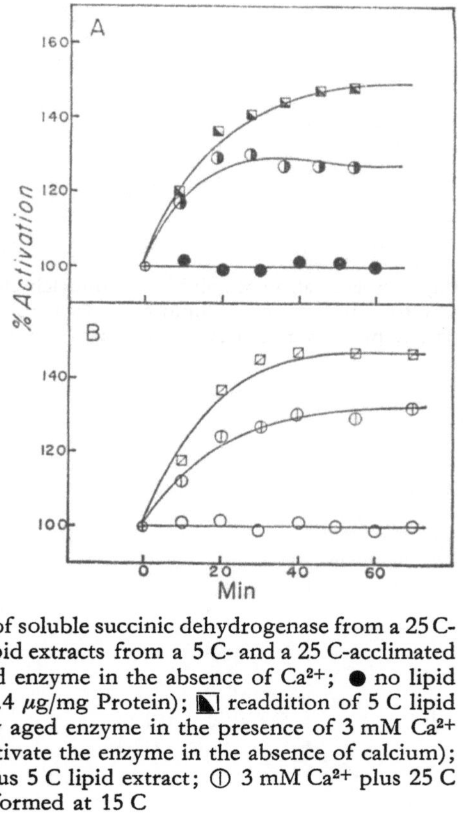

Fig. 1 A and B. The time course of reactivation of soluble succinic dehydrogenase from a 25 C-acclimated goldfish by total mitochondrial lipid extracts from a 5 C- and a 25 C-acclimated goldfish. A Reactivation with freshly prepared enzyme in the absence of Ca^{2+}; ● no lipid added; ◑ readdition of 25 C lipid extract (15.4 μg/mg Protein); ◪ readdition of 5 C lipid extract (18 μg/mg protein). B Reactivation by aged enzyme in the presence of 3 mM Ca^{2+} (aged preparations had lost the ability to reactivate the enzyme in the absence of calcium); ○ 3 mM Ca^{2+} and no lipid; □ 3 mM Ca^{2+} plus 5 C lipid extract; ⦶ 3 mM Ca^{2+} plus 25 C lipid extract. Incubation and assays were performed at 15 C

gardless of the source of the protein (warm or cold acclimated) a total mitochondrial lipid extract from 5 C-acclimated goldfish was more effective (in terms of restoring V_{max}) in reactivating the enzyme than a comparable extract from 25 C-acclimated goldfish (Fig. 2); levels of reactivation were 142 $\pm$ 3.3 and 128 $\pm$ 2.3 per cent respectively. Studies with highly purified phospholipids indicated that the affinity of the enzyme for lipid was largely dependent upon the acidity of the phosphatide species, whereas the magnitude of the reactivation was dependent primarily upon the degree of unsaturation of the fatty acyl residues (Fig. 3); the observed reactivating effectiveness in terms of the concentration required for one-half maximal reactivation according to phosphatide species was PI $>$ CI $>$ PS $>$ PE $>$ PC, whereas for a

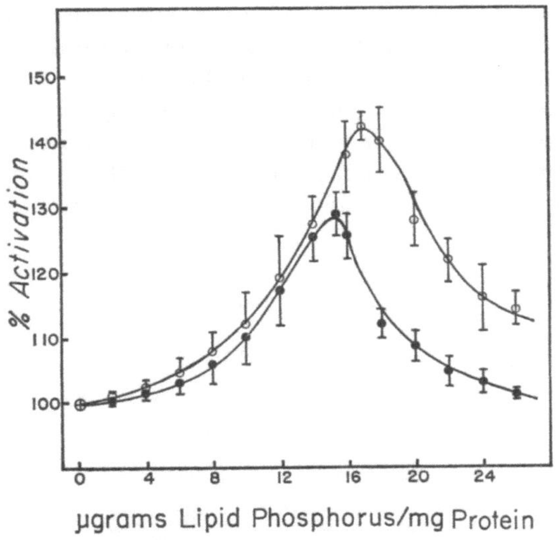

Fig. 2. Reactivation of soluble succinic dehydrogenase by total mitochondrial lipid extracts. The total mitochondrial lipid extracts were from goldfish acclimated to 5 C (○) and 25 C (●). Represented as mean ± standard error for four experiments

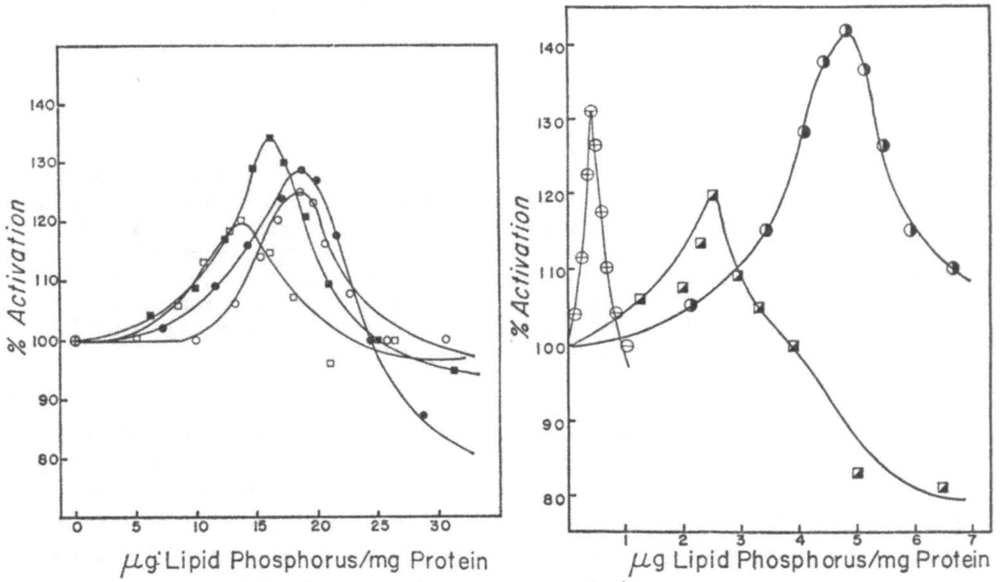

Fig. 3. Right: Reactivation of soluble succinic dehydrogenase by acidic phospholipids. ⊖ phosphatidyl inositol (P_I) and 5 C protein; ◪ cardiolipin (Cl) and 5 C protein; ◑ phosphatidyl serine (PS) and 25 C protein. Left: Reactivation of soluble succinic dehydrogenase by neutral phospholipids. □ Phosphatidyl ethanolamine (PE); ■ dioleyl lecithin; ● egg lecithin; ○ dipalmitoyl lecithin. Soluble enzyme was from 25 C-acclimated goldfish

series of synthetic lecithins the reactivating effectiveness in terms of restoration of V_{max} was dioleyl lecithin > egg lecithin > dipalmitoyl lecithin (it is interesting to note that the most effective acidic phospholipid tested was bovine brain phosphatidyl serine which is highly unsaturated and contains as much as 60% $C_{22:6}$ in the β-position). The higher lipid concentrations required for optimal reactivation with the total mitochondrial lipid extracts as compared with acidic phospholipids probably reflects the predominance of lecithin and phosphatidyl ethanolamine in these preparations; the concentrations required for optimal reactivation with total mitochondrial lipid extracts corresponded closely to the lipid/protein ratio of the intact mitochondrion. Lipid extraction of the particulate enzyme lead to a 3- to 4-fold drop in specific activity, a reduction in the absolute value of K_m for succinate, and a reduction in the slope of the K_m-temperature plot (Hazel, 1972a); as indicated in Fig. 4, readdition of lipid and subsequent reactivation was accompanied by an increase in both V_{max} and K_m and a restoration of the K_m-temperature characteristics of the particulate enzyme. This is the strongest evidence to suggest that SDH functions *in vivo* as a lipoprotein complex. These results were interpreted as indicating that the increased specific activity of SDH following a period of cold-acclimation may be partially or entirely explained by underlying changes in membrane composition which regulate the activity rather than the quantity of this enzyme. It has been suggested that the fluidity of the fatty acyl chains may provide the required motional freedom, allowing enzymes in these membranes to realize conformational changes and movements associated with their activity; such influences of membrane fluidity on enzyme function have been termed "viscotropic" (Kimelberg and Papahadjopoulos, 1972).

Regulatory Aspects of Lipid Biosynthesis in Ectotherms

Although there have been a considerable number of investigations documenting the nature of temperature-induced changes in membrane lipid composition and enhanced lipid biosynthesis in the cold, there have been no studies to date of the enzymes of fatty acid metabolism or phospholipid metabolism with respect to temperature acclimation. Some investigations have been initiated in microorganisms where the variety of fatty acids present and the nature of the temperature-induced changes is much simpler than in ectothermic animals. *E. coli* adjusts lipid composition to the growth temperature, but does not convert saturated to unsaturated fatty acids, suggesting that the temperature controlled step may be a component of the fatty acid incorporation system. In *E. coli* synthesizing phospholipids almost entirely from exogenous fatty acids (which allows one to ignore possible temperature control effected by the relative rates of saturated and unsaturated fatty acid biosynthesis) the lysophosphatidic acid produced by the action of Acyl CoA:glycerol-3-phosphate transacylase exhibited a rise in the palmitate to palmitoleate ratio as the assay temperature was increased (Sinensky, 1971) in a manner that mimicked the *in vivo* temperature control; the second acylation step was found to be similarly temperature-sensitive. Thus in this case, the substrate specificities of the acyltransferase enzymes appeared to be temperature sensitive; such that acute drops in environ-

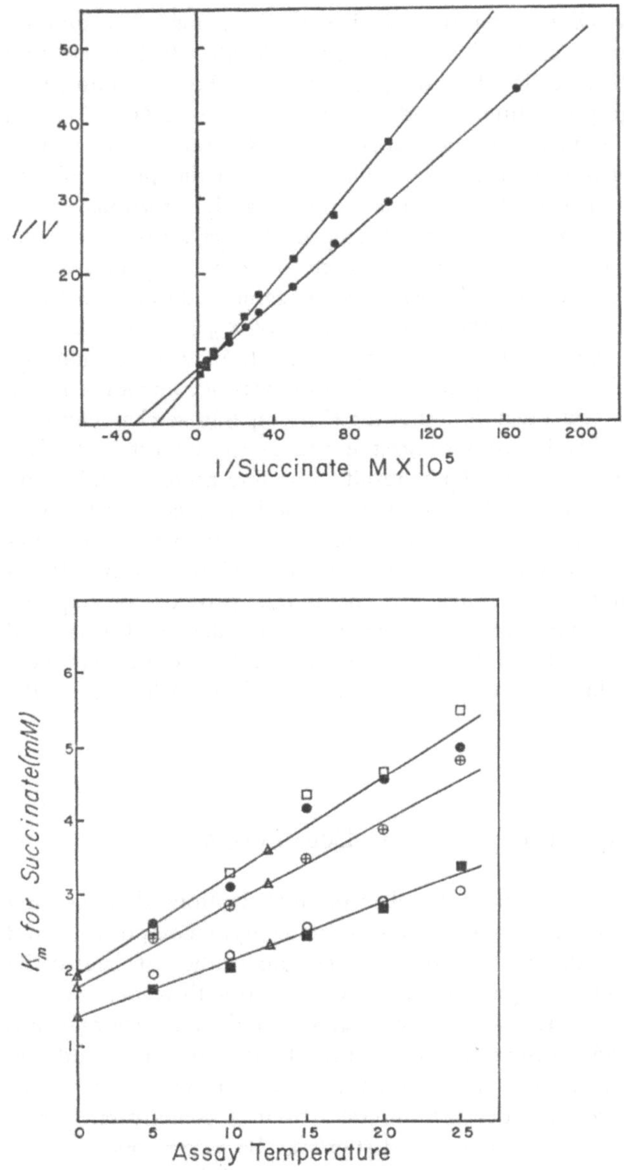

Fig. 4. Top: The effect of extraction and readdition of lipid upon the succinate kinetics of succinic dehydrogenase. Lineweaver-Burk plots for the soluble enzyme from 5 C-acclimated goldfish (●) and the same enzyme preparation reactivated with optimal concentrations of a total mitochondrial lipid extract from 5 C-acclimated foldfish (■). Assay temperature is 25 C. Bottom: The effect of assay temperature upon the K_m for succinate for particulate, soluble, and reactivated succinic dehydrogenase. Particulate enzyme: ● 5 C-acclimated goldfish; □ 25 C-acclimated goldfish. Soluble enzyme: ○ 5 C-acclimated goldfish; ■ 25 C-acclimated goldfish. 5 C soluble protein reactivated with a total mitochondrial lipid extract from 5 C-acclimated goldfish, ⊕; △ points on the regression line

mental temperature would favor incorporation of unsaturated fatty acids into membrane phospholipids. In contrast to this study, *E. coli* that are incorporating endogenous fatty acids into membrane lipid appear to synthesize greater quantities of unsaturated fatty acids in the cold; that is to say that under these conditions environmental temperature influences the relative rates of unsaturated and saturated fatty acid biosynthesis and thus appears to control the type of fatty acid available for incorporation into membrane lipid (Bright-Gaertner and Proulx, 1972). In *B. megaterium* temperature modulation of lipid composition is apparently accomplished by a conversion from saturated to unsaturated fatty acids, for a specific desaturase enzyme is induced at temperatures below 20 C which is totally absent at higher culture temperatures (Fulco, 1970; 1972). Bacilli are unlike *E. coli* but similar to higher organisms in that they produce unsaturated fatty acids directly from saturated analogues by the oxygen dependent removal of two hydrogen atoms. However, information regarding temperature control of lipid metabolism in ectothermic animals is lacking.

References

Baranska, J., Wlodawer, P.: Influence of temperature on the composition of fatty acids and on lipogenesis in frog tissues. Comp. Biochem. Physiol. **28**, 553–570 (1969).

Bright-Gaertner, E., Proulx, P.: Metabolism of phosphoglycerides in *Escherichia coli* during growth at 37 C and during a cold-induced lag phase. Biochem. Biophys. Acta. **27**, 40–49 (1972).

Caldwell, R. S.: Thermal compensation of respiratory enzymes in tissues of the goldfish (*Carassius auratus* L.). Comp. Biochem. Physiol. 31, 79–90 (1969).

Caldwell, R. S., Vernberg, J. F.: The influence of acclimation temperature on the lipid composition of fish gill mitochondria. Comp. Biochem. Physiol. **34**, 179–191 (1970).

Chang, S. B., Matson, R. S.: Membrane stability (thermal) and nature of fatty acids in yeast. Biochem. Biophys. Res. Commun. **46**, 1529–1535 (1972).

Cullen, J., Phillips, M. C., Shipley, G. G.: The effects of temperature on the composition and physical properties of the lipids of *Pseudomonas fluorescens*. Biochem. J. **125**, 733–742 (1971)

Cunningham, C. C., Hager, L. P.: Crystalline pyruvate oxidase from *Escherichia coli* III. phospholipid as an allosteric effector for the enzyme. J. Biol. Chem. **246**, 1583–1589 (1971).

Endo, A., Rothfield, L.: Studies on a phospholipid requiring bacterial enzyme II. the role of phospholipid in the uridine diphosphate galactose: lipopolysaccharide β-3 galactosyl transferase reaction. Biochem. **8**, 3508–3516 (1969).

Farkas, T., Herodek, S.: The effect of environmental temperature on the fatty acid composition of crustacean plankton. J. Lipid. Res. **5**, 369–373 (1964).

Fleischer, S., Brierley, G., Klouwen, H., Slautterback, D. B.: Studies of the electron transfer system XLVII. The role of phospholipids in electron transfer. J. Biol. Chem. **237**, 3264 bis 3272 (1962).

Fulco, A. J.: Temperature-mediated hyper-induction of fatty acid desaturating enzyme. Biochem. Biophys. Acta. **218**, 558–560 (1970).

Fulco, A. J.: The biosynthesis of unsaturated fatty acids by bacilli IV. temperature mediated control mechanisms. J. Biol. Chem. **247**, 3511–3519 (1972).

Haest, C. W. M., DeGier, J., VanDeenen, L. L. M.: Changes in the chemical and barrier properties of the membrane lipids of *E. coli* by variation of the temperature of growth. Chem. Phys. Lipids. **3**, 413–417 (1969).

Hazel, J.: The effect of temperature acclimation upon succinic dehydrogenase activity from the epaxial muscle of the common goldfish (*Carassius auratus* L.): properties of the enzyme and the effect of lipid extraction. Comp. Biochem. Physiol. **43 B**, 837–861 (1972a).

Hazel, J.: The effect of temperature acclimation upon succinic dehydrogenase activity from the epaxial muscle of the common goldfish (*Carassius auratus* L.): lipid reactivation of the soluble enzyme. Comp. Biochem. Physiol. **43 B**, 863–882 (1972b).

Hazel, J., Prosser, C. L.: Interpretation of inverse acclimation to temperature. Z. Vergleich. Physiol. **67**, 217–228 (1970).

Hoar, W. S., Cottle, M. K.: Some effects of temperature acclimatization on the chemical constitution of goldfish tissues. Can. J. Zool. **30**, 49–54 (1951).

Johnston, P. V., Roots, B. I.: Brain lipid fatty acids and temperature acclimation. Comp. Biochem. Physiol. **11**, 303–309 (1964).

Kemp, P., Smith, M. W.: Effect of temperature acclimatization on the fatty acid composition of goldfish intestinal lipids. Biochem. J. **117**, 9–15 (1970).

Kimelberg, H. K., Papahadjopoulos, D.: Phospholipid requirements for ($Na^+ + K^+$)-ATPase activity: head group specificity and fatty acid fluidity. Biochem. Biophys. Acta **282**, 277–292 (1972).

Knipprath, W. G., Mead, J. F.: Influence of temperature on the fatty acid pattern of mosquitofish (*Gambusia affinis*) and guppies (*Lebistes reticulatus*). Lipids **1**, 113–117 (1966).

Knipprath, W. G., Mead, J. F.: The effect of environmental temperature on the fatty acid composition and on the in vivo incorporation of 1-^{14}C-acetate in goldfish (*Carassius auratus* L.). Lipids **3**, 121–128 (1967).

Kuiper, P. J. C., Livne, A., Meverstein, N.: Changes in lipid composition and osmotic fragility of erythrocytes of hamster induced by heat exposure. Biochem. Biophys. Acta **248**, 300–305 (1971).

Lyons, J. M., Raison, J. K.: A temperature-induced transition in mitochondrial oxidation: contrast between cold and warm-blooded animals. Comp. Biochem. Physiol. **37**, 405–411 (1970a).

Lyons, J. M., Raison, J. K.: Oxidative activity of mitochondria isolated from plant tissues sensitive and resistant to chilling injury. Plant Physiol. **45**, 386–389 (1970b).

Morris, R. W., Schneider, M. J.: Brain fatty acids of an Antarctic fish *Trematomus bernacchii*. Comp. Biochem. Physiol. **28**, 1461–1464 (1969).

Mumma, R. O., Fergus, G. L., Sekura, R. D.: The lipids of thermophilic fungi: lipid composition comparisons between thermophilic and mesophilic fungi. Lipids **5**, 100–103 (1970).

Patterson, G. W.: Effect of culture temperature on fatty acid composition of *Chlorella sorokiniana*. Lipids **5**, 597–600 (1970).

Proudlock, J. W., Haslam, J. M., Linnane, A. W.: Specific effect of unsaturated fatty acid depletion on mitochondrial oxidative phosphorylation in *Saccharomyces cerevisiae*. Biochem. Biophys. Res. Commun. **37**, 847–852 (1969).

Raison, J. K., Lyons, J. M., Melhorn, R. J., Keith, A. D.: Temperature-induced phase changes in mitochondrial membranes detected by spin labelling. J. Biol. Chem. **246**, 4036–4040 (1971a).

Raison, J. K., Lyons, J. M., Thomson, W. W.: The influence of membranes on the temperature-induced changes in the kinetics of some respiratory enzymes of mitochondria. Arch. Biochem. Biophys. **142**, 83–90 (1971b).

Richardson, T., Tappel, A. L.: Swelling of fish mitochondria. J. Cell Biol. **13**, 43–53 (1962).

Roots, B. I.: Phospholipids of goldfish (*Carassius auratus* L.) brain: the influence of environmental temperature. Comp. Biochem. Physiol. **25**, 457–466 (1968).

Roots, B. I., Johnston, P. V.: Plasmalogens of the nervous system and environmental temperature. Comp. Biochem. Physiol. **26**, 553–560 (1969).

Rothfield, L., Finkelstein, A.: Membrane biochemistry. Ann. Rev. Biochem. **37**, 463–496 (1968).

Sekuzu, I., Jurtshuk, D., Green, D. E.: Studies on the electron transfer system LI. – Isolation and characterization of the D-(-)-β-hydroxybutyric apodehydrogenase from beef heart mitochondria. J. Biol. Chem. **238**, 975–982 (1963).

Sinensky, M.: Temperature control of phospholipid biosynthesis in *Escherichia coli*. J. Bacteriol. **106**, 449–455 (1971).

Smith, M. W., Kemp, P.: Parallel temperature-induced changes in membrane fatty acids and in the transport of amino acids by the intestine of goldfish (*Carassius auratus* L.). Comp. Biochem. Physiol. **39**, 357–366 (1971).

Steim, J. M., Tourtellote, M. E., Reinert, J. C., McElhaney, R. N., Rader, R. L.: Calorimetric evidence for the liquid-crystalline state of lipids in a biomembrane. Proc. Natl. Acad. Sci. **63**, 104–109 (1969).

Tourtellotte, M. E., Branton, D., Keith, A.: Membrane structure: spin labeling and freeze etching of *Mycoplasma laidlawii*. Proc. Natl. Acad. Sci. **66**, 909–916 (1970).

Van Deenen, L. L. M.: Membrane lipids and lipophilic proteins. In: Tosteson, D. C. (Ed.): The Molecular Basis of Membrane Function, p. 47–78. Englewood Cliffs, New Jersey: Prentice Hall, Inc. 1969.

Verkleij, A. J., Ververgaert, P. H. J., Van Deenen, L. L. M., Elbers, P. F.: Phase transistions of phospholipid bilayers and membranes of *Acholeplasma laidlawii* visualized by freeze fracturing electron microscopy. Biochem. Biophys. Acta **288**, 326–332 (1972).

Williams, M. A., Stancliff, R. C., Packer, L., Keith, A. D.: Relation of unsaturated fatty acid composition of rat liver mitochondria to oscillation period, spin label motion, permeability and oxidative phosphorylation. Biochem. Biophys. Acta **267**, 444–456 (1972).

Wilson, G., Fox, C. F.: Biogenesis of microbial transport systems: evidence for coupled incorporation of newly synthesized lipids and proteins into membranes. J. Mol. Biol. **55**, 46–60 (1971).

Basic Strategies and Mechanisms of Enzyme Adaptation to Temperature

P. W. HOCHACHKA

With 5 Figures

Strategies of Biochemical Adaptation

Since the advent of evolutionary theory, environmental adaptations have been charted at many levels – at physiological, anatomical, and behavioural levels – of biological organization. Less well understood are "biochemical adaptations" to the environment. In part, this lack of information is an historic outcome of the development of adjacent fields of science and of an appropriate technology. But a more important reason for the hiatus in this field stems from the nature of the subject itself: biochemical adaptations usually are not apparent macroscopically, hence the biologist may be aware neither of their existence, nor of their mechanism. To illustrate what I mean, consider the apparent macroscopic attributes of two hypothetical species of fish – one, a surface species, living at the top of the water column at high temperatures and low pressures; the second, an abyssal species, living at low temperatures and high pressures. In much of their anatomy, physiology, and even behaviour the two species may be much alike. Both display comparable hydrodynamic properties and hence comparable swimming velocities; both display comparable growth, comparable osmoregulatory abilities and so forth. Above the biochemical level we may find little reason to regard one species as, say barophilic and stenothermic, and the other as barophobic and eurythermic. Only when we begin to "dissect" the biochemical machinery of the two species do we learn that the two organisms *can conduct the same basic functions at comparable rates despite widely disparate physical conditions because of the existence of basic biochemical differences between them*. It is indeed the biochemical adaptations to their respective environments which makes the two species so apparently similar in the things they can do. Adjustments which occur at the chemical level, and which can be observed only indirectly, then, can obliterate observable differences at higher levels of biological organization.

Further, as I shall illustrate in this paper, biochemical adaptations can largely be grouped into two categories: those which are "compensatory" and those which are "exploitative". Compensatory adaptations are essentially homeostatic in nature, and can be looked upon as corrective or restorative mechanisms. Thus, the organism responds to environmental parameters such as temperature which adversely affect biochemical reaction rates by mobilizing mechanisms for returning those rates to their previous levels.

Other biochemical changes may give the organism a wholly new potential for making use of its environment or for invading a new environment. We term such changes "exploitative adaptations". In contrast to biochemical adjustments which are compensatory or restorative, exploitative adaptations are, strictly speaking, unnecessary. The species could survive without these new potentials, but with them it may do significantly better in its native habitat and more importantly it may enter a new habitat previously unavailable to it.

How have these two strategies been harnessed by ectotherms during enzymic adaptation to temperature? It is perhaps useful to preface our current formulations of this problem by considering the basic effects of temperature on the biochemistry of ectothermic organisms. As it turns out, most of the fundamental effects of temperature on biochemical systems can be categorized into one of two classes, which recently (Hochachka and Somero, 1973) we have termed (1) Rate Effects and (2) Weak Bond – Structural Effects. We shall consider each of these in turn.

The Effect of Temperature on Biochemical Reaction Rates

To an ectothermic organism, the most obvious functional consequence of changes in external temperature is a change in the average kinetic energy of constituent atoms and molecules; such a change in turn is reflected in altered rates at which chemical reactions occur. We are all familiar with these effects and I am sure that we will see many examples of them during this meeting. For the purposes of this paper, suffice to mention that under saturating conditions of substrate, cofactors, etc., rate effects on enzyme catalyzed reactions depend upon the free energy ot activation ($\Delta G^{\ddagger}$), a term that itself consists of activation enthalpy and activation entropy values. Compensatory strategies of adaptation would be expected to lead to a reduction of $\Delta G^{\ddagger}$ for homologous enzyme reactions, since a low $\Delta G^{\ddagger}$ value for a given reaction indicates (a) a catalyst of high efficiency, and (b) a catalyst which is in consequence less dependent upon temperature. In terms which we are more used to, compensatory strategies of low temperature adaptation would tend to lead to (a) catalysts of high turnover number, and (b) catalysts displaying low Q_{10} values. There are some good examples available of this adaptational strategy. Thus, enzymes from midwater organisms display lower values of activation energy than do homologous enzymes from surface or abyssal species, and this property presumably compensates to some extent for the varying temperatures that are encountered during daily vertical migrations through the water column (Hochachka *et al.*, 1970; Moon *et al.*, 1971; Mustafa *et al.*, 1971). Similarly, enzymes of ectotherms often have higher turnover numbers than homologous enzymes from endotherms. By being more efficient, these catalysts compensate for reduced thermal energy; by having a lower Q_{10} (i.e., reduced thermal dependence), they compensate for the inherent thermal instability of their environment (Table 1 and 2).

In contrast to saturating or ideal conditions, under conditions of *non-saturating substrate levels* enzyme-substrate (E-S) affinity has an important influence upon reaction rates: the lower the substrate concentration the more important E-S affinity is in determining the reaction rate. This effect can be so pronounced that for some enzymes (displaying a direct K_m-temperature relationship) low-temperature rates may be actually higher than high-temperature rates. That is, the enzyme

Table 1. The relative efficiencies of 3 enzymes from endothermic organisms compared to the homologous enzymes from ectothermic species. The more efficient a catalyst is, the higher the turnover number. Turnover numer is defined as μ moles substrate converted to product μ moles enzyme/second. It is evident that by this criterion each of the ectothermic enzymes is a considerably more effective catalyst (particularly at low temperatures) than is its homologue in endotherms. (Data from Hochachka and Somero 1973)

Enzyme/Organism	*Assay temperature*	*Turnover number* [a]
phosphorylase b		
rabbit	0	2.4
lobster	0	13.5
triose-P dehydrogenase		
rabbit	5	14.7
lobster	5	55.0
lactate dehydrogenase		
duck	25	82
tuna	25	160

[a] micromoles substrate converted to product/micromole enzyme/sec

Table 2. Q_{10} and activation energy values for 2 enzymes from several species of fishes. Lowest Q_{10} values (i.e. lowest temperature dependencies) are displayed by enzymes from the midwater species (*Stomias* and the midwater sea bass, *Ectreposebastes*), which encounter large temperature change during daily vertical migrations. The low temperature dependence presumably compensates for the varying external temperature. (Data from Hochachka *et al.*, 1970 and Moon *et al.*, 1971)

Enzyme/Organism	*Thermal range*	*Activation energy*	Q_{10}
FDPase			
trout	5–10	12.0	2.4
lungfish	5–15	11.0	2.0
rattail	3– 9	8.1	1.69
Stomias	3–10	3.8	1.28
Pyruvate Kinase			
trout	3–15	24.2	4.63
sea bass	3–12	9.8	1.87

reaction can display Q_{10} values of 1.0 or even less than 1.0. For other enzymes (displaying an inverse K_m-temperature relationship), Q_{10} values may become very high. For isolated enzyme systems, we have observed Q_{10} values of over 25 when the reaction rate is determined at low (K_m) substrate concentration (see Hochachka and Somero, 1973), and these could rise to over 100 if substrate concentration were dropped by an order of magnitude.

Were it not for an essential fact – *that physiological substrate concentrations are almost always 2–10 fold lower than required to fully saturate enzymes involved in their metabolism* – the role of E-S affinity in controlling reaction rates would be of little biological significance. Given the *in vivo* condition, however, reaction rates become highly sensitive to changes in E-S affinity, and it is not surprising that most regulatory metabolites exert their (+ or —) influence by increasing or decreasing E-S affinity.

The key to enzyme regulation is simply control of E-S affinity, and in regard to temperature adaptation this is an important clue. Why? For the simple reason that under physiological conditions the Q_{10} for any given enzyme reaction can be regulated by changing

1. the absolute E-S affinity,
2. the substrate concentration, or
3. both (Fig. 1).

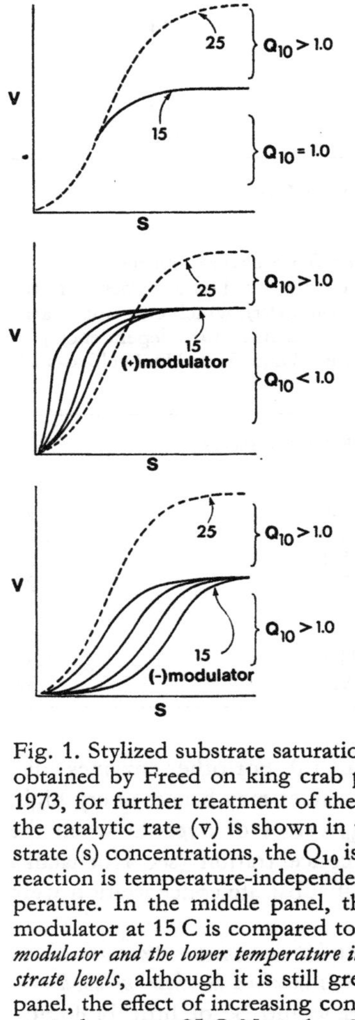

Fig. 1. Stylized substrate saturation curves for an ectothermic enzyme, patterned after data obtained by Freed on king crab phosphofructokinase (see Hochachka and Somero, 1971, 1973, for further treatment of these data). In the upper panel, the effect of temperature on the catalytic rate (v) is shown in the absence of regulatory metabolites. At saturating substrate (s) concentrations, the Q_{10} is greater than 1.0, but at subsaturating substrate levels, the reaction is temperature-independent because of an increase in E-S affinity at the lower temperature. In the middle panel, the effect of increasing concentrations of a positive (+) modulator at 15 C is compared to the control saturation curve at 25 C. *Because both the* (+) *modulator and the lower temperature increase E–S affinity, the Q_{10} can be far less than 1.0 at low substrate levels*, although it is still greater than 1.0 at saturating substrate levels. In the lower panel, the effect of increasing concentrations of a negative (–) modulator is compared to a control curve at 25 C. Now the effect of low temperature (which increases E–S affinity) and the effect of the negative modulator (which decreases E–S affinity) work against each other. Under these conditions, the effective Q_{10} can assume very high values. However, the Q_{10} under saturating substrate concentrations is unchanged

Since E-S affinity is a tightly regulated characteristic, it supplies the organisms with a means of controlling the Q_{10} for any given reaction or reaction pathway. When, as in ectotherms, the E-S affinity is itself dependent upon the thermal range considered, then temperature *per se* contributes to determining the Q_{10} for the reaction. Thus, the design of enzymes displaying a direct K_m-temperature relationship is fundamentally an exploitative strategy of temperature adaptation: it relies upon, or better, takes advantage of, thermal change in order to control the Q_{10} of the reaction, thereby facilitating the organism's survival in its particular thermal regime.

The above statements are tantamount to suggesting that Q_{10} values observed for metabolism of whole animals are not determined by the obvious thermodynamic situation, but by the kinetic properties of enzyme catalysts. A number of interesting corollaries from this premise:

1. Not all metabolic pathways need show the same Q_{10} values.

2. The same metabolic pathway may show different Q_{10} values under different conditions (for e.g. under different substrate levels).

3. The same metabolic pathway may show different Q_{10} values through different regions of the biological temperature range.

We have previously marshalled the experimental data which convincingly demonstrate each of the above corollaries (see Hochachka and Somero, 1971), and our arguments need not be repeated here.

Just as control of E-S affinity is the key to much of enzyme regulation *in vivo*, so is it the key to much of enzymic adaptation to physical parameters like temperature. But what is the basis for the thermal dependence of E-S affinity? To get at this question, we must briefly digress into a consideration of a second major effect of temperature on biochemical systems.

Weak Bond – Structural Effects of Temperature

These effects arise from the action of temperature on biochemical structures and consequently on biochemical functions which are dependent for their fidelity upon "weak bonds" or "weak interactions". As indicated in Table 3, virtually all of the higher orders of biochemical structure are critically if not entirely dependent upon weak bonding interactions. These include the higher levels of protein structure, membrane structure, and nucleic acid structure, as well as most of the biochemical interactions which demand a high degree of stereochemical specificity. The binding of substrate to enzyme is but one such interaction which depends upon *bonding forces only somewhat greater than the thermal energies present in the environment*. In other words, interactions such as E-S interactions are dependent upon bonds, that, as their nam- accurately denotes, are *individually* weak. Normally, their life time is a fleeting one; they are continuously being broken and reformed, and often show critical temperaturee dependence. I will attempt to illustrate the nature of this temperature-dependence by briefly considering the effects of temperature on enzyme-substrate interactions for the enzyme, acetylcholinesterase.

Table 3. A partial list of biological functions in which the underlying structure is stabilized by weak ionic bonds, by hydrogen bonds, by hydrophobic interactions, or by a combination of these

1. enzyme-substrate binding
2. enzyme-cation binding
3. enzyme-cofactor binding
4. enzyme-modulator binding
5. hemoglobin-oxygen binding
6. protein-protein binding
7. protein-lipid binding
8. protein-nucleic acid recognition
9. tRNA-amino acid recognition
10. nucleic acid – nucleic acid interactions
11. hormone-receptor binding
12. chlorophyll-lipid interactions

The Binding Site of Acetylcholinesterase

Acetylcholinesterase (AChE) catalyzes the following reaction:

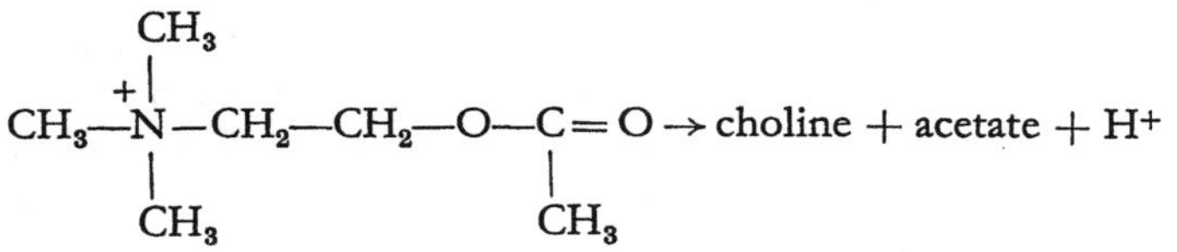

The enzyme has been intensively studied because of its fundamental role in neural transmission and because of the large potential for practical use implicit in any modification of its activity (a large fraction of this so-called practical use, unfortunately, has been directed towards such areas as chemical warfare). As a result, we know a great deal about the binding of its substrate through studies with crude enzyme preparations – more, in fact, that we know about many enzymes which are available as crystalline preparations in relatively large amounts (see Froede and Wilson, 1972).

Given an enzyme such as AChE, how does one proceed to define the mechanism and forces involved in binding substrate? One way is to alter the enzyme so as to alter the binding of substrate – a way that is probably utilized in evolutionary adaptation of the enzyme for function in particular thermal regimes. Another way is to alter the chemical structure of the substrate, then see what happens to enzyme activity or to binding. Experimentally, the latter approach has been by far the most productive, at least to date.

From such studies it has been clear for some time now that the active site of AChE consists of two subsites, an anionic site and an esteratic site (Fig. 2). The first is involved largely *in binding and proper positioning of the substrate, but does not play any role in catalysis.* The esteratic site, on the other hand, is the true catalytic site, the site at which hydrolysis occurs.

As its name implies, the anionic site is the locus of an electrically negative potential which attracts the positive quaternary ammonium head of acetylcholine. A

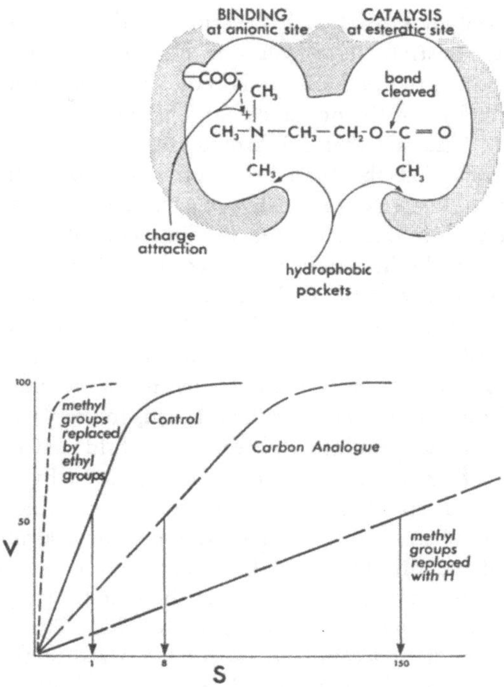

Fig. 2. Upper panel: a diagrammatic representation of the active site of AChE clearly indicating the structural separation of two essential enzyme functions – binding and catalysis. The binding function is largely restricted to the anionic subsite and depends in part on a charge attraction between the quaternary ammonium head and the anionic site on the enzyme, *but mainly upon hydrophobic interactions between the enzyme's binding site and the methyl groups of acetylcholine.* Lower panel: hypothetical "saturation" curves are shown for (1) a control in which the substrate (s) is acetylcholine, (2) a modified substrate which is made more hydrophobic by replacing methyl groups (on N) with ethyl groups, (3) an uncharged carbon analogue of acetylcholine in which the quaternary N is replaced by C, and (4) a hypothetical substrate in which all the methyl groups (on N) are replaced by hydrogen. The curves indicate that, *once bound*, any one of these substrates would be hydrolyzed equally well (i.e. the maximum velocity of the reaction is unchanged); in contrast, the effective E–S affinity is greatly altered by changing the charge properties or the hydrophobicity of the "substrate". See text for further discussion

variety of experimental data support this, but by far the most convincing arises from comparisons of acetylcholine with its carbon analogue, 3,3-dimethylbutyl acetate:

$$\begin{array}{ccccccccc} & CH_3 & & & & & & & \\ & | & & & & & & & \\ H_3C— & C & —CH_2— & CH_2— & O— & C & =O. \\ & | & & & & | & \\ & CH_3 & & & & CH_3 & \end{array}$$

This compound has a shape almost identical to that of acetylcholine, but it has no positive charge. Once the enzyme is saturated by this compound, the maximum catalytic rate (V_{max}) is essentially identical to the V_{max} obtained with acetylcholine as substrate. However, the K_m for the carbon analogue is nearly 10 fold greater; that is, the E-S affinity is reduced by nearly a factor of 10. It is widely accepted that the loss of the positive charge is the cause of this weaker binding.

The fact that 3,3-dimethylbutyl acetate is such a good substrate, however, suggests that the branched methyl groups themselves contribute greatly to binding with the enzyme, an hypothesis that is in fact strongly supported by experimental data. In this case, the binding contributions of the branched methyl groups in acetylcholine can be evaluated from a number of inhibitors including a series obtained by progressively replacing the hydrogen atoms of ammonium ion by methyl groups. Replacement of each H by CH_3 increases binding by a factor of 7, except for the last replacement, which contributes only a factor of 3. As more and more CH_3 groups are added, that is, the more hydrophobic the molecule becomes, the more tightly it is bound to the enzyme. Combined, the replacement of all H groups with CH_3 groups increases binding by nearly 150 fold! AChE affinity for substrate increases even further if the CH_3 groups are replaced by ethyl groups or by longer hydrocarbon chains, further attesting to the critical role of hydrophobic forces in the binding of substrate to this enzyme. From these kinds of studies, it is widely held that most of the affinity of AChE for substrate is accounted for (1) by coulombic and (2) by hydrophobic interactions at the anionic site, and of these, *the sum of the hydrophobic forces is greater at least by several fold.* The question which I now want to raise is this: how would we expect temperature to affect AChE-substrate affinity? If we assume that hydrophobic forces constitute the major contribution to E-S affinity at all temperatures, the question can be effectively re-worded: what is the thermal dependence of hydrophobic interactions?

Low-temperature Lability of Systems Stabilized by Hydrophobic Forces

Much of our current information about the contribution of hydrophobic bonding to the stabilization of various biochemical systems arises from studies of model compounds and of higher levels of protein structure. From such studies the conclusion has emerged that interactions between hydrophobic residues will stabilize the system at all temperatures but considerably more so at high than at low temperatures (see Brandts, 1967, for a rigorous thermodynamic treatment of this problem). In the case of native protein structure, for example, non-polar side chains are surrounded locally by other non-polar groups (Fig. 3). If the overall contribution of these hydrophobic interactions is large enough, the protein may display a cold inactivation or cold lability (Table 4), probably because the preferred mode of accommodation of hydrophobic side-chains in water at low temperatures is by clathrate formation (Fig. 3). The exact physical appearance of these clathrate or "ice-berg" structures is not known, but the tendency towards their formation apparently reflects an effort on the part of water molecules to maximize H bonding in the presence of a non-hydrogen bonding solute. This can be achieved only at a considerable entropy cost, since the hydrophobic residues must be accommodated by an ordered solvation shell of water molecules at low temperatures. These clathrate structures are only marginally stable and are quickly disrupted as the temperature is raised. At elevated temperatures, the native conformational state of the system is the thermodynamically preferred state, and in the case of many cold-labile enzymes and proteins, high temperatures favor a return to native conformation and to normal (catalytic) activity.

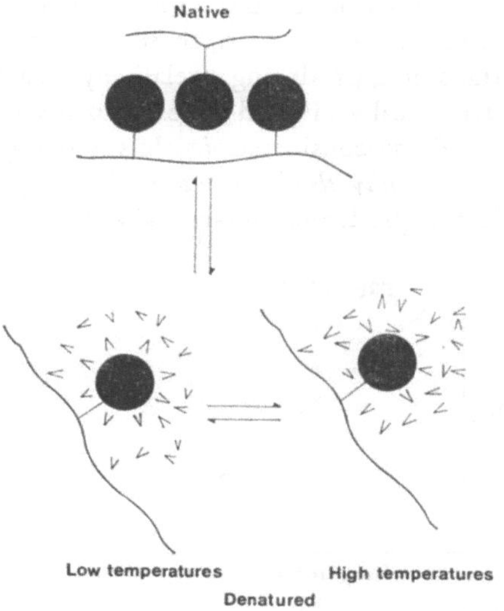

Fig. 3. A diagrammatic representation of hydrophobic bonding in maintenance of native protein structure and of the clathrate melting transition. In the native state (upper middle), quaternary structure is stabilized by non-polar side chains becoming surrounded locally by other non-polar groups (from the same or from a different chain) in such a way as to largely exclude water. In the denatured state, the exposed non-polar groups will be accomodated by an ordered solvation shell of water molecules (represented as V's) at low temperatures (lower left) but by a disordered solvation shell at high temperatures (lower right). In certain cases (Table 4), such proteins may display a cold inactivation or a cold lability. (After Brandts, 1967)

Table 4. A partial list of "cold sensitive" mammalian enzymes. Evidence obtained from independent lines of study (mainly from dissociation-reassociation experiments) indicates that hydrophobic forces are critically involved in the stabilization of oligomeric structure. Because of the low-temperature lability of such structure, low temperature leads to a loss of quaternary structure and hence of catalytic function. Homologous enzymes from ectotherms do not (cannot) possess such „cold sensitivity" either (1) because of increased hydrophobicity of key "contact zones" between or within polypeptide chains, or (2) because stabilization of oligomeric structure is less dependent upon hydrophobic interactions. (Data taken from Beyer, 1972 and Brandts, 1967)

pyruvate carboxylase
glyceraldehyde-3-P dehydrogenase
arginosuccinase
acetylCoA carboxylase
glycogen phosphorylase
mitochondrial ATPase
lipase
phosphopyruvate hydratase (yeast)
glutamate decarboxylase

We would anticipate an identical series of events in the case of AChE-substrate interactions. In the simplest case (which ignores other properties of the molecules such as its charge) the thermodynamically preferred state involves E-S complex formation, for the hydrophobic residues of both the binding site and the substrate interact so as to exclude water. Low temperatures, in contrast, would favor the accomodation of both sets of hydrophobic residues by clathrate formation. As temperature increased, these clathrates would "melt" in a solvent "order-disorder" transition, producing a relatively unstable state. Under these conditions, the thermodynamically favored state again involves E-S complex formation.

These considerations thus allow a fairly precise prediction (Fig. 4): *for AChE, E-S affinity should be low at low temperatures and should increase as temperature rises.* Is this prediction in fact realized?

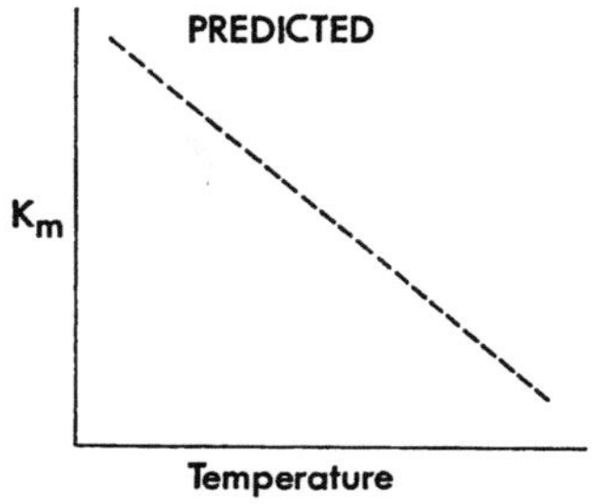

Fig. 4. The predicted effect of temperature on the K_m for AChE

K_m-temperature Relationships for Fish Brain Acetylcholinesterases

Baldwin and Hochachka (1970) and Baldwin (1971) have examined this question in kinetic studies of acetylcholinesterases from the central nervous systems of various species of fishes. In all cases examined, the K_m for acetylcholine remains relatively constant at the "optimal" biological temperature range for each species. In all cases, as the temperature drops below this range, the K_m increases; for some enzymes, the increase is indeed dramatic (Fig. 5). These then are precisely the results that we are predicting. The situation, however, becomes more complex when we consider the upper limits of the biological temperature range.

As temperature increases in these ranges, again in all species examined except tuna, the K_m for substrate increases, whereas our model would predict a continued decrease in the K_m.

We are therefore faced with two important questions concerning AChE-substrate interactions:

1. What factors account for a plateau region in the K_m-temperature relationship?
2. What factors account for an increasing K_m in upper biological temperature ranges?

At the outset, let me stress that we do not know the anwers to these questions. However, Fig. 2 allows some instructive speculation. Thus, *if we assume that hydrophobic interactions predominantly contribute to E-S affinity at all biological temperatures,*

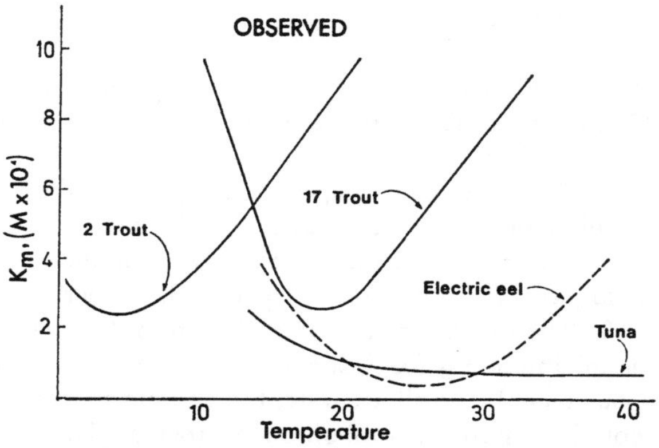

Fig. 5. The observed effects of temperature on the K_m for AChE from cold (2 C) acclimated trout, from warm (17 C) acclimated trout, from the electric eel, and from tuna. (Data from Baldwin and Hochachka, 1970 and Baldwin, 1971)

it takes little chemical-engineering insight to realize that for any given AChE-substrate interaction an increased affinity could be obtained by increasing the hydrophobicity

1. of the anionic site,
2. of the substrate itself, or
3. of both.

As we have already stressed, such an effect can be convincingly demonstrated by replacing the CH_3 groups on acetylcholine with more hydrophobic groups (i.e. with longer hydrocarbon chains). Whereas such a "modification" can be made by the chemist in his laboratory, nature in contrast is, so to speak, "stuck" with acetylcholine as its substrate. Any significant modification of the hydrophobicity of E-S interactions then would have to involve modifications

1. in the number of hydrophobic residues in the anionic site,
2. in the kind of non-polar amino acids present (some being more hydrophobic than others), or
3. in both.

Through such "modifications" we can well envisage that for each AChE there will be an optimal temperature where the strength of hydrophobic E-S interactions are optimal. The precise temperature range will depend upon the hydro phobic nature of the anionic site, and this will presumably be different in each species. At lower temperatures, the E-S affinity will drop because of the low-temperature lability of hydrophobic E-S interactions. The rate of change with temperature again could be species-specific.

Unfortunately, we cannot as readily account for the rise in K_m at temperatures above the "optimal" range. The role of hydrophobic forces in binding and positioning substrate at these temperatures presumably is accentuated because of their unique temperature-dependence. The decrease in E-S affinity presumably could involve shape-changes in the enzyme, possibly because of increased stability of interactions between hydrophobic residues *within the anionic site*. But that is about as far as we can speculate on the matter at this time.

Predictions

Model building is a useful exercise only if it yields predictive, testable premises. That is in fact the purpose of the above discussion, and from it a number of predictions arise:

1. The more cold adapted the organism, the more hydrophobic should be the anionic binding site of AChE.

2. The more cold adapted the organism, the greater the AChE affinity for carbon analogues of acetylcholine (such as 3,3-dimethylbutyl acetate).

3. The minimum K_m for AChE from "winter" trout occurs at about 2 C, while the minimum K_m for AChE from "summer" trout occurs at about 15–20 C. Hence, if our model is at all accurate, the anionic site of the "cold" variant of trout AChE should be more hydrophobic than that of the "warm" form of the enzyme.

All three predictions are experimentally useful.

Summary and Conclusions

Under conditions of saturating substrate levels, the temperature-dependence of enzyme catalyzed reactions is determined strictly by thermo-dynamic parameters. Compensatory strategies of enzymic adaptation to temperature take this fact "into account" and lead to the elaboration of enzymes with altered (a) turnover numbers, and (b) Q_{10}. Enzymes of ectotherms, for example, display higher turnover numbers and lower Q_{10} values than do enzymes of endotherms. The higher turnover numbers compensate for the reduced thermal energy encountered in their environments. The lower Q_{10} values characteristic of ectothermic enzymes presumably compensate for the inherently greater thermal instability of their environment compared to that of endotherms.

Under subsaturating conditions of substrate, enzymic reaction rates are not determined by the thermodynamic situation, but rather by the kinetic properties of the catalyst, in particular by E-S affinity. Significantly, many enzymes from ectotherms display a direct relationship between the K_m and temperature: for such enzymes, as temperature rises, E-S affinity drops. Since E-S affinity is under tight regulation by the organism, this enzyme design allows the organism to "gain" control of the Q_{10} for the reaction. Thus under physiological conditions, the Q_{10} for any given enzyme reaction *can be regulated by changing (1) the E-S affinity,* (2) the *substrate concentration, or (3) both.* The design of enzymes displaying particular K_m-temperature relationships, then, is viewed as an exploitative strategy of adaptation that presumably facilitates survival of the organism in its particular thermal regime.

The physical-chemical basis for temperature-dependent E-S affinities is explored by a brief examination of the binding site and binding functions of acetylcholinesterases.

During AChE function, hydrophobic bonds predominantly contribute to the binding and positioning of substrate at the active site. Since hydrophobic inter, actions are cold-labile, E-S affinity should be reduced at low temperatures. Moreover- the more cold-adapted the organism, the more hydrophobic should be the substrate binding site on the enzyme and the lower the temperature of optimal E-S affinity. A comparison of these and a number of other predictions with available experimental data suggest directions for future studies.

Acknowledgements

My colleague, George Somero, and I considered many of the problems raised in this paper during the preparation of a recent textbook. In addition, I wish to acknowledge the countless numbers of discussions and exchanges between myself and my former students: Hans Behrisch, John Baldwin, Tom Moon, Tariq Mustafa, and Heidi Somero. Frank Marcus and Paul Dreizen helped to sort out "chemical similarities between enzymes" from true "biological adaptations of enzymes". Support for the experimental work summarized in this paper derives from the National Research Council of Canada and the University of British Columbia.

References

Baldwin, J.: Adaptation of enzymes to temperature: acetylcholinesterases in the central nervous system of fishes. Comp. Biochem. Physiol. **40B**, 181–187 (1971).

Baldwin, J., Hochachka, P. W.: Functional significance of isoenzymes in thermal acclimatization. Biochem. J. **116**, 883–887 (1970).

Brandts, J. F.: Heat effects on proteins and enzymes. In: Rose, A. H. (Ed.): Thermobiology, p. 25–72. London: Academic Press 1967.

Froede, H. C., Wilson, I. B.: Acetylcholinesterase. In: Boyer, P. D. (Ed.): The Enzymes, Vol. 5, 89–114. New York: Academic Press 1972.

Hochachka, P. W., Schneider, D. E., Kuznetsov, A.: Interacting pressure and temperature effects on enzymes of marine poikilotherms. Marine Biol. **4**, 285–293 (1970).

Hochachka, P. W., Somero, G. N.: Biochemical adaptation to the environment. In: Hoar, W. S., Randall, D. J. (Eds.): Fish Physiology, Vol. 6, 99–156. New York: Academic Press 1971.

Hochachka, P. W., Somero, G. N.: Biochemical Strategies in Environmental Adaptation. Philadelphia: Saunders Publ. 1973.

Moon, T. W., Mustafa, T., Hochachka, P. W.: The adaptation of enzymes to pressure. II. A comparison of muscle pyruvate kinase from surface and midwater fishes with the homologous enzyme from an off-shore benthic species. Amer. Zool. **11**, 491–502 (1971).

Mustafa, T., Moon, T. W., Hochachka, P. W.: Effects of pressure and temperature on the catalytic and regulatory properties of muscle pyruvate kinase from an off-shore benthic fish. Amer. Zool. **11**, 451–466 (1971).

Acute Temperature Change and the Oxidation Rates of Ectotherm Mitochondria

VERONICA PYE

With 3 Figures

Introduction

There have been many recent comparative studies on the influences of temperature on the oxidative metabolism of whole organisms (Newell and Pye, 1968, 1970a, b; Newell, Ahsanullah and Pye, 1972; Newell, 1969) and of the oxidation rates of mitochondria, Newell and Walkey (1966), Newell and Pye (1971), Davison (1971), Lyons and Raison (1970), Raison, Lyons and Thomson (1971). Recent work has shown that in many species of ectotherms the resting or basal metabolism is relatively independent of acute temperature changes to a maximum temperature which was correlated with the maximum experienced by the organism in its environment (Newell and Northcroft, 1967; Newell and Pye, 1970). Such findings have also been supported by a study of the rates of oxygen consumption of cell-free homogenates and crude mitochondrial suspensions isolated from ectotherms, Newell and Pye (1970, 1971), Davison (1971).

However, Lyons and Raison (1970a, b) have shown that a discontinuous Arrhenius plot for succinate oxidation occurs in chilling sensitive plant tissues and endotherms which is not evident in chilling resistant plant tissues and ectotherms. From this they concluded that the discontinuity and change in activation energy is a unique property of succinate oxidation in endotherms not observed in ectotherms.

In the publication to follow the effect of acute temperature changes on the oxidation rates of mitochondria from mung beans, *Phaseolus aureus* var. *jumbo*, potatoes, *Solanum tuberosum*, and frogs, *Rana pipiens* in the presence of different substrates is examined. Mung beans were chosen as a plant providing uniform etiolated material which could be grown under different thermal regimes in a short period of time. Frogs provided a source of ectotherm tissue and were acclimated to different temperatures under laboratory conditions. Potato tubers were used as a source of chilling-resistant plant tissue.

Materials and Methods

Medium sized specimens of *Rana pipiens*, northern variety, were obtained from West Jersey Biological Farms, Wenonah, N.J. The animals were washed on arrival in a 1% solution of $K\,MnO_4$ and rinsed thoroughly in tap water. The frogs were then kept in aquaria maintained at constant temperatures for at least two weeks prior to sacrifice. The aquaria contained very dilute amphibian ringer solution of the following composition, NaCl 0.1 mM, KCl 0.02 mM, CaCl 0.007 mM, $NaHCO_3$ 0.035 mM at pH 7.8. Live meal worms, *Tenebrio sp*, were provided ad lib as food at regular intervals. After the animals had been acclimated, mitochondria were isolated from approximately 30–40 gms wet weight of hind limb skeletal muscle obtained from nine frogs. The animals were killed by severing the spinal cord just beneath the cranium and pithed. The procedure used for the preparation of mitochondria from this source is a modification of that described by Makinen and Lee (1967).

The muscle, in ice-cold isotonic KCl solution, was freed from connective and fatty tissue, especially from the blood vessels which are pigmented with melanin. It was then transferred to the following Chappell-Perry medium: 100 mM KCl, 50 mM Tris HCl pH 7.5, 1 mM ATP, 5 mM $MgCl_2$ and 1 mM EDTA, and minced for 10 seconds with a Moulinex 66 blender (Alençon, France). The coarsely minced tissue was suspended in the above medium (10 ml/g tissue) and to this was added *B. subtilis* proteinase (3 mg/g tissue) which had been dissolved in a little of the medium. The mixture was incubated at pH 7.5 for seven minutes. Additional medium was added (10 ml/g tissue) and the muscle was homogenised using a Polytron (Kinematica, Lucerne) for 10–15 secs. After readjusting to pH 7.3, the homogenate was centrifuged as described by Makinen and Lee (1967). However, the mitochondrial pellet was resuspended in the following medium, 100 mM KCl, 50 mM Tris HCl pH 7.5, 0.2 mM ATP, 1 mM $MgCl_2$, 0.2 mM EDTA and 0.1 % Bovine Serum albumen (BSA). This medium was used for all further washings.

Mung beans were grown from seed on perforated sheets of aluminium suspended in trays containing tap water. White light was excluded from the growth chambers, which were maintained at a constant temperature. Mitochondria were isolated from the hypocotyls of such dark-grown, five day old beans. Mitochondria isolated from mung beans and also those isolated from potatoes, were purified on a discontinuous sucrose gradient. The procedure for the preparation and the purification of plant mitochondria from these two sources of tissue was that described by Douce, Christensen and Bonner (1972).

Oxygen uptake was measured polarographically using a Clarke oxygen electrode (Yellow Springs Instrument Company) inserted into a 2 ml Plexiglass chamber, the temperature of which could be varied by a thermostatically controlled water jacket. The reaction mixture was stirred by means of a small magnetic stirrer and reagents were introduced through a small hole in a moveable well. The electrode medium contained 0.3 M mannitol, 0.01 M KCl, 0.005 M $MgCl_2$ and 0.01 M phosphate buffer pH 7.2. The oxygen content of this medium at a variety of different temperatures was calculated from the solubility tables of air saturated water in the "Handbook of Chemistry and Physics". After use of 1799 (bis-hexafluoroacetonylacetone) the chamber was soaked for one minute in 50 % solution of dimethyl formamide to remove all trace of the uncoupler.

Reduced Nicotine adenine dinucleotide (NADH) cytochrome c reductase activity was measured optically using a Zeiss spectrophotometer using the millimolar extinction coefficient at 550 nm of 18.5 $mM^{-1} \times cm^{-1}$ for cytochrome c.

The protein content of the mitochondria was determined by a modification of the phenol method described by Miller (1959) using crystalline BSA as a standard.

The concentrations of solutions of adenosine-5'-diphosphate (ADP) and NADH were determined optically at 260 nm and 340 nm on the basis of millimolar extinction coefficients of 15.4 and 62.2 respectively.

Reagents

Crystallised lyophilised *B. subtilis* proteinase (Nagarse) was obtained from the Enzyme Development Corp. N.Y.C. ADP, adenosine-5'-triphosphate (ATP), NADH and cytochrome c were obtained from Boehringer.

All other chemicals used were of analytical reagent grade. Glass distilled water which had been passed through an ion exchange column was used throughout. All substrates and cofactors were adjusted to a pH of approximately 7.2 before use.

Effect of Experimental Temperature on the Oxidation Rates of Frog Mitochondria

Frog skeletal muscle mitochondria were prepared in the manner described above from animals which had been acclimated to 15 C for two weeks. The oxidation rates were measured in the presence of the following substrates: Glutamate 17.5 mM

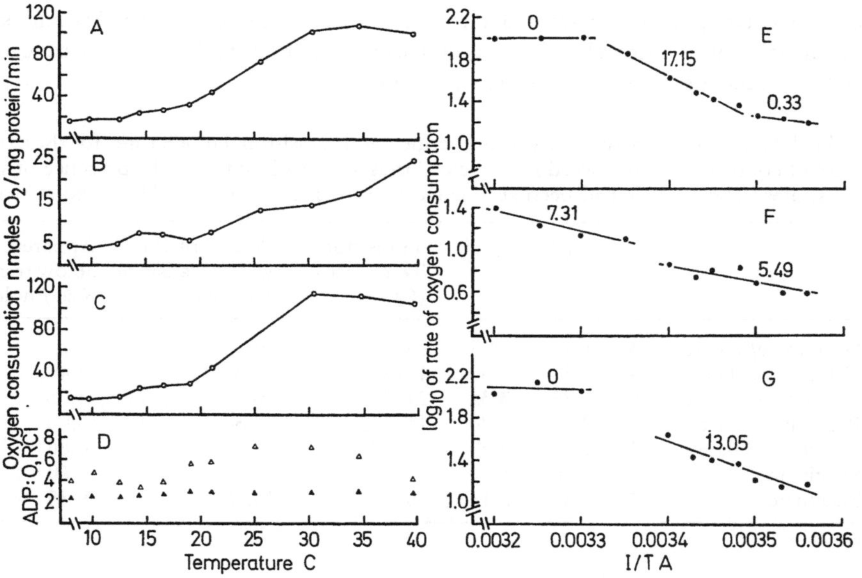

Fig. 1 A–G. The effect of experimental temperature on the rates of oxygen consumption, measured in n moles O_2/mg mitochodrial protein/min, of frog skeletal muscle mitochondria in the presence of glutamate 17.5 mM and malate 1.0 mM. A State 3 respiration. B State 4 respiration. C Uncoupled respiration. D Effect of temperature on ADP : O ratios indicated by ▲ and respiratory control indices (RCl) indicated by △. E, F and G Arrhenius plots of the data presented in A, B and C respectively. The figures beside the graphs denote the apparent activation eneriges (Ea) in kcals above and below the transition temperatures. The frogs had been acclimated to 15 C

with 1 mM malate, pyruvate 17.5 mM with 1 mM malate, succinate 17.5 mM, the latter preceded by a one minute soak with 0.25 mM ATP. ADP-stimulated and ADP-limited respiration (states 3 and 4, respectively, Chance and Williams, 1956) were measured at a variety of temperatures. The rate of oxidation of uncoupled mitochondria was measured after the addition of 0.05 mM 1799. The results for glutamate and malate oxidation are shown in Fig. 1. The mitochondria were tightly coupled over the range of experimental temperatures, and showed good respiratory control. Similar results were obtained when pyruvate and malate were used as substrates. However, when succinate was used as a substrate there was no discontinuity in the Arrhenius plot of state 3 respiration, and only a slight break in state 4. The Q_{10} values over various temperature ranges are shown in Table 1. From this table it can be seen that the Q_{10} values are lowest for the state 4 rates, and that over the range 5–15 C i.e. up to the maximum temperature experienced by the animals during acclimation, the Q_{10} values reach a minimum. Table 1 also shows the transition temperature or temperatures at which a discontinuity occurs in the Arrhenius plots, and apparent activation energies for the oxidation of various substrates above and below such transition temperatures. It is evident that the apparent activation energies for state 3 and uncoupled respiration, except in the presence of succinate differ by 12–17 Kcals above and below the transition temperature, while that of state 4 respiration differ by 2–6 Kcals. Also the apparent activation energies for state 4 are higher above the transition temperature than below – the converse being true for state 3 rates. The discontinuities in the Arrhenius plots occur at temperatures above the acclimation temperature in all cases save that of state 3 succinate respiration, where none occurred.

Table 1. Q_{10} values, apparent activation energies, and transition temperatures for the oxidation rates of coupled and uncoupled frog skeletal muscle mitochondria in the presence of various substrates. The animals had been acclimated to 15 C for two weeks. Measured at pH 7.2

	State	Q_{10}^{5-15}	Q_{10}^{15-25}	Transition Temperature (C)	Apparent Ea above transition Kcal/mole	Apparent Ea below transition Kcal/mole
Glutamate/Malate	3	1.50	3.85	26–30	0	17.15
Glutamate/Malate	4	1.50	2.00	19–21	7.31	5.49
Glutamate/Malate	1799	1.75	3.32	21–30	0	13.05
Pyruvate/Malate	3	2.15	4.20	15–17.5	3.88	16.00
Pyruvate/Malate	4	1.33	3.30	15–17.5	12.3	6.86
Succinate	3	2.50	1.97	—	13.6	13.60
Succinate	4	1.33	2.13	18–21	14.85	9.15

Effect of Experimental Temperature on the Oxidation Rates of Purified Mung Bean Mitochondria

The oxidation rates of mung bean mitochondria were measured in the presence of 50 mM malate and 17.5 mM succinate as substrates, the latter preceded by a one minute soak with 0.25 mM ATP. The results for malate oxidation are shown in Fig. 2. Again the mitochondria were tightly coupled over the whole experimental

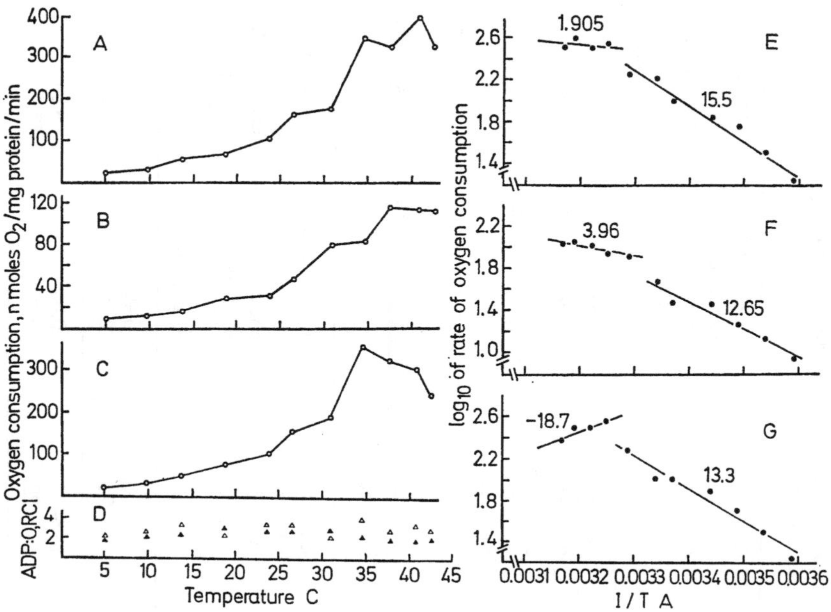

Fig. 2 A–G. The effect of experimental temperature on the rates of oxygen consumption, measured in n moles O_2/mg mitochondrial protein/min, of purified mung bean mitochondria in the presence of 50 mM malate as a substrate. A State 3 respiration. B State 4 respiration. C Uncoupled respiration. D ADP : O ratios ▲ and RCl △. E, F und G, Arrhenius plots of the data presented in A, B and C respectively. The figures beside the graphs denote the apparent activation energies (Ea) in kcals above and below the transition temperatures. The mung beans had been grown at 28 C

range and showed good respiratory control. Similar results were obtained for succinate oxidation. The Q_{10} values, transition temperatures and apparent activation energies are shown in Table 2. Again, it is noticeable that the lowest Q_{10} values occur in state 4 rates over a temperature range up to the maximum temperature experienced by the beans during growth. The state 3 apparent activation energies above and below the transition temperature differ from each other by 12–13 Kcals, whereas those in state 4 differ by approximately 4–8 Kcals. Unlike the frog data, the activation energies were higher below the transition temperature for all rates. However the transition temperatures, both for malate and succinate oxidation rates, occur at temperatures just above that at which the beans were grown.

Effect of Experimental Temperature on the Oxidation Rates of Purified Potato Mitochondria

The oxidation rates of coupled and uncoupled potato mitochondria were measured in the presence of 17.5 mM glutamate and 1 mM malate, 50 mM malate preceded by a one minute soak in 0.05 % of the thiamine pyrophosphate (TPP) and 17.5 mM succinate preceded by a one minute soak in 0.05 mM ATP. The result for succinate

Table 2. Q_{10} values, apparent activation energies, and transition temperatures for the oxidation rates of coupled and uncoupled purified mung bean mitochondria in the presence of malate and succinate as substrates. The beans had been grown at 28 C

Substrate	State	Q_{10}^{15-25}	Q_{10}^{20-30}	Q_{10}^{25-35}	Transition temperature (C)	Apparent Ea above the transition temperature Kcal/mole	Apparent Ea below the transition temperature Kcal/mole
Malate	3	1.77	2.54	2.82	31 –34.5	1.91	15.5
Malate	4	1.50	3.48	2.72	26.5–31	3.96	12.65
Malate	1799	1.69	2.30	2.44	31 –34.5	–18.70	14.30
Succinate	3	2.26	2.25	1.06	26 –30	0.86	13.30
Succinate	4	1.38	1.44	—	26 –30	7.01	10.79

Table 3. Q_{10} values, apparent activation energies and transition temperatures for the oxidation rates of coupled and uncoupled purified potato mitochondria in the presence of glutamate with malate, malate with TPP, and succinate as substrates

Substrate	State	Q_{10}^{10-20}	Q_{10}^{20-30}	Q_{10}^{30-40}	Transition temperature (C)	Apparent activation energy above transition temperature Kcal/mole	Apparent activation energy below transition temperature Kcal/mole
Glutamate/malate	3	2.50	2.29	1.54	17 –21	6.40	20.10
	4	1.55	1.64	1.70	17 –21	12.80	16.90
Malate/TPP	3	1.92	1.22	—	18.5–22	0.46	18.30
	4	1.70	1.29	—	18.5–22	0	13.70
Succinate	3	2.00	1.43	1.41	15.5–20	4.57	17.80
	4	1.28	1.60	2.49	20 –26	13.7	7.32
	1799	1.92	1.40	1.42	15.5–20	2.74	14.60

oxidation are shown in Fig. 3. Similar results were obtained in the presence of malate and glutamate. The Q_{10} values, transition temperatures and apparent activation energies are shown in Table 3. These results support those obtained with frogs and mung beans in that the lowest Q_{10} values are usually obtained in the state 4 rates. It is not known what temperature the potatoes were kept at prior to their use, but from the Q_{10} values and the transition temperatures obtained, and the correlation between them and acclimation temperature formed for frogs and mung beans one would expect them to have experienced temperatures about 20 C. Discontinuities do occur in Arrhenius plots of the oxidation rates in the presence of all the substrates used, even succinate. Furthermore such discontinuities occur in the state 3 and state 4 rates of coupled mitochondria and in uncoupled mitochondria. The state 3 apparent activation energies above and below the transition temperature differ from each other by 13–18–Kcals whereas those in state 4 differ by 4–13 Kcals. For uncoupled mitochondria the difference was 12 Kcals.

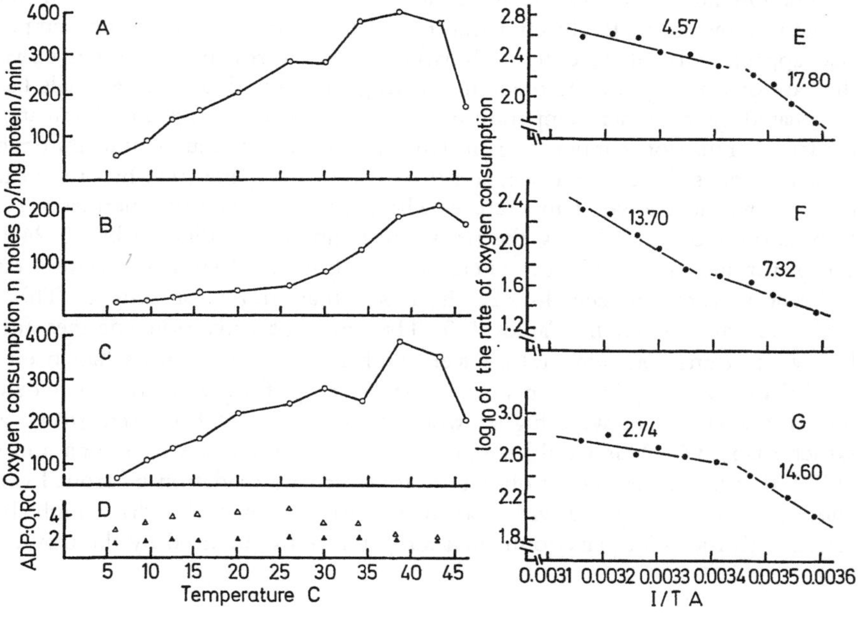

Fig. 3 A–G. The effect of experimental temperature on the rates of oxygen consumption, measured in n moles O_2/mg mitochondrial protein/min of purified potato mitochondria in the presence of 17.5 mM succinate as a substrate. A State 3 respiration. B State 4 respiration. C Uncoupled respiration. D ADP : O ratios ▲, RCl △. E, F und G Arrhenius plots of data presented in A, B and C respectively. The figures beside the graphs denote the apparent activation energies (Ea) in kcals above and below the transition temperatures

Variation of the Q_{10} Values and Discontinuities in Arrhenius Plots with Acclimation Temperature

Frogs were kept in aquaria maintained at 27 C for at least two weeks. Mitochondria were prepared from them in the manner described above and their oxidation rates were measured in the presence of 17.5 mM pyruvate with 1 mM malate, and 17.5 mM glutamate with 1 mM malate as substrates. Mung beans were grown at 18 C for ten days. Mitochondria were prepared from the etiolated hypocotyls and their oxidation rates in the presence of 50 mM malate and 17.5 mM succinate were measured. The Q_{10} values for the oxidation rates over various temperature ranges, the transition temperatures in the Arrhenius plots and the apparent activation energies for both the frog and mung bean mitochondria are shown in Table 4.

When the Q_{10} values for frog mitochondria are compared with those in Table 1 it will be noticed that although the lowest values still occur for the state 4 rates, the range over which the low Q_{10}'s occur extends up to 28 C, the maximum temperature experienced by the animals during acclimation. When the values for mung beans are compared with those in Table 4 the lowest Q_{10} values are again for state 4 respiration and extend only up to 18 C, the temperature at which the beans were grown. This supports the suggestion put forward by Newell and Pye (1970), that acclimation affects the temperature range over which the oxidative metabolism of ectotherms is relatively temperature independent. The transition temperatures also appear to be affected by acclimation temperatures, being shifted upwards in the case of malate/pyruvate oxidation in frog mitochondria when animals had been acclimated to a higher temperature. There was little change in glutamate/malate oxidation transition temperatures in frog. In the case of mung bean mitochondrial oxidation rates there was a decrease in transition temperature when the beans had been grown at a lower temperature. The apparent activation energies for state 3 respiration above and below the transition temperatures differed by 10–24 Kcals – except for succinate oxidation where no transition was found. For state 4 rates the differences were between 4–12 Kcals, lower than those for state 3. This agrees with the data shown in Tables 1–3. The apparent activation energies for frogs below the transition temperature were higher when the animals had been stored at 27 than when they had been stored at 15 C, except for glutamate/malate oxidation. Similarly such values were higher when the mung beans had been grown at 28 C rather than 18 C. This would support the suggestion made by Lyons and Raison (1970) that the lower activation energy for succinate oxidation by trout liver mitochondria, when compared with that for cat fish liver mitochondria, might be associated with the colder environmental temperature experienced by the trout.

Intactness of Mung Bean Mitochondrial Membranes at Different Temperatures

The cause of the discontinuities in Arrhenius plots of oxidation rates and the increase in Q_{10} values that occurs above the maximum environmental temperature is not known. However, it seemed interesting to investigate the possibility that the integrity of the outer mitochondrial membranes might be influenced by experimental temperature and whether such an influence, if present, differed at temperatures above and below the transition temperature. Therefore the following integrity test was used to determine the intactness of purified mung bean mitochondria (Douce *et al.*, 1973). NADH cytochrome *c* reductase activity was measured in electrode medium containing 0.067 mM cytochrome *c*, 0.075 mM KCN and approximately 0.1 mg mitochondrial protein. The reaction was initiated by 1.67 mM NADH.

Table 4. The effect of acclimation temperature of the Q_{10} values for the oxidation rates, apparent activation energies and transition temperatures for the oxidation rates of coupled and uncoupled frog and mung bean mitochondria in the presence of various substrates. The frogs were acclimated to 27 C and the mung beans were grown at 18 C

Tissue Source	Substrate	State	Q_{10}^{8-18}	Q_{10}^{18-28}	Q_{10}^{28-38}	Transition Temperature (C)	Apparent Activation energy above transition temperature, Kcal/mole	Apparent Activation energy below transition temperature, Kcal/mole
Frogs kept at 27 C	Pyruvate/malate	3	3.00	2.02	0.84	28 –31.5	—2.51	20.15
		4	2.1	1.26	1.50	21.5–25	4.11	13.95
		1799	3.58	2.16	—	28 –31.5	—2.28	20.6
	Glutamate/malate	3	2.95	2.84	—	28 –31	—9.9	15.3
		4	1.20	1.64	6.72	21 –25	20.2	7.99
Mung beans grown at 18 C	Malate	3	3.12	1.64	—	18 –22	19.4	9.14
		4	1.22	2.00	—	18 –22	3.66	9.14
	Succinate	3	1.68	1.66	—	None	8.56	8.56
		4	1.20	1.80	—	18 –22	2.74	6.85

The final volume was 3 ml. The activity was also measured in a bursting medium, similar to that described above, except that the electrode medium was replaced with 5 mM phosphate buffer pH 7.2. Plant mitochondria are able to oxidise externally added NADH by means of two dehydrogenases. One is located on the outer membrane and is completely antimycin A insensitive, the other is located on the inner membrane and is antimycin A sensitive. In suspensions which have their outer membranes intact the cytochrome *c* is unable to penetrate the outer membrane and therefore is reduced only by the external dehydrogenase, the rate of reduction being unaffected by the addition of antimycin A, 0.0013 mM. If the outer membranes are broken, the cytochrome *c* can reach the inner membrane and be reduced by the internal dehydrogenase too. On the addition of antimycin A in these circumstances the internal dehydrogenase is inhibited and the rate is greatly reduced to that shown by the external dehydrogenase only. Knowing the rates of reduction it is possible to calculate the % intactness of the mitochondrial membranes at different experimental temperatures. The results are shown in Table 5. In Table 2 it was shown that the transition temperatures of the oxidation rates of mitochondria isolated from mung beans grown at 28 C were between 26–34 C. The values shown in Table 5 do not show that there is a sudden change in mitochondrial intactness at the transition temperature and therefore it is unlikely that changes in Q_{10} values and discontinuities in Arrhenius plots are caused by disruption of the membranes.

Table 5. The effect of experimental temperature on the intactness of the outer membranes of mitochondria isolated from mung beans grown at 28 C, measured by NADH cytochrome c reductase activity

Experimental temperature, (C)	28	32	34	36	38	41.5
Intactness % of outer membranes	87.8	89.3	91.2	87.8	93.7	91.4

These data suggest that the mitochondria are capable of retaining their integrity over a wide range of experimental temperatures and are remarkably resistant to breakage – a fact which is borne out by unpublished data (Douce and Pye) on the effect of the osmolarity of the suspending medium on the intactness of plant and frog mitochondria.

Discussion and Conclusion

The data which are presented above confirm and amplify those reported earlier (Newell and Pye 1968, 1970a, b; Davison 1970). The state 4 oxidation rates of mitochondria from mung beans, potatoes and frogs, in the presence of high levels of substrates, are relatively independent of temperature. The Q_{10} values are between 1.20–1.60 over a temperature range which extends up to the maximum environmental temperature experienced by the organism prior to the experiment. The Q_{10} values over a similar temperature range for the state 3 and uncoupled rates are not so low, being in the range of 1.7–3.0, and show relative temperature dependence. Work on sarcosomes from *Calliphora erythrocephala* has shown that low Q_{10} values are obtained for state 4 rates with high levels of substrates (Davison, 1970) but differs from the results above in that the state 3 rates also show relative temperature independence. Both for the results cited above, and also for those cited by Davison (1970) the range over which low Q_{10} values are obtained is related to acclimation or growth temperature.

It had previously been reported that substrate concentration influenced metabolic homeostasis in the rates of oxidation of crude mitochondrial preparations isolated from whole *Littorina littorea* (Newell and Pye, 1971), temperature dependence being shown at high concentrations of pyruvate. This suggested that temperature-induced variations in enzyme-substrate affinity in the presence of low substrate concentrations may form the basis for control of metabolic homeostasis. Such changes have been described for LDH (Somero and Hochachka, 1969) and pyruvate kinase (Somero, 1969). However this present work indicates that for frog, mung bean and potato mitochondria state 4 respiration is relatively independent of temperature even at high substrate levels, and therefore might indicate the presence of one or more different control mechanisms in these cases. The temperature independence of the state 4 rates and the temperature dependence of the state 3 rates are similar to the effect of acute temperature change on the quiescent and active rates respectively of oxygen consumption of whole organisms (Newell and Pye, 1970a, b). It is tempting to regard the active rate of oxygen consumption of a whole organism as a state where, amid a multitude of metabolic processes, ATP is converted to ADP, which then might be compared with state 3 respiration in an isolated system. Similarly the state 4 respiration might be compared with the quiescent rate in intact organisms. Whether or not such direct comparisons are valid, the effect of acute temperature change on the rates in the two systems is similar. From the data presented above it would appear that tightly coupled ADP-limited oxidation is relatively temperature independent, and that ADP-stimulated oxidative phosphorylation and uncoupled oxidation is relatively temperature dependent, over a range extending up to the maximum experienced by the organism in its environment.

Discontinuities in Arrhenius plots for the oxidation rates of mitochondria from chill resistant plants and ectothermic animals used in this study do occur when substrates other than succinate are used, in the state 4 rates of succinate oxidation and occasionally in the state 3 rates of succinate oxidation also. This does not agree with the findings of Lyons and Raison (1970a, b) who put forward that discontinuities in succinoxidase activities only occur in endotherms and chill-sensitive plants. Moreover the discontinuities described above occur at the upper end of the temperature range over which low Q_{10} values are obtained for the state 4 rates, and therefore are influenced by the acclimation or growth temperature of the organisms. This would not agree with the findings of Lyons and Raison (1970a, b) which showed that discontinuities occurred at temperatures below which endotherms and chilling-sensitive plants suffered damage, 24 and 12 C respectively, the lowest end of the organisms' tolerated range. Kumamoto *et al.* (1971) concluded from thermodynamic arguments that a phase change must occur in mitochondrial membrane systems at the transition temperature. Such conclusions were supported by experimental observation of such changes by Steim *et al.* (1969) and Raison *et al.* (1971).

However the latter did not detect any phase changes in the membranes of trout liver and potato mitochondria.

In all cases the difference in apparent activation energies above and below the transition temperatures were lower for the state 4 rates than for the state 3 and uncoupled rates. Acclimation temperature influences the apparent activation energies, lower values being obtained below the transition temperature at lower acclimation or growth temperatures. Such findings agree with those of Lyons and Raison (1970). There was no significant difference in the intactness of the membranes of mung bean mitochondria above and below the transition temperatures, therefore the discontinuities in the Arrhenius plots are unlikely to be due to an increase in the permeability of substrates due to breakage of the membranes. Raison, Lyons and Thomson (1971) also found that substrate permeability was not important in the occurrence of discontinuities by showing them to be present in mitochondrial suspensions which had been sonically and osmotically disrupted.

In conclusion, it may be said that the results presented above indicate the presence of discontinuities in Arrhenius plots in the oxidation rates of mitochondria from chill-resistant plants and ectotherms. The transition temperatures of such discontinuities, and the apparent activation energies are influenced by the acclimation or growth temperature of the organism. Moreover, regions of relative temperature independence occur in the state 4 oxidation rates of such mitochondria in the presence of high concentrations of substrates. However, it should be pointed out that ADP is at a low concentration in state 4 respiration and therefore the temperature independence of this state may reflect a similar phenomenon to the substrate limited respiration described for other ectotherm and enzyme preparations (Newell and Pye, 1971). Acclimation or growth temperature also effects the temperature range over which such metabolic homeostatis occurs.

Summary

Discontinuities occur in the Arrhenius plots of the oxidation rates of mitochondria isolated from chill-resistant plants and an ectothermic animal. The transition temperatures of such discontinuities, and the apparent activation energies are influenced by the acclimation or growth temperature of the organism. Moreover regions of relative temperature independence occur in the state 4 oxidation rates of such mitochondria in the presence of high concentrations of substrates. Acclimation or growth temperature also affects the temperature range over which such metabolic homeostasis occurs.

Acknowledgements

I would like to thank Prof. W. D. Bonner, Jr. and Dr. R. Douce for their helpful and constructive criticisms. This work was carried out on leave of absence from the Dept. of Zoology and Comparative Physiology, Queen Mary College, University of London, London, E. 1. made possible by an International Fellowship from the American Association of University Women.

References

Chance, B., Williams, G. R.: The respiratory chain and oxidative phosphorylation. Advan. in Enzymol. **XVII**, 65–134 (1956).
Davison, T. F.: The effect of temperature on oxidative phosphorylation in isolated flight muscle sarcosomes. Comp. Biochem. Physiol. **38 B**, 21–34 (1971).
Douce, R., Christensen, E. L., Bonner, W. D., Jr.: Preparation of intact plant mitochondria. Biochim. Biophys. Acta **275**, 148–160 (1972).
Douce, R., Mannella, C. A., Bonner, W. D., Jr.: The external NADH dehydrogenases of intact plant mitochondria. Biochim. Biophys. Acta **292**, 105–116 (1973).
Handbook of Chemistry and Physics, 42nd ed. Cleveland, Ohio: Chemical Rubber Publ. Co. 1960–61.
Kumamoto, J., Raison, J. K., Lyons, J. M.: Temperature "Breaks" in Arrhenius plots: A thermodynamic consequence of a phase change. J. Theoret. Biol. **31**, 47–51 (1971).
Lyons, J. M., Raison, J. K.: Oxidative activity of mitochondria isolated from plant tissues sensitive and resistant to chilling injury. Plant Physiol. **45**, 386–389 (1970a).
Lyons, J. M., Raison, J. K.: A temperature-induced transition in mitochondrial oxidation: contrasts between cold and warm blooded animals. Comp. Biochem. Physiol. **37**, 405–411 (1970b).
Makinen, M. W., Lee, C. P.: Biochemical studies of skeletal muscle mitochondria I. Microanalysis of cytochrome content, oxidative and phosphorylative activities of mammalian skeletal muscle mitochondria. Arch. Biochem. Biophys. **126**, 75–82 (1968).
Miller, G. L.: Protein determinations for large numbers of samples. Anal. Chem. **31**, 964 (1959).
Newell, R. C.: The effect of temperature fluctuation on the metabolism of intertidal invertebrates. Am. Zool. **9**, 293–307 (1969).
Newell, R. C., Ahsanullah, M., Pye, V. I.: Aerial and aquatic respiration in the shore crab *Carcinus maenas* (L) Comp. Biochem. Physiol. **43 A**, 239–252 (1972).
Newell, R. C., Northcroft, H.: A re-interpretation of the effect of temperature on the metabolism of certain intertidal invertebrates. J. Zool. **151**, 277–298 (1967).
Newell, R. C., Pye, V. I.: Seasonal variations in the effect of temperature on the respiration of certain intertidal algae. J. Marine Biol. Assoc. U.K. **48**, 341–348 (1968).
Newell, R. C., Pye, V. I.: Seasonal changes in the effect of temperature on the oxygen consumption of the winkle *Littorina littorea* (L) and the mussel *Mytilus edulis* (L) Comp. Biochem. Physiol. **34**, 367–383 (1970a).
Newell, R. C., Pye, V. I.: The influence of thermal acclimation on the relation between oxygen consumption and temperature in *Littorina littorea* (L) and *Mytilus edulis* (L). Comp. Biochem. Physiol. **34**, 385–397 (1970b).
Newell, R. C., Pye, V. I.: Temperature induced variations in the respiration of mitochondria from the winkle *Littorina littorea* (L). Comp. Biochem. Physiol. **40 B**, 249–261 (1971).
Newell, R. C., Walkey, M.: Oxidative activity of mammalian liver mitochondria as a function of temperature. Nature (London) **212**, 428–429 (1966).
Raison, J. K., Lyons, J. M., Mehlhorn, R. J., Keith, A. D.: Temperature-induced phase changes in mitochondrial membranes detected by spin labelling. J. Biol. Chem. **246**, 4036–4040 (1971).
Raison, J. K., Lyons, J. M., Thomson, W. W.: The influence of membranes on the temperature-induced changes in the kinetics of some respiratory enzymes of mitochondria. Arch. Biochem. Biophys. **142**, 83–90 (1971).
Somero, G. N.: Pyruvate kinase variants of the Alaskan King Arab. Evidence for a temperature dependent interconversion between two forms having distinct and adaptive kinetic properties. Biochem. J. **114**, 237–241 (1969).
Somero, G. N., Hochachka, P. W.: Isoenzymes and short term compensation in poikilotherms: Activation of lactate dehydrogenase isoenzymes by temperature decreases. Nature (London) **233**, 194–195 (1969).
Steim J. M., Tourtellotte, M. E., Reinert, J. C., McElhaney, R. N., Rader, R. L.: Calorimetric evidence for the liquid crystalline state of lipids in a biomembrane. Proc. Natl. Acad. Sci. U.S. **63**, 104–109 (1969).

Effects of Acclimation Temperature on Aerobic Energy Production in Eel Liver: Oxidative Phosphorylation in Isolated Mitochondria

Ekkehart Wodtke

With 3 Figures

Introduction

Various acclimation patterns of enzymes and of enzyme systems involved in electron transport in fish liver have been published (Reviews: Precht, 1968; Prosser, 1967; Hazel and Prosser, 1970). Measurements of the maximal rate of enzyme activity and of respiration rate have been made under different assay conditions (temperature, incubation time, composition of assay medium) with different test systems (liver slices, liver homogenates, liver mitochondrial fractions) based on different units (wet weight, dry weight, protein, nitrogen).

It has often been suggested that changes in quantity of oxidative enzymes are involved in temperature compensation of respiration rates. However, the investigations carried out so far do not justify conclusions of this nature. For example, isolation of mitochondria as reported by Kanungo and Prosser (1959) or by Caldwell (1967) may lead to significant lysosomal contamination which is known to induce partial uncoupling of oxidative phosphorylation (Mellors *et al.*, 1967). In addition, mitochondria may contain free fatty acids – derived from the homogenate – critical to mitochondrial function in reducing respiratory control, elevating state 4 respiration and impairing coupling reactions. Obviously this is of great significance when considering the influence of acclimation temperature on mitochondrial oxidative phosphorylation in vertebrates (see Patkin and Masoro, 1960; Gumbmann and Tappel, 1962; Kadenbach, 1966; van den Bergh, 1967).

Some Considerations on Factors Influencing Respiration Rates

Changes of respiration rate in tissues and tissue homogenates may be due to:

1. Changes in the amount of mitochondrial protein;

2. changes in the amount of respiratory assemblies per mitochondrial protein, perhaps mediated by hormones (Hoch, 1971), or some other "Langzeiteffekte" (Precht, 1965);

3. altered activity of respiratory units based on conformational changes of protein not necessarily involved in dehydrogenation and oxidation, perhaps due to substrates (e.g. free fatty acids), hormones (Hoch, 1971), or other "Direkteffekte" (Precht, 1965);

4. kinetic control by allosteric mechanisms of regulatory enzymes, especially in glycolysis (Hochachka and Somero, 1971) resulting in an increased or decreased hydrogen input into the electron transport chain.

5. Thermodynamic control of activity of enzymes involved in mitochondrial electron transport by means of the phosphorylation potential, resulting in an altered specific activity.

Changes in respiration rates as mentioned under points 1. to 5. may be divided in

a) *indirect effects* of acclimation temperature AT on respiratory units (systemic changes) (1, 2): in such cases regulators affect the architecture of the respiratory units at the level of their biosynthesis;

b) *direct effects* of AT on respiratory units (functional changes) (3, 4, 5): in such cases regulators affect the mode of operation of respiratory units.

Both types of effects may be superimposed in bringing about adaptive changes of respiratory rates *in vivo* and in complex test-systems. In experimental systems direct actions of regulators on respiratory units (b) can be excluded if carefully isolated mitochondria are used.

Some Considerations on Factors Influencing Phosphorylation Rates

The phosphorylation rate of mitochondria, i.e. moles of ATP synthetized per minute, depends on the respiration rate and on the efficiency of oxidative phosphorylation. Efficiency of oxidative phosphorylation can be measured by the ADP/O ratio, i.e. moles of ATP synthetized per gram atom oxygen consumed.

However, it should be kept in mind that energy derived from water production is fixed at first in the form of a non-phosphorylated energy-rich intermediate (Slater, 1953; Storey, 1970). This intermediate can serve to generate ATP, or promote intramitochondrial ion transport, or combust to heat. As a consequence, efficiency of oxidative phosphorylation need not necessarily equal efficiency of energy fixation.

Measurements of ADP/O ratios in isolated cells from fish acclimated to different temperatures can show whether or not altered efficiency occurs *in situ*. If it does this might be due to

a) *indirect effects* of acclimation temperature on mitochondrial biosynthesis, resulting in structural changes of the phosphorylation machinery;

b) *direct effects* of extrinsic cytoplasmatic components on mitochondria.

Again a) and b) may be superimposed *in vivo* and in complex test-systems. Direct effects of regulators (b) can be eliminated in experiments with isolated, intact, and well washed mitochondria.

Experimental

Eels show partial compensation of respiration *in vivo* (Precht, 1951, 1961; Schultze, 1965). The studies reported below were carried out in order to establish whether or not acclimation temperature affects respiratory properties of liver mitochondria of eels and whether the rates of oxidative phosphorylation strictly follow the pattern of respiration rates. As experimental animals adult eels acclimated in tap water to 7 C (21 days and more) and to 25 C (12–14 days) were used.

Preparation and Assay of Mitochondria

Mitochondria were isolated in a sucrose-mannitol medium containing albumin (see Table 1), from centrifugation fractions between 700 g · 10 min and 4500 g · 10 min. Albumin was present in order to bind critical extrinsic substrates that might have been released during and following cell rupture (see Davis, 1967; Hoch and Motta, 1968). Two additional washes were included; the final suspension contained about 40 mg mitochondrial protein per ml. Mitochondria thus isolated remained tightly coupled for at least 10 hours after preparation.

The mitochondrial protein content of the liver was determined by a biuret assay procedure with bovine serum albumin as standard, following Beisenherz *et al.* (1957) and Szarkowska and Klingenberg (1963).

Mitochondrial respiration in state 4 and state 3 as well as coupled phosphorylation were measured polarographically according to Eastabook (1967) at an experimental temperature of 25 C. Concentrations of ADP test solutions were determined enzymatically according to Adam (1962).

Mitochondrial Protein Content

Fig. 1 demonstrates that acclimation to a high temperature lowers the mitochondrial protein content of livers by 34 % ($p < 0.01$). Since total protein content of the livers of eels acclimated to 7 C and 25 C does not differ significantly (Wodtke, unpublished), Fig. 1 shows that the relative amount of the respiratory cell compartment is influenced by the acclimation temperature (compare Jankowsky and Korn, 1965; Thiessen and Mutchmor, 1967).

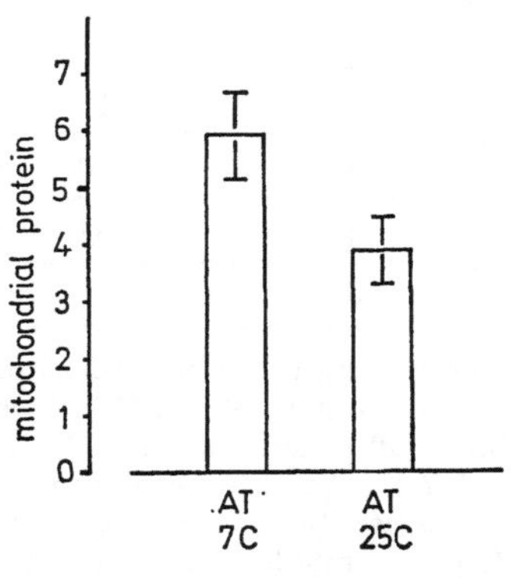

Fig. 1. Mitochondrial protein (in mg/g fresh weight) of the livers of cold- and warm-acclimated eels. The heights of the columns represent mean values, the vertical bars standard deviations

Oxidative Phosphorylation

Respiration of mitochondria was measured in the presence of different substrates in state 4 and state 3 (see Fig. 2). ADP/O ratios were determined for each experiment.

The respiration rates represent activities of mitochondrial systems which differ with respect to the dehydrogenase and the region of the electron transport chain involved. For example, oxidation of malate, succinate, and ascorbate is coupled to phosphorylation reactions at the sites I+II+III, II+III, and III respectively (Fig. 2).

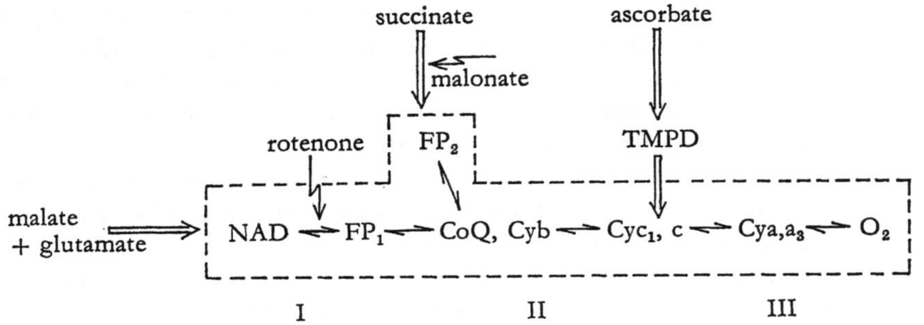

Fig. 2. Schematic representation of mitochondrial functions relevant to respiration and phosphorylation in this study. →: inhibiting; ⇒ „hydrogen (electron) input"; ⇌: possible directions of electron transport (during respiration: to O_2); I, II, III: phosphorylation sites of electron transport chain. NAD = nicotine-amide-dinucleotide. FP = flavoproteins. CoQ = coenzyme Q. Cy = cytochromes

Table 1 clearly demonstrates that acclimation to 25 C increases mitochondrial specific activity of respiration in state 3 with malate + glutamate by more than 100%. In state 4, however, the effect is lower. This is due to the lower respiratory control

Table 1. Influence of acclimation temperatures (7 C and 25 C) on respiration rates of liver mitochondria from eels, with malate + glutamate as substrates, in states 3 and 4

AT	state	z	n	RR ± S.D.	state	$RR_{AT\ 25\ C}$ in % of $RR_{AT\ 7\ C}$	p
7 C	4	8	16	9.39 ± 1.54	4	160	0.02
	3	8	15	23.0 ± 3.22			
25 C	4	9	33	15.0 ± 4.03	3	214	0.001
	3	12	18	50.0 ± 12.9			

About 2 mg of mitochondrial protein in 50 μl (225 mM mannitol, 75 mM sucrose, 50 mM tris/HCl, 0.05% BSA; pH 7.4) were added to the respiration vessel (V = 0.925 ml) containing 0.875 ml assay medium (225 mM mannitol, 75 mM sucrose, 50 mM tris/HCl, 5 mM Pi; pH 7.4). State 4 respiration was started with 10μl of 0.925 M malate + 0.925 M glutamate. State 3 respiration was obtained on addition of determined amounts of ADP (150–300 nmoles). Experimental temperature: 25 C.
AT = acclimation temperature; z = no. of preparations (each containing two livers); n = no. of experiments; RR = mean respiration rate [ng atoms O · min^{-1} · mg^{-1}]; S.D. = standard deviation; p = level of significance.

ratio of liver mitochondria from cold acclimated eels. The effects of acclimation temperature on state 3 respiration with succinate and ascorbate + TMPD as substrates are less pronounced (Wodtke, unpublished).

If it is accepted that state 3 respiration is a measure of the concentration of the functional respiratory chain units in the mitochondria (Kadenbach, 1966; compare: Chance and Williams, 1955), it would appear that cold acclimation of the eel is followed by a decrease of the concentration of the functional respiratory assemblies relative to the total protein content of liver mitochondria.

Table 2. Influence of acclimation temperatures (7 C and 25 C) on ADP/O-ratios of liver mitochondria from eels

substrate	ADP/O-ratio ($\bar{x} \pm$ S.D.) AT 7 C	ADP/O-ratio ($\bar{x} \pm$ S.D.) AT 25 C	p	Δ ADP/O ($\bar{x}$)	phosphorylation sites involved	~ relative Δ ADP/O per coupling site
Mal + Glu	1.71 ± 0.21 (15)	2.07 ± 0.13 (19)	0.001	0.36	I, II, III	2
Succ	0.96 ± 0.15 (13)	1.31 ± 0.11 (19)	0.001	0.35	II, III	2
AscTMPD	0.46 ± 0.19 (3)	0.64 ± 0.08 (7)	> 0.05	0.18	III	1

Mitochondria (about 2 mg in experiments with malate + glutamate and succinate, about 1 mg in experiments with ascorbate + TMPD) were tested as described in Table 1. Experiments with succinate included 5 μM rotenone, experiments with ascorbate + TMPD included 2 μM rotenone + 10 mM malonate. Abbreviations: see Table 1
TMPD = tetramethyl-p-phenylenediamine

The effect of cold acclimation on ADP/O ratios is shown in Table 2. It is evident from these data that the efficiency of oxidative phosphorylation – at least at the experimental temperature of 25 C – increases in liver mitochondria of eels following acclimation to a high temperature. Since the three types of substrate used link with the respiratory chain at different points (Fig. 2) it is possible to assess the contribution of the three coupling sites to the acclimation effect. Table 2 demonstrates that Δ ADP/O is the same irrespective of whether phosphorylation sites II and III (succinate as substrate) or sites I, II and III (malate plus glutamate) are in operation. Thus the acclimation effect appears to be due entirely to sites II and III, site I remaining unaffected by the two acclimation temperatures. It has to be admitted, however, that the interpretation of the data involving site III is somewhat difficult, since due to the small number of experiments the difference in ADP/O ratio, although of the proper magnitude, is not significant at the 5% level (Table 2). Thus whether the acclimation effect is due to sites II plus III, or to site II only, cannot be decided on the strength of the present data.

At any rate, the difference between sites II and III on the one hand, site I on the other may be of significance with regard to modifications of the chemical hypothesis of oxidative phosphorylation recently published (Storey, 1970, 1971; Chance and Erecinska, 1971). Storey postulates that the energy conserved in the Cyt a, a_3-

oxygen step is fed in equal parts to phosphorylation sites II and III, resulting in ADP/O ratios of unity.

Cyt a, a_3 depends strongly on mitochondrial lipids (Rothfield and Romeo, 1971; Fessenden-Raden and Racker, 1971) and these are known to change under the influence of acclimation temperature (Caldwell and Vernberg, 1970; Hazel, this symposium). Thus it may be speculated that the decrease in ADP/O ratio at coupling sites II and, perhaps, III is due to altered properties of the inner membrane phospholipids.

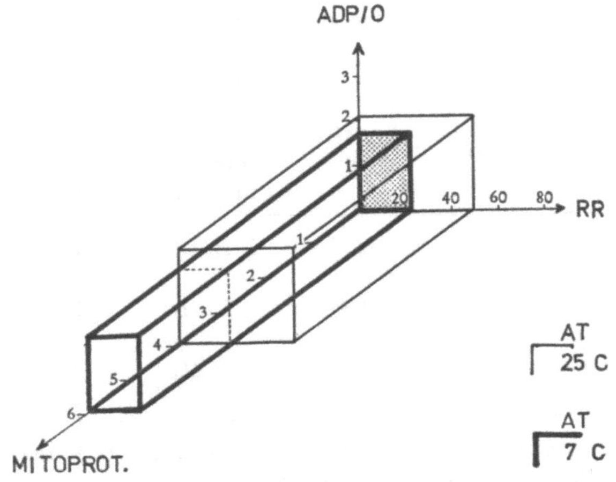

Fig. 3. Influence of acclimation temperatures (7 C and 25 C) on the capacity of phosphorylation of 1 g of liver. The volumes of the prisms represent the capacity of phosphorylation [nM ATP/min] of 1 gram liver with malate + glutamate as substrates. ADP/O = ADP/O-ratio [nM ATP/n g atom O]. RR = respiration rate (st 3) [n g atom O/mg · min]. MITOPROT. = mitochondrial protein content of organ [mg/g]

In Fig. 3 the aerobic phosphorylation capacity of the livers of warm- and cold-acclimated eels is plotted in a rectangular co-ordinate system. The maximum rate of respiration of liver mitochondria (Table 1) and their phosphorylation efficiency (Table 2) result in an increased capacity of the livers of warm-acclimated eels although total mitochondrial protein content is diminished (Fig. 1).

The areas in the plane of (ADP/O · RR) represent the mitochondrial capacities of phosphorylation (= maximum "specific activity of phosphorylation"); the rectangular prisms represent the capacities of phosphorylation of 1 gram of liver. The ratio AT7C/AT25C of the areas equals 0.38, whereas that of the volumes equals 0.58.

Conclusion

Following cold acclimation the mitochondrial protein content of the liver of eels increases, whereas the maximum respiratory activity of liver mitochondria decreases. This effect of low acclimation temperature on mitochondrial specific activity is likely to be due to a decrease in the amount of electron transport chain enzymes per mitochondrial protein.

Diminished efficiency of oxidative phosphorylation following cold acclimation, if measured at 25 C, occurs at coupling sites II and, perhaps, III. This is an intrinsic property of the mitochondria studied and might be due to alterations in the inner membrane.

Fig. 3 summarizes the effects and shows that aerobic energy production in the liver of adult eels increases following warm acclimation. This may be interpreted as adaptive and advantageous to the eel in order to meet the demands of increased metabolic activity in the warm environment.

Summary

1. Silver eels (300–500 g) were acclimated in tap water to 7 C (21 days and more) and to 25 C (12–14 days).

2. The efficiency of oxidative phosphorylation of isolated mitochondria from cold acclimated eels is lower than that from warm-acclimated eels.

3. It is suggested that the ADP/O ratio of mitochondria of cold-acclimated eels is dimished at coupling sites II and, perhaps, III, but not at site I.

4. The capacity of oxidative phosphorylation is calculated by combining phosphorylation efficiency, respiratory rate and protein content of liver mitochondria. It is shown that the phosphorylation *capacity* of the *whole organ* increases with high temperature acclimation but not as much as the *specific activity* of phosphorylation. of mitochondria.

References

Adam, H.: Adenosin-5'-diphosphat und Adenosin-5'-monophosphat. In: Bergmeyer, H. U. (Ed.): Methoden der enzymatischen Analyse, p. 573–577. Weinheim/Bergstraße: Verlag Chemie 1962.

Beisenherz, G., Boltze, H. J., Bücher, Th., Czok, R., Carbade, K. H., Meyer-Arendt, E., Pfleiderer, G.: Diphosphofructose-Aldolase, Phosphoglyceraldehyd-Dehydrogenase, Milchsäure-Dehydrogenase, Glycerophosphat-Dehydrogenase und Pyruvatkinase aus Kaninchenmuskulatur in einem Arbeitsgang. Z. Naturforsch. (B) **8**, 555–577 (1953).

Caldwell, R. S.: Temperature acclimation in the goldfish (*Carassius auratus* L.): Studies on terminal electron transport and the role of lipids. Ph. D. Thesis, Duke University, North Carolina (1967).

Caldwell, R. S., Vernberg, F. J.: The influence of acclimation temperature on the lipid composition of fish gill mitochondria. Comp. Biochem. Physiol. **34**, 179–191 (1970).

Chance, B., Erecinska, M.: Flowflash kinetics of the cytochrome a_3-Oxygen Reaction in coupled and uncoupled mitochondria using the liquid dye laser. Arch. Biochem. Biophys. **143**, 675–687 (1971).

Chance, B., Erecinska, M., Williams, G. R.: Respiratory enzymes in oxidative phosphorylation. III. The steady state. J. Biol. Chem. **217**, 409–427 (1955).

Davis, J. T.: A technique for the isolation of mitochondria from bovine lymphocytes. In: Colowick, S. P., Kaplan, N. O. (Eds): Methods in Enzymology, Vol. X. Oxidation and Phosphorylation, p. 114–117. New York and London: Academic Press 1967.

Estabrook, R. W.: Mitochondrial respiratory control and the polarographic measurement of ADP: O ratios. In: Colowick, S. P., Kaplan, N. O. (Eds.): Methods in Enzymology, Vol. X. Oxidation and Phosphorylation, p. 41–47. New York and London: Academic Press 1967.

Fessenden-Raden, J. M., Racker, E.: Structural and functional organization of mitochondrial membranes. In: Rothfield, L. I. (Ed.): Structure and Function of Biological Membranes, p. 401–438. New York-London: Academic Press 1971.

Gumbmann, M., Tappel, A. L.: The tricarboxylic acid cycle in fish. Arch. Biochem. Biophys. **98**, 262–270 (1962).

Hazel, J., Prosser, C. L.: Interpretation of inverse acclimation to temperature. Z. Vergleich. Physiol. **67**, 217–228 (1970).

Hoch, F.: Energy transformations in animals: regulatory mechanisms, p. 173. Philadelphia-London-Toronto: W. B. Saunders Co. 1971.

Hoch, F., Motta, M. V.: Reversal of early thyroid hormone action on mitochondria by bovine serum albumine in vitro. Proc. Natl. Acad. Sci. U.S. **59**, 118–122 (1968).

Hochachka, P., Somero, G. N.: Biochemical adaptation to the environment. In: Hoar, W. S., Randall, D. J. (Eds.): Fish Physiology, Vol. 6, 99–156 New York and London: Academic Press 1971.

Jankowsky, H. D., Korn, H.: Der Einfluß der Adaptationstemperatur auf den Mitochondriengehalt des Fischmuskels. Naturwiss. **52**, 642–643 (1965).

Kadenbach, B.: Der Einfluß von Thyreoidhormonen in vivo auf die oxidative Phosphorylierung und Enzymaktivitäten in Mitochondrien. Biochem. Z. **344**, 49–75 (1966).

Kanungo, M. S., Prosser, C. L.: Physiological and biochemical adaptation of goldfish to cold and warm temperatures – II. Oxygen consumption and oxidative phosphorylation of liver mitochondria. J. Cellular Comp. Physiol. **54**, 265–274 (1959).

Mellors, A., Tappel, A. L., Sawant, P. L., Desai, I. D.: Mitochondrial swelling and uncoupling of oxidative phosphorylation by lysosomes. Biochem. Biophys. Acta **143**, 299–309 (1967).

Patkin, J., Masoro, E. J.: Effects of cold stress on mitochondrial oxidative phosphorylation. Am. J. Physiol. **199**, 201–202 (1960).

Precht, H.: Der Einfluß der Temperatur auf die Atmung und auf einige Fermente beim Aal (*Anguilla vulgaris* L.). Biol. Zentr. **70**, 71–85 (1951).

Precht, H.: Beiträge zur Temperaturadaptation des Aales (*Anguilla vulgaris* L.) I. Z. Vergleich. Physiol. **44**, 451–462 (1961).

Precht, H.: Über die Bedeutung des Blutes für die Temperaturadaptation von Fischen. Zool. Jb. Physiol. **71**, 313–327 (1965).

Precht, H.: Der Einfluß „normaler" Temperaturen auf Lebensprozesse bei wechselwarmen Tieren unter Ausschluß der Wachstums- und Entwicklungsprozesse. Helgoländer Wiss. Meeresuntersuch. **18**, 487–548 (1968).

Prosser, C. L.: Metabolic and central nervous acclimation of fish to cold. In: Troshin, A. S. (Ed.): The cell and environmental temperature, p. 375–383. London-New York: Pergamon Press 1967.

Rothfield, L. I., Romeo, D.: Enyzme reactions in biological membranes. In: Rothfield, L. I. (Ed.): Structure and Function of Biological Membranes, p. 251–284. New York and London: Academic Press (1971).

Schultze, D.: Beiträge zur Temperaturadaptation des Aales (*Anguilla vulgaris* L.) II. Z. Wiss. Zool. **172**, 104–133 (1965).

Slater, E. C.: Mechanism of phosphorylation in the respiratory chain. Nature **172**, 975–978 (1953).

Storey, B. T.: Chemical hypothesis for energy conservation in the mitochondrial respiratory chain. J. Theoret. Biol. **28**, 233–259 (1970).

Storey, B. T.: Chemical hypothesis for energy conservation in the mitochondrial respiratory chain. II. Redox properties of the coupling site enzymes at sites I and II. J. Theoret. Biol. **31**, 533–552 (1971).

Szarkowska, L., Klingenberg, M.: On the role of ubiquinone in Mitochondria. Biochem. Z. **338**, 674–697 (1963).

Thiessen, C. I., Mutchmor, J. A.: Some effects of thermal acclimation on muscle apyrase activity and mitochondrial number in *Periplaneta americana* and *Musca domestica*. J. Insect Physiol. **13**, 1837–1842 (1967).

Van den Bergh, G. S.: Fatty acid activation and oxidation by Mitochondria. In: Colowick, S. P., Kaplan, N. O. (Eds): Methods in Enzymology, Vol. X. Oxidation and Phosphorylation, p. 749–755. New York and London: Academic Press 1967.

Temperature-compensated Oxygen Consumption of Subcellular Preparations from Vertebrate Ectotherm, Endotherm, and Tumor Tissues

Leland G. Johnson and Richard C. Newell

With 7 Figures

Introduction

There are many examples of long term adjustments to environmental temperature which result in fundamental differences in response to temperature change among organisms acclimated to various temperatures (Scholander, Flagg, Walters, and Irving, 1953; Prosser and Brown, 1961). Some of these adjustments are genetic and have taken place over very long time spans; but within the limits of genetically-determined capacities, individual organisms can reorganize physiological functions in response to exposure to new environmental temperatures. Frequently, there are modifications of rate:temperature relationships of various activities so that when comparisons are made at common intermediate temperatures, cold-acclimated animals tend to perform at higher rates than warm-acclimated animals.

Another type of compensatory mechanism results in a relative insensitivity to acute temperature fluctuations within periods of time as short as a few hours or less. Bullock (1955) and Newell (this symposium) have cited a number of studies on a variety of activities in ectothermic animals which show this type of relative temperature independence. Several studies on oxygen consumption in ectothermic organisms have revealed temperature-independent responses in the absence of organismic or even cellular integrity (Newell, this symposium). Relative temperature independence at low substrate levels has been demonstrated for the oxygen consumption of crude, cell-free homogenates prepared in physiological saline solution and for partially purified suspensions of mitochondria in sucrose. Temperature alterations had relatively little effect on oxygen consumption rates of these preparations throughout the organisms' normal environmental temperature ranges. In each case, this stability of oxygen consumption rate was obtained for measurements taken at temperatures up to the highest normally encountered environmental temperature. Beyond this point, a temperature was encountered at which a large increase in the rate occurred in response to a small increase in the measurement temperature. Thus, the rate of reaction:temperature ($R:T$) curve for homogenate or mitochondrial oxygen consumption shows a flat region ("plateau") and a region of sharp increase. Finally, after a peak response, higher experimental temperatures produce a decline in the $R:T$ curve. Davison (1971) has reported similar temperature-independent plateaus in substrate and ADP-stimulated respiration of blowfly flight muscle sarcosomes. This type of temperature independence may be useful for homeostasis in ectotherms because it could provide protection against short-term temperature fluctuations which would supplement other mechanisms such as behavioral thermoregulation. The manner in which the positions of various parts of the *R*:*T* curves are shifted in response to seasonal temperature changes or storage at different temperatures under experimental conditions has been reviewed by Newell (this symposium).

Relative temperature independence also has been reported for the respiratory responses of mitochondria from several endotherms (Newell and Walkey, 1966). The adaptive significance of this temperature-independent respiratory response for animals with centrally regulated, stable body temperatures is not clear. But the *R*:*T* curves for these mitochondrial preparations did contain definite plateaus through a range of temperatures below and up to the normal body temperature. Somewhat above the body temperature there was a sharp increase in the metabolic rate. Horwitz and Nelson (1968) found similar plateaus in the *R*:*T* curves for rat and bat liver mitochondria and they reported that the effect could be abolished by an extra washing with sucrose.

This paper reports the results of our studies on the influence of temperature on the oxygen consumption of cell-free preparations from organisms representing several vertebrate classes. We have examined temperature effects on preparations from three adult ectothermic vertebrates, from various stages of the ectotherm-endotherm transition in chicken development, and from normal and tumor cells in mice. We hope that presentation of these results (some of which are preliminary) and a discussion of hypotheses regarding their significance will stimulate criticism and further investigations of temperature compensated responses in vertebrates.

Adult Ectothermic Vertebrates

Cell-free homogenates of the livers of frogs (*Rana temporaria*), terrapins (*Emys orbicularis*), and trout (*Salmo trutta*) were prepared in phosphate buffered (pH 7.4) physiological saline solutions. In each case liver tissue was minced and suspended in approximately three volumes of saline solution, homogenized over ice in a Virtis homogenizer, and centrifuged for 25 minutes at 1,000 g at 2 C to remove unbroken cells and particulate debris. The supernatants were pipetted off carefully and stored on ice until transferred to 5 ml Warburg flasks for measurement of oxygen consumption in Gilson G. R. 20 respirometers. Equal volumes of homogenate and 2 mM pyruvate in saline were mixed in the flasks to produce a concentration of 1 mM added pyruvate in the final reaction mixture. Standard Gilson respirometry techniques were used in the oxygen consumption measurements and results were expressed per unit total TCA-precipitable material in the homogenates.

A cell-free homogenate (in Young's fresh water teleost saline: 5.50 g NaCl, 0.14 g KCl, 0.12 g $CaCl_2$ per liter) was prepared from livers of 29 trout which had been held in aquaria at 7.2 C. Measurements of oxygen consumption were made at several temperatures above and below the storage temperature (Fig. 1A). Similar measurements were made on cell-free homogenates (in Amphibian Ringer solution) from livers of 32 frogs (Fig. 1B) which had been held in a tank with a water temperature of 7.2 C. However, these frogs had opportunity to crawl up and remain out of the water at will. The air temperature above the water was ap-

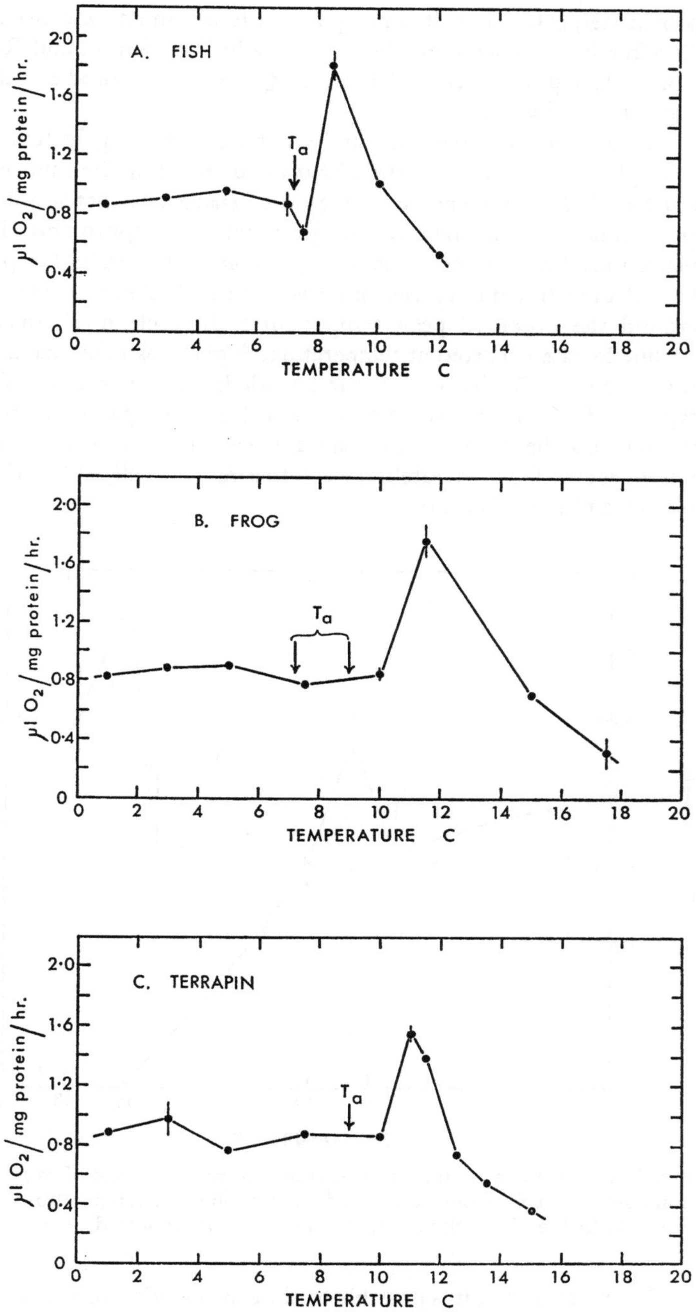

Fig. 1. A–C. Graphs showing the effect of temperature on the rate of oxygen consumption of cellfree liver homogenates. Bars indicate standard errors. The largest standard error plus several other representative ones are included in each curve. T_a indicates acclimation temperature. A Fish *(salmo trutta)*; B frogs *(Rana temporaria)*; C terrapins *(Emys orbicularis)*

proximately 9 C. Thus, it is not possible to assign an exact acclimation temperature. Cell-free homogenates (in the "turtle saline" of Simon and Robin, 1970) of livers from 3 terrapins which had been held at 9 C were used for the measurements represented in Fig. 1 C.

In all cases, oxygen consumption rates were largely temperature independent through a range of temperatures below and including the environmental temperature, but the *R*:*T* curves showed a region of sharp rate increase just above the highest environmental temperature. The peak rate was approximately twice the average plateau-level rate in each case. It is possible that even higher peak rates might have been demonstrated if a series of measurements had been made at very close intervals around the observed peak temperatures. In each *R*:*T* curve a region of sharp decline in rate occurred at temperatures higher than the one at which the peak rate was obtained. To the best of our knowledge, this is the first demonstration of this type of *R*:*T* curve discontinuity in cell-free preparations from vertebrate ectotherms and these *R*:*T* curve patterns are strikingly similar to those reported for homogenates from invertebrate ectotherms (Newell, 1967, 1969, this symposium; Newell and Pye, 1971 a, b).

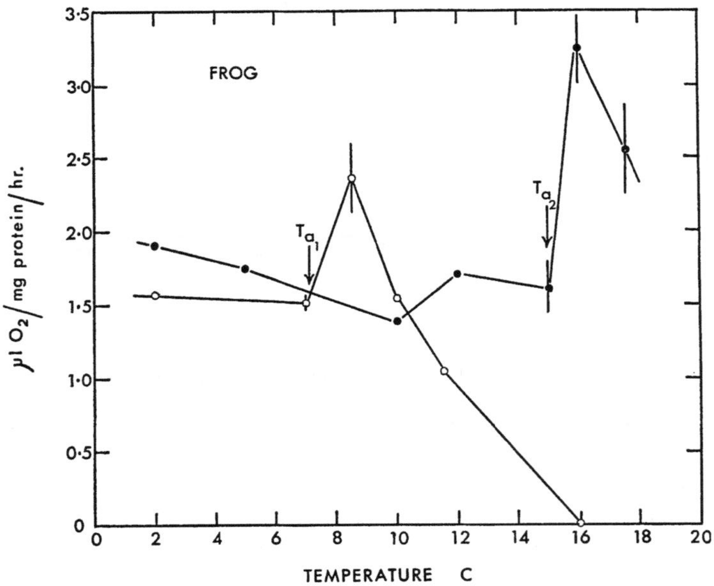

Fig. 2. Graphs showing the effect of temperature on the rate of oxygen consumption of liver homogenates from frogs acclimated to two different temperatures. Bars indicate standard errors as in Fig. 1. Acclimation temperatures are indicated (T_a)

In invertebrates, it is possible to shift the position of various areas of *R*:*T* curves, including the range of the plateau, with thermal acclimation (Newell, this symposium; Newell and Pye, 1970b, 1971 a, b). In order to test the possibility that such shifting might occur in vertebrate ectotherms, a batch of frogs was divided into two samples which were held at different temperatures for 30 days. One group was stored in water at 7.2 C and the other was stored in water at 15 C. The *R*:*T* curves for liver homogenates from the two groups are shown in Fig. 2. On the basis of other experiments, it seems unlikely that the actual peak rate was measured for the frogs acclimated to the lower temperature. Nevertheless, there are a number of obvious differences between the two *R*:*T* curves and the results indicate that frogs show the same sort of *R*:*T* curve shifting with thermal acclimation which has been reported for invertebrates. These results have been confirmed and extended with purified mitochondrial preparations under more rigorously defined experimental measurement conditions (Pye, this symposium).

Avian Developmental Transition

Endotherms undergo a developmental transition from ectothermism to endothermism. This fairly abrupt alteration in temperature relationships apparently involves, in its final stages, completion of establishment of necessary central regulatory mechanisms. Transient endothermic responses are seen in the domestic chicken *(Gallus domesticus)* by about the nineteenth day of incubation and characteristically endothermic responses in both metabolism and maintenance of stable body temperature are seen by about a week after hatching (Freeman, 1964, 1966, 1967; Wekstein and Zolman, 1967, 1968, 1969). Because of this developmental transition, the temperature relationships of chicken mitochondrial oxygen consumption at various stages were of interest.

Mitochondrial suspensions in sucrose were used for measurements on endotherms because the activity of chicken (and mouse) homogenates in saline deteriorated rather quickly. Liver mitochondria were extracted by homogenization of minced liver tissue in ice-chilled, phosphate-buffered, 0.88 M sucrose followed by differential centrifugation. The final pellet obtained by high-speed centrifugation was resuspended in 0.88 M sucrose. Oxygen consumption was measured in Gilson respirometers after equal volumes of this suspension and a reaction medium were mixed in the reaction flasks to give final concentrations of sucrose 0.44 M, KCl 0.0158 M, $MgSO_4$ 0.00158 M, ATP 0.0002 M, and Cytochrome *c* 0.4×10^{-5} M. Different concentrations of added succinate were provided in various experiments as described below. Further details regarding techniques can be obtained from Johnson, Newell, and Hanson (1972).

Fig. 3 shows an *R*:*T* curve for oxygen consumption of a mitochondrial preparation from a young White Leghorn adult measured in the presence of 1 mM succinate. This "plateau" of apparent temperature compensation is found only in the presence of lower added substrate concentrations and temperature dependence is observed in similar preparations from adult chickens at higher substrate levels. The results of a series of measurements of oxygen consumption at various added succinate concentrations and temperatures are presented in Fig. 4. It is evident that at lower added substrate levels the rates are relatively temperature independent between 30 and 40 C, but that with 4.0 mM added succinate the rates are very temperature dependent between 30 and 40 C. These data are quite comparable to results presented elsewhere by Newell and Pye and others which have repeatedly demonstrated that temperature-compensated responses usually are expressed only

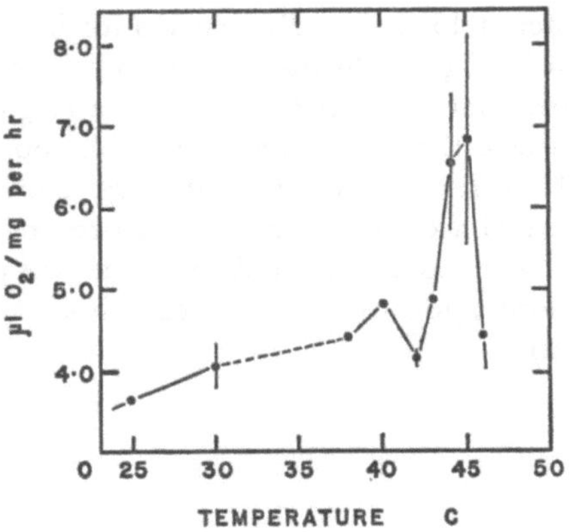

Fig. 3. Graph showing the effect of temperature on the rate of oxygen consumption of an adult chicken liver mitochondrial suspension in sucrose measured in the presence of 1.0 mM added succinate. Bars indicate standard errors as in Fig. 1. (For detailed explanation, see Johnson, Newell and Hanson, 1972)

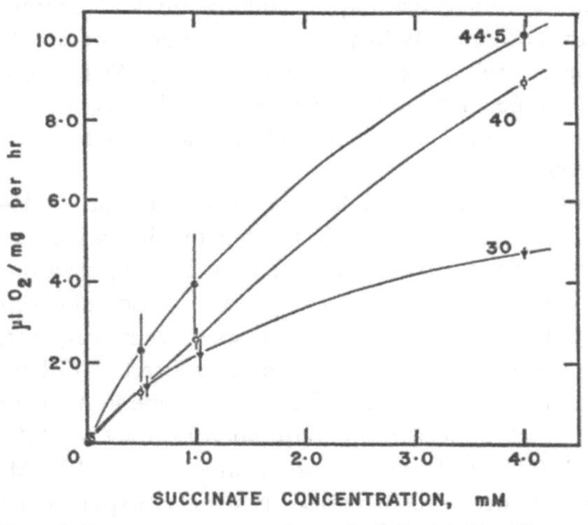

Fig. 4. Oxygen consumption of adult chicken liver mitochondrial suspensions measured in sucrose at several temperatures with various added succinate concentrations. Bars indicate standard errors

in the presence of relatively lower added substrate concentrations and that they are replaced by temperature dependence with higher added substrate concentrations.

The chick embryo responds essentially as an ectotherm throughout most of the incubation period. For purposes of comparison with the endothermic adult, mitochondrial preparations were made from embryonic liver and their temperature relationships were measured under the same conditions used in experiments with the adult material. Even at low added substrate levels, only temperature-dependent $R:T$ curves were obtained (e.g., Fig. 5A). Progressively later stages were studied and temperature-dependent responses likewise were obtained for early post-hatching chicks. For example, when measurements were made on liver mitochondria from

5-day post-hatching chicks at several added substrate levels, we obtained confirmation of temperature dependence at all substrate concentrations measured (Fig. 5B). A series of experiments in which measurements were made in the presence of 1.0 mM succinate at various stages revealed that there was an apparent rotation of the portion of the $R:T$ curve between 30 and 40 C so that the Q_{10} for 30 to 40 C

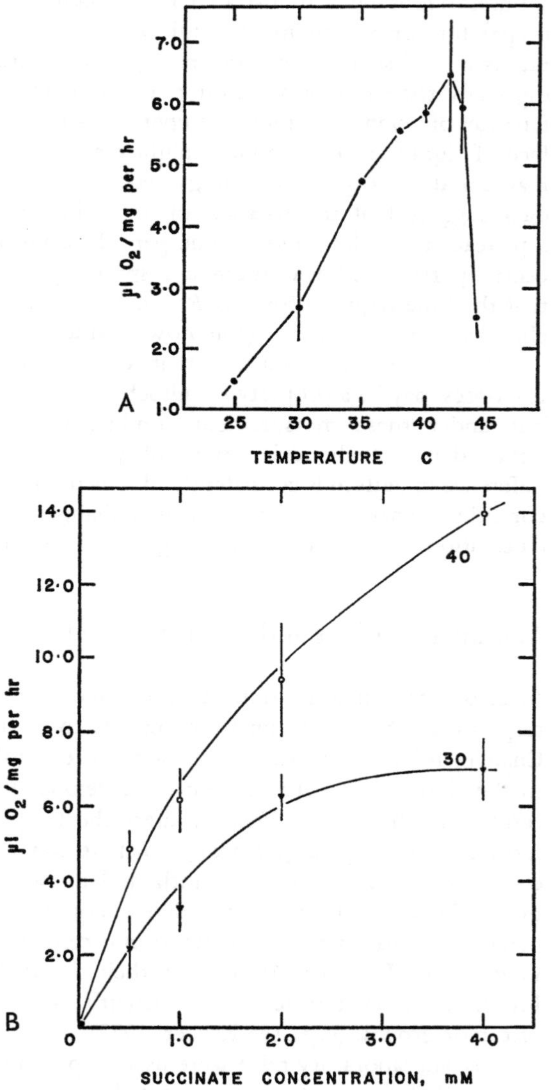

Fig. 5. A Graph showing the effect of temperature on the rate of oxygen consumption of a liver mitochondrial suspension from 17-day chick embryos in sucrose with 1.0 mM added succinate. Bars indicate standard errors as in Fig. 1. B Oxygen consumption of a liver mitochondrial suspension from 5-day post-hatching chickens measured in sucrose at two temperatures with various added succinate concentrations. Bars indicate standard errors

interval fell from 1.90 in the five-day post-hatching to 1.17 in the fourteen-day post-hatching chick.

Because the occurrence of temperature-compensated oxygen consumption at low substrate levels appears to be phylogenetically widespread, the temperature-dependent responses of mitochondria from embryos and very young post-hatching chickens appear to be exceptional results. Possibly, temperature-compensated responses might be demonstrated for mitochondria from these stages under other preparatory or measurement conditions, but assuming that the oxygen-consumption rate is indeed strictly temperature dependent during these stages, an adaptive value especially for embryos with limited metabolic reserves might be proposed. Maintenance of even a minimal temperature-compensated $R{:}T$ curve plateau in the face of significant egg cooling might deplete excessive amounts of stored nutrient material at the expense of energy requirements of developmental processes. These data suggest that the rotation of a portion of the $R{:}T$ curve which establishes a plateau takes place over a time period during and after the establishment of the centrally regulated endothermic response pattern and, perhaps, it is significant that the time required for this $R{:}T$ curve change is in the range of proposed mitochondrial turnover times (Rabinowitz and Swift, 1970).

In conclusion, it appears that the possibility for temperature-compensated responses may be a property of mitochondria from many or most mature organisms, but under these measurement conditions, temperature compensation was not obtained in mitochondria from embryonic and early post-hatching chicks. If this difference in pattern is consistently obtained under other extraction and measurement conditions, analysis of developmental changes in this system might prove useful in elucidation of mechanisms of temperature compensated responses.

Mammalian Liver and Tumor Mitochondria

In ectotherms, modification of mitochondrial oxygen consumption $R{:}T$ curves in response to thermal acclimation appears to be a normal physiological occurrence. Changes following warm acclimation, for example, include translation of the curve to the right and an altered heat-lethal temperature for the system. In at least one ectotherm, thermal acclimation alters the temperature at which mitochondria are most tightly coupled (Davison, 1971). It was of interest to compare temperature effects on tumor cell mitochondria and mitochondria from other tissues to determine whether such $R{:}T$ curve differences are associated with the differential temperature sensitivity of both intact tumor cells (Cavaliere *et al.*, 1967; Vollmar, 1941; Levine and Robbins, 1971) and their isolated mitochondria (Christiansen and Kvamme, 1969). Further, there currently is considerable interest in the energetics of tumor growth (e.g., see Racker, 1972).

P-strain laboratory mice were used in preliminary experiments to investigate the existence of temperature-compensated responses in mouse mitochondria. Minced liver was homogenized in a Virtis homogenizer in 9 volumes of phosphate buffered (pH 7.4) 0.68 M sucrose and all subsequent procedures were carried out at 1 to 4 C. The homogenate was centrifuged at 1,000 g for 25 min. The supernatant was spun at 17,000 g for 30 min and pellets were resuspended gently in sucrose. Initial experiments showed that the presence of enough endogenous substrate to produce measurable respiration without added substrate always resulted in temperature dependent responses when substrate was added under experimental conditions. Therefore, a second high-speed centrifugation was used in most experiments including the experiments comparing tumor and liver mitochondria. Although a second centrifugation usually was necessary, some preparations contained sufficiently low endogenous substrate levels to produce the temperature-compensated response after only one high-speed centrifugation (e.g., data presented in Fig. 6). The final pellet was resuspended in a small volume of sucrose and stored over ice until transferred to the reaction flasks. Equal volumes of mitochondrial suspension and mixed reagents were placed in flasks so that the final concentrations in the reaction mixtures were: sucrose 0.34M, KCl 0.0158M, $MgSO_4$ 0.00158M, ATP 0.0002M, Cyt. *c* 0.4×10^{-5}M and succinate 1.0mM. The pH of the reaction medium had been adjusted to 7.4 at 38 C. Oxygen consumption was measured in Gilson differential respirometers (model G. R. 20).

On the 5th day of implantation, twenty four mice (strain C57 × DBA2) were sacrificed to provide Ehrlich ascites tumor cells and liver. Ascites cells were sedimented at 1000 g for 10 min at 2 C, made up to 30 ml in 0.68M sucrose in phosphate buffer (pH 7.4), homogenized over ice with a Virtis 23 homogenizer, and diluted to 150 ml in sucrose. Liver homogenates were prepared from the same mice as outlined above. Mitochondrial isolation, reaction media, and measurement techniques were as described. Results were expressed in terms of dry weight of total TCA-precipitable material in the mitochondrial suspensions.

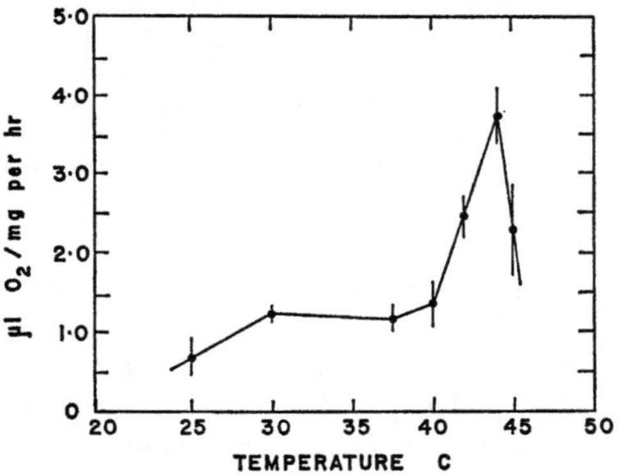

Fig. 6. Graph showing the effect of temperature on the rate of oxygen consumption of a crude mitochondrial preparation from livers of *P*-strain laboratory mice measured in the presence of 1.0 mM succinate. Bars indicate standard errors

Fig. 6 shows the *R*:*T* curve for oxygen consumption of a crude mitochondrial preparation from the liver of a P-strain laboratory mice. It is evident that the rate is quite temperature dependent between 25 and 30 C and very markedly so between 39 and 44 C, but that the rate appears essentially insensitive to temperature change between 30 and approximately 39 C. *R*:*T* curves for mitochondria from the livers and from ascites tumors of C57 × DBA2 mice in the presence of 1 mM succinate are superimposed in Fig. 7. It will be noticed that in each case there is a region of temperature insensitivity followed by a marked increase in metabolism at higher temperatures. The *R*:*T* curve for the ascites mitochondria is, however, to the left of that for liver mitochondria. The region of marked temperature dependence thus occurs at lower temperatures in the ascites mitochondria and the point of thermal decline occurs just above 40.5 C rather than at temperatures above 43 C as in the liver mitochondria. No measurable uptake occurred at 44 C in the tumor mitochondria whereas the rate was 7.8 μl/mg/hr in the liver mitochondria.

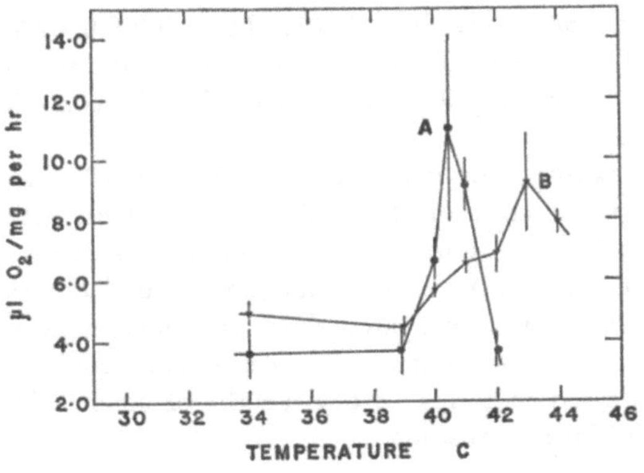

Fig. 7. Graphs showing the effect of temperature on the rate of oxygen consumption of suspensions of mitochondria extracted from *A* Ehrlich ascites tumor cells and *B* from livers of the same mice (strain C57 × DBA2) measured in the presence of 1.0 mM added succinate. Bars indicate standard errors. [The difference in level from the *P*-strain mouse mitochondria may be attributable to preparative procedures (see text), strain differences, or other factors.] No measurable oxygen uptake occurred at 44 C in the tumor mitochondria. Dr. V. I. Pye participated in these experiments (see acknowledgment)

We do not know yet whether the rate:temperature curves are the same for the metabolism of mitochondria from other normal tissues or how this finding relates to other aspects of tumor cell metabolism (e.g. Racker, 1972; Koobs, 1972). Nevertheless, if this type of relationship proves to be of general occurrence in comparisons between tumor cell and normal tissue mitochondria, it may be related to the enhanced metabolism and growth of tumor cells compared with normal cells at body temperatures. Further, the increased sensitivity to temperature-induced uncoupling and membrane damage in ascites mitochondria compared with those from other tissues (Christiansen and Kvamme, 1969) may be analogous to the translation of the metabolic *R*:*T* curve which accompanies thermal acclimation in many ectotherms. The data also would suggest that differential inhibition of Ehrlich ascites tumor metabolism could be accomplished by raising the body temperature to a carefully selected point without thermal damage to other mitochondria. Possibly, examination of temperature effects under these conditions might be used to select appropriate inhibitory temperatures for other types of tumors.

Discussion

The data presented here demonstrate that the metabolism of cell-free preparations of a variety of vertebrates, including both ectotherms and endotherms, is nearly independent of temperature over at least part of the range of tolerance of the organism. In this respect the data are similar to those obtained on many invertebrates (for review see Newell, 1969, 1970) and on some endotherms (Newell and Walkey, 1966; Horwitz and Nelson, 1968). Further, the data obtained with liver homogenates of frogs acclimated to different temperatures show that the plateau, or region of metabolic homeostasis, varies in its temperature range according to acclimation temperature. Such results also closely resemble those obtained on intertidal invertebrates (Newell and Pye, 1970a, b, 1971a, b; Newell, this symposium; Pye, this symposium) and on insect flight muscle sarcosomes (Davison, 1971).

The possible biochemical mechanisms underlying such compensation for acute temperature fluctuation have received considerable attention recently, and have been reviewed by Atkinson (1966), Hochachka (this symposium) and Hochachka and Somero (1971). There are, for example, many alternative and cross-linked metabolic pathways which may be differently affected by temperature and by means of which carbon flows through glycolysis and the Krebs cycle is maintained constant and independent of temperature. They have also emphasized the importance of variations in the enzyme-substrate affinity of regulatory enzymes in maintenance of temperature-compensated metabolism. Further, different isozymes each capable of contributing to thermal compensation over a particular temperature range are known to occur and to be activated in response to thermal acclimation. Such properties may thus account not only for the maintenance of temperature-independent respiration over a particular temperature range, but also for the alterations in the range of such metabolic homeostasis in response to thermal acclimation.

Davison (1971) has demonstrated that the rapid increase in the metabolism of *Calliphora* sarcosomes, which occurs above the temperature tho which the flies were acclimated, is accompanied by the uncoupling of respiration. This result is in accordance with the suggestion of Newell and Walkey (1966) that the steep increase in the metabolism of mammalian liver mitochondria at experimental temperatures above the normal body temperature was associated with uncoupling. Since the temperature at which this occurred in *Calliphora* varied with the temperature to which the flies had been acclimated, it follows that variations in the susceptibility to uncoupling may be implicated in the variations of temperature range over which metabolic homeostasis occurs.

Thus, the control of metabolism in response to temperature change is complicated and is likely to involve several different mechanisms according to the organization of the system and the experimental conditions involved. The variations in K_m with temperature reported for LDH and a variety of other enzymes (see Somero, 1969; Hochachka and Somero, 1971; Hochachka, this symposium) would satisfactorily account for the high Q_{10} values obtained at low experimental temperatures (called "phase 1" of the $R:T$ curve by Davison, 1971) and also for the presence of a region of temperature independence (called "phase 2" by Davison). However, the fact that we have obtained low Q_{10} values with succinate as well as pyruvate (see also Newell and Pye, 1971a, b) suggests that variations in the enzyme-substrate affinity of some intramitochondrial enzyme system rather than LDH may be implicated in the maintenance of temperature-independent metabolism in response to acute temperature change. Alternatively, the system may be subject to overall control by the coupling enzymes. The maximal rates of respiration, with their associated fall in respiratory control (Davison, 1971) and high Q_{10} values for small temperature changes above the plateau range, may be due to thermal damage of the mitochondrial membranes (Christiansen and Kvamme, 1969; Raison, Lyons and Thomson, 1971) since respiratory control is affected by even slight damage of this kind (Chance, 1959). It is also possible that the very temperature-dependent portion of the $R:T$ curve is associated with a temperature-induced phase change in the mitochondrial membranes which may cause configurational changes in the enzyme proteins as is discussed by Raison, Lyons and Thomson (1971) and Hazel (this symposium).

The biological significance for ectotherms of compensation for acute temperature change and hypotheses relating such compensation to metabolic regulation and thermal acclimation have been treated extensively by Newell (this symposium). Clearly, the range of thermal responses observed in subcellular preparations might contribute to the variety of thermal responses of ectothermic organisms which range from the more commonly reported temperature-dependent responses to virtual temperature independence (Bullock, 1955; Newell, 1969; Aleksiuk, 1971). It has been suggested that the range between the plateau level of activity seen in temperature-compensated responses and the temperature-dependent maximal responses observed, for instance, in the presence of higher added substrate concentrations represents a "scope for activity". Because of the apparently simultaneous translation of both types of $R:T$ curves with thermal acclimation, the maximal "scope for activity" is adjusted to the organism's environmental temperature range. Thus, it is proposed that the temperature range permitting the greatest possibilities for adjustment of metabolism with changes in general activity, hormonal state, nutritional condition, or other systemic factors is rather closely attuned to the environmental temperature range which the organism experiences.

While the significance of temperature-compensated responses for the temperature relationships of centrally regulating endotherms is obscure, some hypothetical comments might help to put the endotherm results in perspective. It appears that at least under certain measurement conditions, it is possible to elicit a variety of patterns of response to temperature change from endotherm preparations which is similar to that observed in ectotherm preparations. Normally, of course, the temperature range of the cellular environment of a typical endotherm varies only to the extent of the diurnal fluctuations of the organism's body temperature. It is interesting that the body temperatures of the endotherms which we have studied appear to bear the same sort of relationship to the $R:T$ curves for oxygen consumption of subcellular preparations which can be observed for acclimation temperatures in ectotherms. Thus, the maximum "scope for activity" is again expressed at or around the endothermic organism's normal body temperature.

It might be argued, therefore, that the presence of $R:T$ curve plateaus in endotherms under certain measurement conditions simply represents one evidence of a general basic metabolic regulatory pattern. Normally this "scope for activity" would be utilized only within a very narrow range of temperatures (i.e., at the organism's regulated body temperature). If variation of available intracellular or even intramitochondrial substrate level (or the concentration of other substances whose level influenced the rate of substrate utilization) were key factors in determining the oxidative activity level, a number of regulators might act through controlling release or transmembrane penetration of these materials. In this manner, they would influence the degree of utilization of the "scope of activity". For instance, the calorigenic effect of thyroid hormone might be one example of such a regulatory effect. Further, this would lead to the proposal that the variable results obtained in studies attempting direct stimulatory effects of thyroid hormone on mitochondria *in vitro* may be related to choice of substrate levels utilized in test systems. Possibly, this problem should be re-explored under a variety of test conditions with an emphasis on use of relatively lower substrate concentrations than are commonly chosen for such studies. Certainly, it would seem worthwhile to examine the action of thyroid hormone and various other metabolic regulators as well in light of these proposals.

The $R:T$ curve characteristics of a variety of normal and malignant or diseased tissues should also be explored further. Possibly, the model of $R:T$ curve translation available as a result of studies on ectotherm systems (see Newell, this symposium) would prove valuable in interpretation of results gained in studies on endotherm material. Should "shifted" $R:T$ curves prove to be common, information gained from their study might well prove useful for diagnosis or even for therapy.

Finally, these results must be regarded as preliminary and tentative and there is an urgent need for more data gained under conditions in which the physiological condition of the mitochondria could be better assessed (see, for example, Pye, this symposium). But is is hoped that the hypothetical proposals made in this paper will serve as a stimulus to further studies of the occurrence and significance of temperature-compensated metabolic responses in vertebrates.

Acknowledgements

We worked with Dr. V. I. Pye on the mouse tumor mitochondria experiments and we gratefully recognize her contribution to that portion of the work reported here.

This work was supported by equipment grants from the Royal Society of London and the S. R. C. L. G. Johnson was a N. S. F. Science Faculty Fellow and received support at Augustana College from Research Corporation and the Augustana Research and Artist Fund. We are very grateful to Mr. B. C. Michley and Dr. A. M. Neville of the Department of Experimental Pathology, Chester Beatty Research Institute for the mouse tumor material and also to Mr. C. Fowler and Miss V. Patchen for technical assistance in all of the experiments.

References

Aleksiuk, M.: Temperature-dependent shifts in the metabolism of a cool temperate reptile, *Thamnophis sirtalis parietalis*. Comp. Biochem. Physiol. **39 A**, 495–503 (1971).

Atkinson, D. E.: Regulation of enzyme activity. Ann. Rev. Biochem. **35**, 85–124 (1966).

Bullock, T. H.: Compensation for temperature in the metabolism and activity of poikilotherms. Biol. Rev. **30**, 311–342 (1955).

Cavaliere, R., Ciotto, E. C., Giovanella, B. P., Heidelberger, C., Johnson, R. O., Margottini, M., Mondovi, B., Moricca, G., Rossi-Fanelli, A.: Selective heat sensitivity of cancer cells. Cancer **20**, 1351–1381 (1967).

Chance, B.: Quantitative aspects of the control of oxygen utilization. In: Wolstenholme, G. E. W., O'Connor, C. M. (Eds.): Ciba Foundation Symposium on the Regulation of Cell Metabolism, p. 91–129. London: J. &. A. Churchill Ltd. 1959.

Christiansen, E. N., Kvamme, E.: Effects of thermal treatment on mitochondria of brain, liver and ascites cells. Acta Physiol. Scand. **76**, 472–484 (1969).

Davison, T. F.: The effect of temperature on oxidative phosphorylation in isolated flight muscle sarcosomes. Comp. Biochem. Physiol. **38 B**, 21–34 (1971).

Freeman, B. M.: The emergence of the homeothermic-metabolic response in the fowl *(Gallus domesticus)*. Comp. Biochem. Physiol. **13**, 413–422 (1964).

Freeman, B. M.: The effects of cold, noradrenaline and adrenaline upon the oxygen consumption and carbohydrate metabolism of the young fowl *(Gallus domesticus)*. Comp. Biochem. Physiol. **18**, 369–382 (1966).

Freeman, B. M.: Some effects of cold on the metabolism of the fowl during the perinatal period. Comp. Biochem. Physiol. **20**, 179–193 (1967).

Hochachka, P. W., Somero, G. N.: Biochemical adaptation to the environment. In: Hoar, W. S., Randall, D. J. (Eds.): Fish Physiology, Vol. VI, p. 99–156. New York: Academic Press 1971.

Horwitz, B. A., Nelson, L.: Effect of temperature on mitochondrial respiration in a hibernator *(Myotis austroriparius)* and a non-hibernator *(Rattus rattus)*. Comp. Biochem. Physiol. **24**, 385–394 (1968).

Johnson, L. G., Newell, R. C., Hanson, R. D.: Temperature relationships of mitochondrial respiration during chicken development. Comp. Biochem. Physiol. **42 B**, 693–701 (1972).

Koobs, D. H.: Phosphate mediation of the Crabtree and Pasteur effects. Science **178**, 127–133 (1972).

Levine, E. M., Robbins, E. B.: Differential temperature sensitivity of normal and cancer cells in culture. J. Cell. Physiol. **76**, 373–380 (1971).

Newell, R. C.: Oxidative activity of poikilotherm mitochondria as a function of temperature. J. Zool. (London) **151**, 299–311 (1967).

Newell, R. C.: Effects of fluctuations in temperature on the metabolism of intertidal invertebrates. Am. Zoologist **9**, 293–307 (1969).

Newell, R. C.: Biology of Intertidal Animals. London: Logos Press 1970.

Newell, R. C., Pye, V. I.: Seasonal changes in the effect of temperature on the oxygen consumption of the winkle *Littorina littorea* (L.) and the mussel *Mytilus edulis* L. Comp. Biochem. Physiol. **34**, 367–383 (1970a).

Newell, R. C., Pye, V. I.: The influence of thermal acclimation on the relation between oxygen consumption and temperature in *Littorina littorea* (L.) and *Mytilus edulis* L. Comp. Biochem. Physiol. **34**, 385–397 (1970b).

Newell, R. C., Pye, V. I.: Quantitative aspects of the relationship between metabolism and temperature in the winkle *Littorina littorea* (L.) Comp. Biochem. Physiol. **38 B**, 635–650 (1971a).

Newell, R. C., Pye, V. I.: Temperature-induced variations in the respiration of mitochondria from the winkle *Littorina littorea* (L.). Comp. Biochem. Physiol. **40 B**, 249–261 (1971b).

Newell, R. C., Walkey, M.: Oxidative activity of mammalian liver mitochondria as a function of temperature. Nature (London) **212**, 428–429 (1966).

Prosser, C. L., Brown, F. A., Jr.: Comparative Animal Physiology, 2nd ed. Philadelphia: W. B. Saunders Co. 1961.

Rabinowitz, M., Swift, H.: Mitochondrial nucleic acids and their relation to the biogenesis of mitochondria. Physiol. Rev. **50**, 376–427 (1970).

Racker, E.: Bioenergetics and the problem of tumor growth. Am. Scientist **60**, 56–63 (1972).

Raison, J. K., Lyons, J. M., Thomson, W. W.: The influence of membranes on the temperature-induced changes in the kinetics of some respiratory enzymes of mitochondria. Arch. Biochem. Biophys. **142**, 83–90 (1971).

Scholander, P. F., Flagg, W., Walter, V., Irving, L.: Climatic adaptation in arctic and tropical poikilotherms. Physiol. Zool. **26**, 67–92 (1953).

Simon, L. M., Robin, E. D.: Changes in heart and skeletal muscle cytochrome oxidase activity during anaerobiosis in the freshwater turtle, *Pseudemys scripta elegans*. Comp. Biochem. Physiol. **37**, 437–443 (1970).

Somero, G.: Enzymic mechanisms of temperature compensation: immediate and evolutionary effects of temperature on enzymes of aquatic poikilotherms. Am. Naturalist **103**, 517–530 (1969).

Vollmar, H.: Über den Einfluß der Temperatur auf normales Gewebe und auf Tumorgewebe. Z. Krebsforsch. **51**, 71–99 (1941).

Wekstein, D. R., Zolman, J. F.: Homeothermic development of the young chick. Proc. Soc. Exptl. Biol. Med. **125**, 294–297 (1967).

Wekstein, D. R., Zolman, J. F.: Sympathetic control of homeothermy in the young chick. Am. J. Physiol. **214**, 908–912 (1968).

Wekstein, D. R., Zolman, J. F.: Ontogeny of heat production in chicks. Federation Proc. Am. Soc. Exptl. Biol. **28**, 1023–1027 (1969).

Molecular Mechanisms of Temperature Adaptation in Arctic Ectotherms and Heterotherms

HANS W. BEHRISCH

With 8 Figures

Introduction

The dispersion into and exploitation of the Arctic by northern organisms has been made possible by the evolutionary development of metabolic characteristics which permit the animal to maintain and control its metabolic processes in an environment of low, and sometimes widely changing, temperatures. The molecular mechanisms of adaptation to temperature appear to be exceedingly complex, and a large literature on the subject has accumulated (see Hochachka and Somero, 1971, for a recent review). One of the more outstanding observations in recent years is that numerous physiologic functions in ectotherms are insensitive to temperature (Newell, 1966, 1967; Newell and Northcroft, 1967) and thus there must be mechanisms which permit the animal's metabolism to remain independent of thermal change. There is a steadily growing literature on the mechanisms possibly responsible.

The present discussion will be limited to a selected group of arctic and subarctic marine ectotherms, seals and hibernators, animals with which we have been working in this laboratory for the last two years. The hibernators belong to a group of animals called heterothermic (see Bartholomew, 1968) since their temperature relations alternate between endothermy and ectothermy. However, at tissue levels the term heterothermy may be applied even to non-hibernating mammals if it is considered that their core tissues are endothermic whereas the temperature of some of their peripheral tissues is determined, to a varying degree, by the external temperature. The reason for including these mammals in the present communication is the question of whether animals as diverse as trout, seal and ground squirrel might perhaps confront a common problem, low temperature, with similar methods. Thus the central question of this paper might be phrased in the following manner: how do arctic ectotherms and heterotherms alter their biochemistry to permit continued physiologic function at very low temperatures?

It is generally known that temperate fishes, when acclimated to different temperatures, exhibit a rate compensation of metabolism in such a way as to offset the effect of a decrease in temperature (Kanungo and Prosser, 1959; Ekberg, 1958; Precht, 1964, 1965; Dean, 1969). Thus the numerous studies on metabolic compensation in eurythermal fishes such as *Carrassius carrassius*, *Anguilla vulgaris* and *Salmo gairdneri* have established rather firmly that as a rule, the metabolic rate of cold-acclimated members of the same species is higher than that of warm-acclimated animals, when metabolic rate measurements are carried out at a temperature intermediate to the two acclimation temperatures. Further, when comparisons are made of the metabolic rates of ectotherms that have adapted – in evolutionary time – to widely different temperatures, an almost identical pattern of metabolic compensation is observed (Scholander *et al.*, 1953; Somero, Giese and Wohlschlag, 1968; Wohlschlag, 1964). However, it is becoming apparent that temperature affects not only the extent of metabolism, but also the relative contributions of the various pathways which contribute to the total metabolic rate of the animal. The mechanisms by which this other aspect of metabolic compensation, namely "metabolic reorganization" is effected, is essentially unknown, although a number of possibilities have been suggested (Hochachka and Hayes, 1962; Hochachka, 1968; Caldwell and Vernberg, 1970).

It is instructive to analyze the metabolic chemistry of organisms against an evolutionary and ecological background by discussing theoretical mechanisms of thermal adaptation in ectotherms in teleological terms. In this context the factors which govern the direction and rate of metabolic flow in the cell are considered in terms of the "demands" and "supply" of the cell. The application of this functionalistic philosophy has contributed very significantly to the recognition of integrated reaction sequences as metabolic units with special function and hence selective value. In these terms, just as the individual reaction in the cell has meaning only in light of the sequence in which it occurs, the sequence is useful only in the light of its integration into the overall metabolism of the organism. It is the primary function of adaptation to maintain this integration of metabolism.

"Strategies" of Adaptation to Temperature

There appear to be three ways in which an organism could maintain integration of metabolism over a wide thermal range; (a) the concentrations of intracellular enzymes could be varied in a manner compensatory to thermal change, (b) enzymes with different degrees of "catalytic efficiency" could be synthesized in response to a chronic change in environmental temperature, and (c) the nature of the intracellular milieu could be altered in such a way as to render regulation of metabolism stable over a certain physiological range of temperatures. All three mechanisms appear to have at least some support in fact.

The first mechanisms would appear to be the most attractive as it is known that changes in enzyme activity follow alterations in diet (Schimke, 1962; Weber *et al.*, 1965) or hormonal stimulation (Exton and Park, 1965; Schimke, 1969); it is assumed that a change in environmental temperature would cause a similar result. The difficulty with this idea is that there appears to be no reliable information on intracellular concentrations of enzymes as opposed to measures of enzyme activity. Therefore, although the literature is replete with data on changes in enzyme activity in response to a change in some factor, it is not at all clear whether these changes are based on changes in enzyme concentration.

One of the most important parameters in the regulation of intermediary metabolism is enzyme-ligand affinity. It was originally observed, in endothermic systems, that affinity of an enzyme for its ligand (substrate, cofactor, or modulator) is markedly sensitive to temperature (Helmreich and Cori, 1964; Taketa and Pogell, 1965). Should this observation be extended to ectotherms which experience marked changes in temperature, it follows that control of metabolism in these animals would be extremely sensitive to thermal change. A generally observed effect of a decrease in temperature is an increase in enzyme-substrate affinity (Fig. 1) which is undoubtedly rooted in a temperature-induced change in the conformation of the enzyme-molecule (Behrisch, 1969b; Iwatsuki and Okazaki, 1967). As a result, at low (and physiological) concentrations of substrate, the rate of enzyme reaction may be equal to or higher than that at higher temperatures. The relationship between temperature and the enzyme-ligand interaction has been investigated for a number of enzymes from fishes, molluscs and crustaceans, and a noteworthy pattern has become discernible. For example, in fructose-1,6-diphosphatase (FDPase) from muscle of Alaskan king crab *(Paralithodes camtschatica)* it appears that the temperature at which maximum

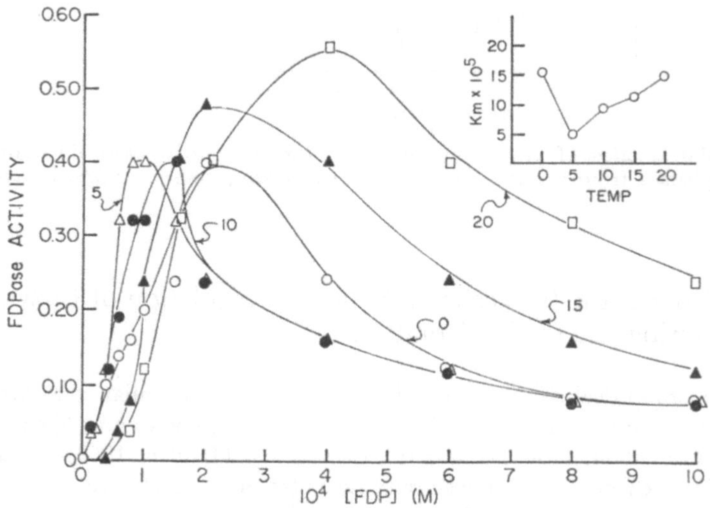

Fig. 1. Substrate saturation curves for Alaskan king crab muscle fructose diphosphatase at a series of temperatures: ○ 0 C; △ 5 C; ● 10 C; ▲ 15 C; □ 20 C. (Behrisch, 1971)

affinity for substrate occurs closely approximates the temperatures of acclimatization (Behrisch, 1971). Similar results have been obtained with lactate dehydrogenase from trout, tuna and lungfish (Hochachka and Somero, 1968), pyruvate kinase from trout, the antarctic fish *Trematomus* (Somero and Hochachka, 1968) and isocitrate dehydrogenase from trout (Moon, 1970). A striking feature of these results is that, at temperatures above and below the acclimatization temperature, values of the apparent K_m increase markedly, resulting a in complex U-shaped temperature-K_m curve (Fig. 2). In many cases, the increase in K_m is so great that at above and below the temperature of minimum K_m (maximum affinity), it would be very difficult for the enzyme to become saturated with substrate *in vivo*, and control would be virtually impossible. Therefore it seems that animals which experience thermal changes on a seasonal basis must be able to bypass this metabolic "lesion" at the new temperature. Thus Alaskan king crab acclimatized to winter (—1 to 1 C) and summer (6–9 C) temperatures have different variants of muscle FDPase with different thermal relationships with respect to K_m for substrate (Fig. 2). In each case minimum K_m (maximum affinity) falls near the acclimatization temperature. Similar behavior has been observed for brain acetylcholinesterase for rainbow trout (Baldwin and Hochachka, 1970).

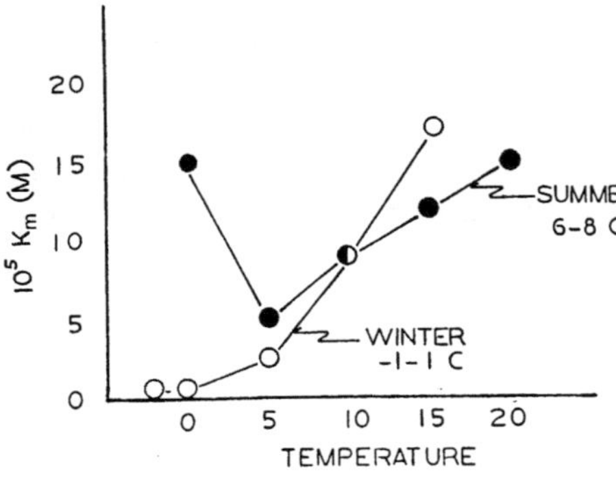

Fig. 2. Effect of temperature on affinity for substrate in seasonal variants of fructose diphosphatase from skeletal muscle of Alaskan king crab *(Paralithodes camtschatica)*

Role of the Intracellular Milieu in Regulation of Enzyme Activity during Thermal Adaptation

A good deal of evidence indicates that the intracellular concentrations of numerous ions vary during adaptation to temperature (Toews and Hickman, 1969; Hickman *et al.*, 1964; Houston, 1959; Houston and Madden, 1968) and a large literature on ionic effects on enzyme activity exists (for a comprehensive list see Dixon and Webb, 1964). What is the effect of these changes in ion concentration in an ectotherm? An example is provided by the yellowfin sole *Limanda aspera* which inhabits waters of the Bering Sea the year around. Temperatures in these waters are relatively stable and vary from −1.86 C in winter to as high as about 4–5 C in late summer (Johnson and Hartmann, 1971). An examination of specimens collected at various times of the year shows that the animal apparently has the same variants of lactate dehydrogenase (LDH) during all seasons. Thus in partially purified muscle LDH from this fish, minimum K_m for substrate occurs at about 4 C at all seasons (Fig. 3), a temperature quite removed from that experienced during the greater part of the year. It may thus be reasoned that catalytic efficiency of this enzyme would be appreciably curtailed for much of the year, unless there are mechanisms that increase enzyme-substrate affinity at the lower temperatures. Examination of the effects of various cations on LDH activity shows that in the presence of 150 mM K^+ (an approximately physiological value) K_m for pyruvate falls to a minimal value at 0 C and remains insensitive to —2 C (Fig. 3). Clearly then, variations in a key cation can alter quite markedly the thermal relationship of enzyme-substrate affinity. The probable importance of cations in the control of enzyme activity is further underscored by observations that the bivalent cations Mg^{2+} and Mn^{2+} enhance the activities of both glucose-6-phosphate and 6-phosphogluconate dehydrogenase from gill tissue of the arctic tanner crab at low concentrations of substrate, and that K^+ can maintain activity of glyceraldehyde phosphate dehydrogenase independent of temperature (Behrisch, 1972a). These findings are of particular relevance to metabolic control in ectotherms, because in addition to those of a number of bivalent cations, the concentrations of Na^+ and K^+ vary noticeably during thermal acclimatization (Rao, 1962, 1967; Hickman *et al.*, 1964; Houston, Madden and deWilde, 1970; Umminger, 1969, 1970). Changes in the concentrations of these cations and in compartmentation are known to occur and, judging from available evidence (see Behrisch and Hochachka, 1969a, b; Behrisch, 1969a, b, 1971), such changes in cation concentration may well be a major method of regulation of enzyme and metabolic activity during thermal adaptation.

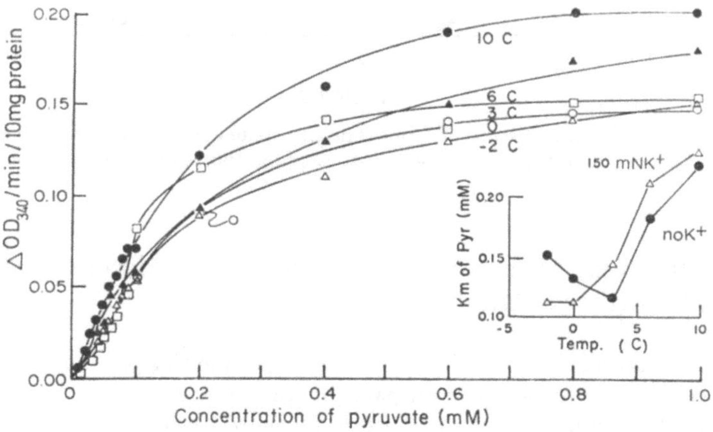

Fig. 3. Effect of temperature and a monovalent ion (K^+) on enzyme-substrate affinity in lactate dehydrogenase from yellowfin sole *(Limanda aspera)*. (After Behrisch, 1972a)

Another factor that is potentially very important is pH. It is known (Robin, 1962; Rahn, 1965; Reeves and Wilson, 1969) that temperature-caused changes in intracellular pH occur in ectotherms. A decrease in temperature results in a rise in the ionization constant of water. Since it appears that most, if not all, ectotherms regulate their H^+/OH^- ratios (at some point above neutrality) at all temperatures, a rise in intracellular pH at low temperature is observed. This temperature-dependent pH change could have profound effects on metabolic control in an ectotherm if the behavior of key enzymes is sharply affected by pH. Thus for example, fructose diphosphatase from liver of rainbow trout shows such a dependence. There is an alkaline pH optimum (Behrisch, 1969a) and the animal can thus take advantage of the described pH-temperature relationship. Similar behavior has also been observed in lactate dehydrogenase from muscle of yellow fin sole (Behrisch, 1972a) and citrate synthase from trout (Hochachka and Lewis, 1971).

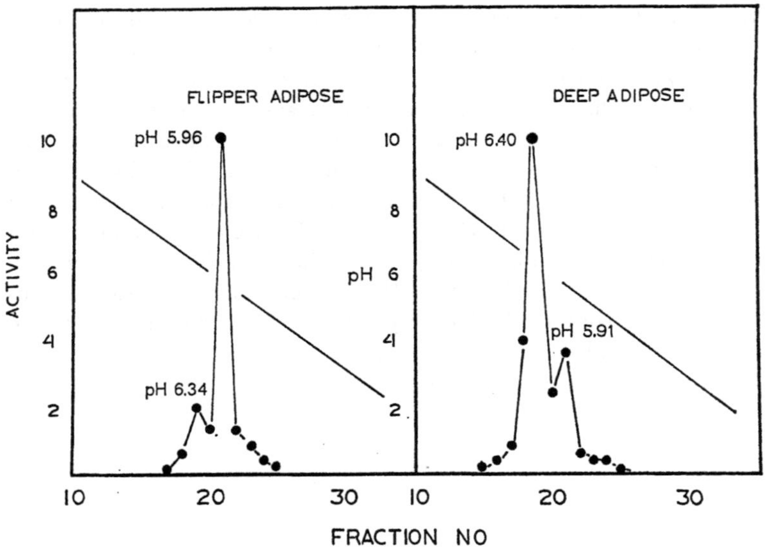

Fig. 4. Electrofocussing of 6-phosphogluconate dehydrogenase from flipper (heterothermic) and deep (endothermic) adipose tissues of the spotted seal *(Phoca vitulina)*

Molecular Aspects of Temperature Adaptation in Ectothermic Tissues of Northern Mammals

The second category of organisms discussed in this paper are those which experience ectothermy in only certain parts of the body. Most, if not all mammals and birds in the Arctic and Subarctic are faced with a large and constant loss of heat from their extremities and peripheral tissues. It is thought that these animals offset this heat drain through the use of a counter-current heat exchange circulation, as has been elegantly described in the porpoise (Scholander and Schevill, 1955). As a result, temperatures in the extremities of many arctic mammals may vary considerably, and at least in the case of arctic marine mammals which reside in a large heat sink, temperatures in the limbs may vary from as low as 0 to 40 C or above. In contrast, temperatures in the tissues from the deep parts of the body are endothermic. Available evidence indicates that a number of physiologic processes from tissues from peripheral parts of the body are less sensitive to low temperature than are those from the core tissues. Caudal nerve preparations from various northern mammals continue to conduct at temperatures at which nervous conduction ceases in deep nerve preparations (Miller, 1967, 1970; Miller and Irving, 1967). Similarly, the rate of cell division in peripheral tissues is thermally less sensitive than that in central tissues (Feltz and Fay, 1966). In addition, fats from the peripheries of high latitude or cold-acclimated "endotherms" contain larger fractions of unsaturated fatty acids than do those from deep adipose tissue depots (Meng, West and Irving, 1970), and the melting point of these lipids is markedly altered by variations in the acclimation temperature (Irving, Schmidt-Nielsen and Abrahamson, 1957). Therefore, are the enzymes from ectothermic regions of the body from, say, a seal *(Phoca vitulina)* the same as those from endothermic regions of the body or are they different? If they do differ, what are the sites at which selection appears to have been active in the evolution of the counter-current heat exchange habit?

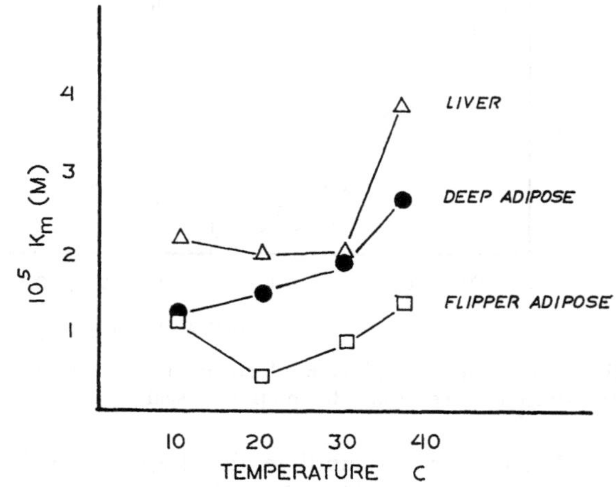

Fig. 5. Relationship between temperature and apparent K_m for substrate in 6-phosphoglucontae dehydrogenase from various sources in spotted seal

Electrofocussing experiments on 6-phosphogluconate dehydrogenase (6-PG deh.) taken from adipose tissue from flipper and the deep part of the body indicate that there are two variants of the enzyme present and that both are present in either tissue (Fig. 4). However, the relative quantities of both variants are clearly different in either source. Kinetic investigations on the effects of temperature on enzyme-substrate affinity yield the results seen in Fig. 5. Thus deep adipose 6-PG deh. shows a decrease to 1/3 of the normothermic value in K_m for substrate as temperature is decreased from 37 to 5 C. In contrast, the 6-PG deh. from flipper adipose tissue is relatively insensitive to temperature in its relationship with substrate. One of the products of the 6-PG deh. reaction, NADPH, is a powerful inhibitor of the enzyme. Affinity of the endothermic 6-PG deh. for NADPH is markedly sensitive to thermal change, so that at low temperature, the operation of this enzyme and the pentose shunt would be virtually "shut off" by very low concentrations of this metabolite. In marked contrast, the sensitivity of the ectothermic 6-PG deh. to its inhibitor NADPH is unaffected by temperature (Fig. 6), thus assuring thermal independence of this chief method of control at this key step in intermediary metabolism.

Studies on Hibernating Animals

In a small hibernating animal such as the arctic ground squirrel *(Citellus undulatus)* the onset of hibernation involves the gradual decrease in metabolic rate (as measured

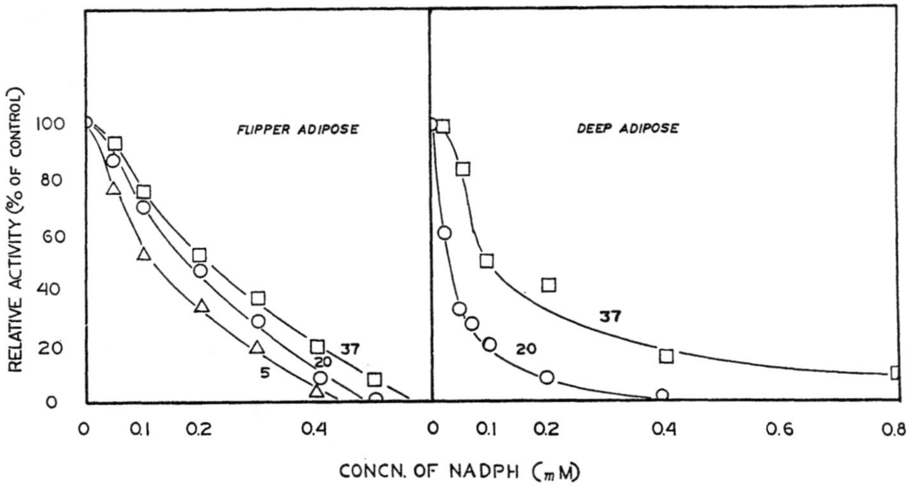

Fig. 6. Inhibition of 6-phosphogluconate dehydrogenase at varying temperatures from flipper and deep adipose tissues from spotted seal

by O_2 consumption) until in the deep hibernation state, metabolic rate may be reduced to as low as 1/200th of the normothermic rate (Burlington, 1972; Hannon *et al.*, 1972). In this state, body temperatures may descend to as low as 0 C, and within limits parallel changes in the hibernaculum temperature. In addition, many hibernators, including the arctic ground squirrel, spontaneously arouse from the deep hibernation state in a cyclic fashion. During this arousal metabolic rate and body temperature rise rapidly to normothermic levels. After a period of several hours or even days the animal descends once more into the deep hibernating state. This alternation between ectothermy and endothermy is typical for what has been called a heterothermic mode of life.

Through isolated studies it appears that some enzymes from ectotherms are more efficient catalysts at low temperatures than are those of endotherms (Assaf and Graves, 1969; Cowey, 1967). An obvious, but seldom asked, question is thus the following: are the enzymes of the hibernating mammal different from those of the non-hibernating one? If they do differ, what parameters of enzyme regulation are likely to be significant in the differentiation between a hibernating and a non-hibernating enzyme? For these studies we chose the enzyme fructose 1,6-diphosphatase from liver of the arctic ground squirrel. The enzyme is the rate-limiting one in the gluconeogenetic sequence (Krebs, 1963; Horecker *et al.*, 1966; Newsholme and Gevers, 1967) and its regulation will affect in large part the flow of carbon at the higher levels of carbohydrate metabolism. One of the first observations on the effects of temperature on enzyme-ligand affinity was made with FDPase from rat liver in which it was seen that a decrease from 42 to 5 C resulted in a 100-fold increase in affinity for the allosteric inhibitor of the enzyme, 5′ AMP (Taketa and Pogell, 1965). Thus FDPase and gluconeogenic activity would be effectively turned off by extremely low concentrations of AMP at temperatures that are commonly experienced by a hibernator. However, indirect evidence (Galster and Morrison, 1970) indicates that gluconeogenesis does occur during hibernation, albeit at a reduced rate, and thus there must be mechanisms that permit the continued operation of this key regulatory enzyme in the hibernating ground squirrel.

As with the ectothermic enzymes (Hochachka and Somero, 1968, 1971), low temperature acts as a positive modulator of ground squirrel FDPase in both the summer (non-hibernating) and winter (hibernating) animals (Fig. 7a, b). It is at once apparent, however, that a decrease in temperature exerts a much larger influence on affinity for substrate in the hibernating enzyme than in the non-hibernating enzyme. Thus the very large increase in FDPase-FDP affinity in the hibernating FDPase is one mechanism that would allow for activity at low temperature in the hibernating enzyme. Moreover, values of the Hill constants (n_H) are stable over the entire range in the non-hibernating enzyme while in the hibernating FDPase a drop in temperature from 37 to 5 C results in a large increase in n_H values from about 2 to almost 4, indicating a marked increase in homotropic interaction between the FDP-binding sites (Fig. 7b). Similar to the findings of Taketa and Pogell (1965) in their classic work on rat liver FDPase, the non-hibernating squirrel FDPase is remarkably sensitive to low concentrations of AMP, and would thus be completely shut down at hiber-

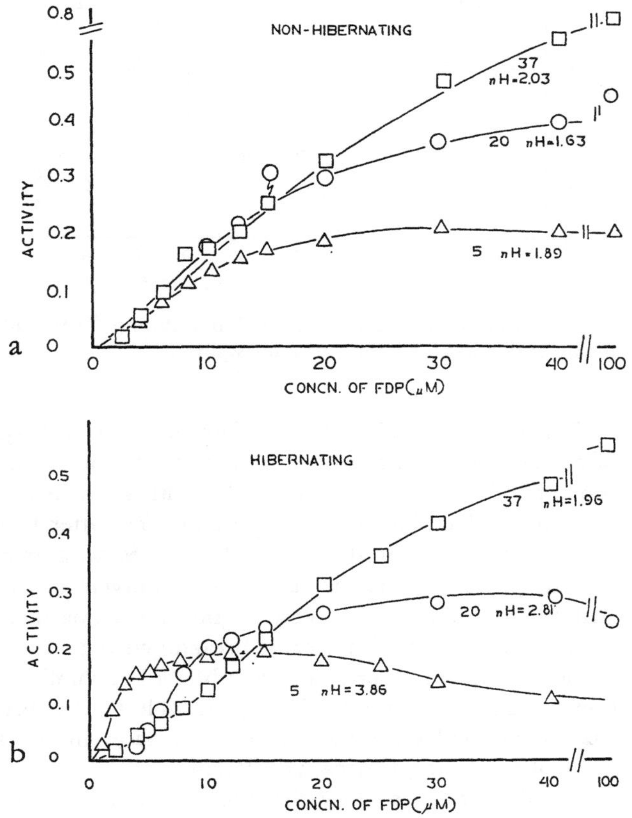

Fig. 7 a and b. Substrate saturation curves of fructose diphosphatase from liver of summer (nonhibernating) and winter (hibernating) arctic ground squirrel *(Citellus undulatus)* at different temperatures

nating temperatures. It would therefore seem necessary to circumvent this powerful inhibition at low temperature, during the hibernation months. Thus the AMP saturation curves in Fig. 8 show that K_i values for AMP are remarkably temperature insensitive in the hibernating (ectothermic) enzyme, in a manner similar to that seen in the FDPase from liver of migrating salmon (Behrisch, 1969b). These purely kinetic findings suggest that there may indeed be different variants of liver FDPase in the hibernating and non-hibernating ground squirrel. In fact, electrofocussing experiments show that the hibernating FDPase variants are replaced by non-hibernating variants of the enzyme presumably in preparation for the summer months. We are finding similar, and equally clear, results in pyruvate kinase in the arctic ground squirrel (Johnson and Behrisch, unpublished findings) and in various aspects of mitochondrial metabolism (Reynafarje, unpublished findings).

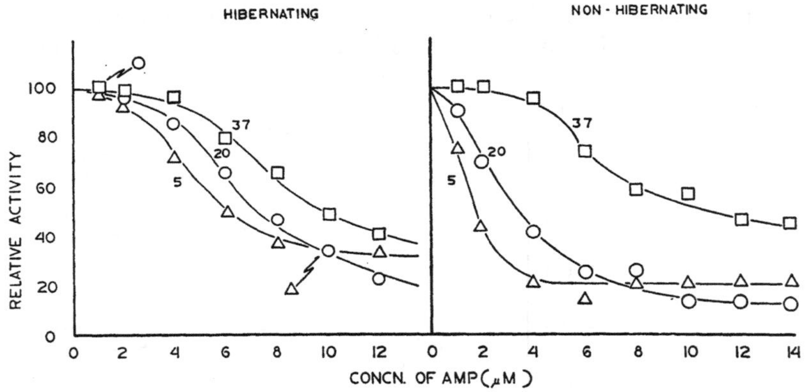

Fig. 8. Effect of temperature on AMP inhibition of liver fructose diphosphatase from hibernating and non-hibernating ground squirrels

It is unclear whether these variants are new isoenzymes (true individual species of proteins) or are instead enzymes, only one (or a small number) of the component subunits of which have been newly synthesized in response to the environmental stimulus. In fact, energetically speaking, the latter would be a more efficient way for the animal to produce "new" enzymes with improved catalytic capabilities. Whichever of these is the case, it is intuitively obvious that when the animal is adapting to a new environmental regime, it must balance nicely the rate of production of the new variant with that of breakdown (or conversion) of the old – and less efficient – variant (Morris and Smith, 1967; Schimke, 1969). The recent identification of enzymes that are solely responsible for the breakdown and metabolism of enzyme molecules provides an exciting facet in the study of the integration of metabolism during thermal adaptation.

In addition, it is likely that the observed alterations in enzyme-ligand affinity are rooted in conformational alterations that are caused by changes in temperature. In this connection, Iwatsuki and Okazaki (1967) were able to demonstrate that observed changes in enzyme-substrate-affinity were linked to temperature-dependent changes in enzyme configuration. This also seems to be the case in FDPase from liver of rainbow trout (Behrisch, 1969a). Sucrose density gradient centrifugation of the trout enzyme at two different temperatures (2 and 18 C) yields symmetrical peaks, each with a specific sedimentation velocity. The trout FDPase has a $S_{20,W}$ value of around 6.0 at 18 C and a value of about 6.8 at 2 C. Whether such changes in sedimentation velocity are due to differences in the degree of subunit interaction or to variations in tightness of coiling of the individual chains is at present an open question. Further experiments in this direction should yield much useful insight into such matters.

The increase in homotropic cooperation paralleling the rise in enzyme-substrate affinity at low temperature (Fig. 7) raises an interesting point. It appears through these and previous studies in this laboratory that an increase in site-site (or subunit-subunit) interaction may in fact underlie the relatively high efficiencies of low temperature ectothermic enzymes. Clearly, as the regulatory properties of these enzymes are based on their allosteric behavior (Gerhart, 1964), any factor influencing the extent of interaction between these subunits will in turn affect the catalytic and regulatory behavior of the molecule (Gerhart, 1964; Changeux, 1964). Thus it is perhaps significant that an allosteric activator of fructose diphosphatase, phosphoenolpyruvate (Behrisch, 1971, 1972b) has the effect of not only increasing enzyme-substrate interaction but also the degree of site-site interaction in the binding of substrate, in a manner identical to that of low temperature which also acts as a positive modulator of the ectothermic enzymes discussed here.

In this connection, it has been suggested (Noble, 1969) that cooperative allosteric interactions, such as the haeme-haeme interactions in haemoglobin, which have been shown to be temperature-sensitive (Monod, Wyman and Jacob, 1965), require that the change in inter-subunit bonding energy accompanying the attachment of a ligand molecule be not uniform but depend rather on the extent to which the protein is already "liganded". Thus the energy of subunit bonding might be the primary cooperative parameter (Noble, 1969). For example, consider the oxygenation of haemoglobin. The ligand reactivity of fully liganded Hb is very similar to that of the liganded isolated chains. Noble then postulates that in unliganded Hb the chains exist in a strained confirmation which is characterized by a low affinity for ligand but permits a higher energy of interchain interaction. The interaction of a molecule of ligand raises the free energy of this conformation and thus reduces the interchain bonding.

It seems not unreasonable to accept that the allosteric properties of enzymes confer selective advantage to the cell or organism only when they are effective under physiological concentration of substrate, modulator and cofactor. Therefore it seems that a parameter governing subunit-subunit interaction might be of importance in the regulation of enzyme activity and it is proposed that a low energy of subunit bonding might be of adaptive advantage to ecto- or heterothermic organism, such as a hibernating mammal, living in an environment of reduced thermal energy. In an ectothermic enzyme, the binding of ligand at low temperature would result in a great increase in the free energy of the conformation (relative to the proposed low bonding energy). This would lead to enhanced subunit-subunit (or site-site) interaction at low temperature, as observed in the increased Hill constants of substrate saturation in the hibernating ground squirrel FDPase at low temperature.

Conclusion

This brief discussion of some of the work from this laboratory has been an attempt to show that organisms as diverse as "true" ectotherms (crabs, fish), a pinniped, and a small hibernator can apparently cope with a common problem – low temperature – in much the same way.

In every case of which I am aware, the regulatory properties of control enzymes are linked to their allosteric behavior, which in turn is founded on the subunit structure of these enzymes. During adaptation to environments of low, and changing, temperature selection appears to have favored the evolution of enzymes whose regulatory characteristics can effectively function in an environment of reduced thermal energy. It is suggested that two sites at which selective pressure has been greatest during evolution to cold are (1) enhanced enzyme-ligand affinity at low temperature, and (2) reduced subunit bonding energy, which permits allosteric behavior at low temperatures.

Acknowledgements

The work described in this manuscript was supported by the National Science Foundation (Grant No. GB 20476) and the National Institutes of Health (Grants No. GM 10402 and 5-S05-RR07001-5). The author wishes to thank Mr. Craig E. Johnson and Drs. Jon A. Percy and Peter W. Hochachka for their valuable discussions during the preparation of this manuscript.

References

Assaf, S. A., Graves, D. J.: Structure and catalytic properties of lobster muscle glycogen phosphorylase. J. Biol. Chem. **244**, 5544–5555 (1969).

Baldwin, J., Hochachka, P. W.: Functional significance of isozymes in thermal acclimation. Acetylcholinesterase from trout brain. Biochem. J. **116**, 883–889 (1970).

Bartholomew, G. A.: Body Temperature and Metabolism. In: Gordon, M. S. (Ed.): Animal Function: Principles and Adaptation, p. 290–354. London: The Macmillan Comp. Ltd. 1968.

Behrisch, H. W.: Temperature and the regulation of enzyme activity in poikilotherms. Regulatory properties of fish fructose diphosphatase. Ph. D. Thesis, p. 125. University of British Columbia 1969a.

Behrisch, H. W.: Temperature and the regulation of enzyme activity in poikilotherms. Fructose disphosphatase from migrating salmon. Biochem. J. **115**, 687–696 (1969b).

Behrisch, H. W.: Temperature and the regulation of enzyme activity in poikilotherms. Regulatory properties of fructose diphosphatase from muscle of the Alaskan king crab. Biochem. J. **121**, 399–409 (1971).

Behrisch, H. W.: Molecular mechanisms of adaptation to low temperature in marine poikilotherms. Some regulatory properties of dehydrogenases from two arctic species. Mar. Biol. **13**, 267–275 (1972a).

Behrisch, H. W.: Regulation by phosphoenolpyruvate of fructose 1,6-diphosphatase in skeletal muscle: evidence for an allosteric activator of the enzyme. Can. J. Biochem. 50, 710–713 (1972b).

Behrisch, H. W., Hochachka, P. W.: Temperature and the regulation of enzyme activity in poikilotherms. Properties of trout fructose-diphosphatase. Biochem. J. **111**, 287–295 (1969a).

Behrisch, H. W., Hochachka, P. W.: Temperature and the regulation of enzyme activity in poikilotherms. Properties of lungfish fructose-diphosphatase. Biochem. J. **112**, 601–607 (1969b).

Burlington, R. F.: Recent advances in intermediary metabolism of hibernating mammals. In: Hibernation-Hypothermia: perspectives and challenges, p. 3–16. Amsterdam: Elsevier Publ. Co. 1972.

Bygrave, F. L.: Studies on the interaction of metal ions with pyruvate kinase from Ehrlich ascites tumor cells and from rabbit muscle. Biochem. J. **101**, 488–494 (1966).

Bygrave, F. L.: The ionic environment and metabolic control. Nature (London) **214**, 667 to 671 (1967).

Caldwell, R. S., Vernberg, F. J.: The influence of acclimation temperature on the lipid composition of fish gill mitochondria. Comp. Biochem. Physiol. **34**, 179–191 (1970).

Changeux, J. P.: Allosteric interactions interpreted in terms of quaternary structure. Brookhaven Symp. Biol. **17**, 232–249 (1964).

Cowey, C. P.: Comparative studies on the activity of D-glyceraldehyde-3-phosphate dehydrogenase from cold- and warm-blooded animals with reference to temperature. Comp. Biochem. Physiol. **23**, 696–976 (1967).

Dean, J. M.: The metabolism of tissues of thermally acclimated trout *(Salmo gairdneri)*. Comp. Biochem. Physiol. **29**, 185–196 (1969).

Dixon, M., Webb, E. D.: The Enzymes. 2nd Ed., p. 950. London: Longmans, Green and Co. 1964.

Ekberg, D. R.: Respiration in tissues of goldfish adapted to high and low temperatures. Biol. Bull. **114**, 308–316 (1958).

Ekberg, D. R.: Respiration in tissues of goldfish adapted to high and low temperatures. Biol. Bull. **114**, 308–316 (1958).

Ekberg, D. R.: Aerobic and anaerobic metabolism in gills of the crucian carp adapted to high and low temperatures. Comp. Biochem. Physiol. **5**, 123–128 (1962).

Exton, J. H., Park, C. R.: Gluconeogenesis in liver: effects of adrenalectomy, cortisol, and glucagon. Federation Proc. **24**, 537a (1965).

Feltz, E. T., Fay, F. H.: Thermal requirements *in vitro* of epidermal cells from seals. Cryobiology **3**, 261–264 (1966).

Galster, W. A., Morrison, P. R.: Cyclic changes in carbohydrate concentrations during hibernation in the arctic ground squirrel. Am. J. Physiol. **218**, 1228–1232 (1970).

Gerhart, J. C.: Subunits for control and catalysis in aspartate transcarbamylase. Brookhaven Symp. Biol. **17**, 222–231 (1964).

Hannon, J. P., Beyer, R. E., Burlington, R. F., Somero, G. N., Bland, J. H.: A guide for future studies of low temperature metabolic function. In: Hibernation-Hypothermia: perspectives and challenges, p. 99–119. Amsterdam: Elsevier Publ. Co. 1972.

Haschemayer, A. E. V.: Compensation of liver protein synthesis in temperature-acclimated toadfish, *Opsanus* tau. Biol. Bull. **135**, 130–140 (1968).

Helmreich, E., Cori, C.: The effects of pH and temperature on the kinetics of the phosphorylase reaction. Proc. Natl. Acad. Sci. U.S. **52**, 647–654 (1964).

Hickman, C. P., McNab, R. A., Nelson, J. S., Van Breemen, E. D., Comfort, D.: Effect of cold acclimation on electrolyte distribution in rainbow trout *(Salmo gairderni)*. Can. J. Zool. **42**, 577–597 (1964).

Hochachka, P. W., Hayes, F. R.: The effect of temperature acclimation on pathways of glucose metabolism in the trout. Can. J. Zool. **40**, 261–270 (1962).

Hochachka, P. W., Lewis, J. S.: Temperature and regulation of citrate synthase variants. J. Biol. Chem. **264**, 1544–1552 (1971).

Hochachka, P. W., Somero, G. N.: The adaptation of enzymes to temperature. Comp. Biochem. Physiol. **27**, 659–668 (1968).

Hochachka, P. W., Somero, G. N.: Biochemical adaptation to the environment. In: Hoar, W. S., Randall, D. J. (Eds.): Fish physiology, Vol. 6, p. 100–156. London: Academic Press 1971.

Horecker, B. L., Pontremoli, S., Rosen, S., Rosen, O.: Structure and function in fructose diphosphatase. Federation Proc. **25**, 1521–1528 (1966).
Houston, A. H.: Osmoregulatory adaptation of steelhead trout *(Salmo gairdneri)*. Can. J. Zool. **37**, 729–748 (1959).
Houston, A. H., Madden, J.: Environmental temperature and plasma electrolytic regulation in the carp *Cyprius carpio*. Nature (London) **217**, 969–970 (1968).
Houston, A. H., Madden, J., deWilde, M. A.: Environmental temperature and the body fluid system of the fresh water teleost IV. Water-electrolyte regulation in thermally acclimated carp, *Cyprius carpio*. Comp. Biochem. Physiol. **34**, 805–818 (1970).
Irving, L., Schmidt-Nielsen, K., Abrahamson, N. S. B.: On the melting point of animal fats in cold climates. Physiol. Zool. **30**, 93–105 (1957).
Iwatsuki, N., Okazaki, R.: Mechanisms of regulation of deoxythymidine kinase *Escherichia coli*. II. Effect of temperature on the enzyme's activity and its kinetics. J. Mol. Biol. **29**, 155–165 (1967).
Johnson, P. R., Hartmann, C. W.: Environmental atlas of Alaska, p. 111. College, Alaska: Univ. of Alaska Press 1971.
Kanungo, M. S., Prosser, C. L.: Physiological and biochemical adaptation of goldfish to cold and warm temperatures. I. Standard and active oxygen consumption of cold- and warm-acclimated goldfish at various temperatures. J. Cellular Comp. Physiol. **54**, 265–273 (1959).
Krebs, H. A.: Renal gluconeogenesis. Advan. Enzyme Regulation **1**, 385–400 (1963).
Meng, M. S., West, G. C., Irving, L.: Fatty acid composition of caribou bone marrow. Comp. Biochem. Physiol. **30**, 187–191 (1969).
Miller, L. K.: Caudal nerve function as related to temperature in some Alaskan mammals. Comp. Biochem. Physiol. **21**, 679–686 (1967).
Miller, L. K.: Temperature-dependent characteristics of peripheral nerves exposed to different thermal conditions in the same animal. Can. J. Zool. **48**, 75–81 (1970).
Miller, L. K., Irving, L.: Temperature-related function in warm and cold-acclimated muskrats. Am. J. Physiol. **231**, 1295–1298 (1967).
Monod, J. P., Wyman, J., Changeux, J. P.: On the nature of allosteric transitions: a plausible model. J. Mol. Biol. **12**, 88–118 (1965).
Moon, T. W.: Isozymes in temperature acclimation: trout $NADP^+$ linker isocitrate dehydrogenase. Federation Proc. **29**, 400a (1970).
Morris, D., Smith, M. W.: Protein synthesis in the intestine of goldfish acclimatized to different temperatures. Biochem. J. **102**, 648–653 (1967).
Newell, R. C.: Effect of temperature on the metabolism of poikilotherms. Nature (London) **212**, 426–428 (1966).
Newell, R. C.: Oxidative activity of poikilotherm mitochondria as a function of temperature. J. Zool. (London) **151**, 299–311 (1967).
Newell, R. C., Northcroft, H. R.: A re-interpretation of the effect of temperature on the metabolism of certain marine invertebrates. J. Zool. (London), **151**, 277–298 (1967).
Newsholme, E. A., Gevers, W.: Control of glycolysis and gluconeogenesis in liver and kidney cortex. Vitamins and Hormones **25**, 1–87 (1967).
Noble, R. W.: Relation between allosteric effects and changes in the energy of bonding between molecular subunits. J. Mol. Biol. **39**, 479–491 (1969).
Precht, H.: Ergänzende Versuche zur Bedeutung des Blutes für die Temperaturadaptation bei Fischen. Zool. Anz. **175**, 301–310 (1964).
Precht, H.: Über die Bedeutung des Blutes für die Temperaturadaptation von Fischen. Zool. Jahrb. Abt. Allg. Zool. Physiol. **71**, 313–328 (1965).
Rahn, H.: Gas transport from the external environment to the cell. In: de Reuck, A. V. S., Porter, R. (Eds.): Ciba Foundn. Symp. on the development of the lung, p. 3–23. Boston: Little, Brown and Co. 1965.
Rao, K. P.: Physiology of acclimation to low temperature in poikilotherms. Science, N. Y. **137**, 682–683 (1962).
Rao, K. P.: Biochemical correlates of temperature acclimation. In: Prosser, C. L. (Ed.): Molecular mechanisms of temperature adaptation, p. 227–244. Washington, D. C.: AAAS Press 1967.

Reeves, R., Wilson, T.: Intracellular pH in bullfrog striated and cardiac muscle as a function of body temperature. Federation Proc. **28**, 782a (1969).

Robin, E.: Relationship between temperature and plasma pH and carbon dioxide tension in the turtle. Nature (London) **195**, 249, 251 (1962).

Schimke, R. T.: Differential effects of fasting and protein-free diets on levels of urea enzymes in rat liver. J. Biol. Chem. **237**, 1921–1924 (1962).

Schimke, R. T.: Roles of synthesis and degradation in regulation of enzyme levels in mammalian tissues. Current Topics Cell. Regul. **11**, 77–124 (1969).

Scholander, P. F., Flagg, W., Waters, V., Irving, L.: Climatic adaptation in arctic and tropical poikilotherms. Physiol. Zool. **26**, 67–92 (1953).

Scholander, P. F., Schevill, W. E.: Counter-current heat exchange in the fins of whales. J. Appl. Physiol. **8**, 279–282 (1955).

Somero, G. N., Giese, A. C., Wohlschlag, D.: Cold adaptation in the Antarctic fish *Trematomus bernachii*. Comp. Biochem. Physiol. **26**, 223–233 (1968).

Somero, G. N., Hochachka, P. W.: The effect of temperature on catalytic and regulatory function of pyruvate kinase of the rainbow trout and the Antarctic fish *Trematomus bernachii*. Biochem. J. **110**, 395–400 (1968).

Taketa, K., Pogell, B.: Allosteric inhibition of rat liver fructose diphosphatase by adenosine 5' monophosphate. J. Biol. Chem. **240**, 651–662 (1965).

Toews, D., Hickman, C. P.: The effect of cycling temperatures on electrolyte balance in skeletal muscle and plasma of rainbow trout *(Salmo gairdneri)*. Comp. Biochem. Physiol. **29**, 905–918 (1969).

Umminger, B. L.: Physiological studies on super-cooled killifish *(Fundulus heteroclitus)*. I. Serum inorganic constituents in relation to osmotic and ionic regulation at subzero temperatures. J. Exptl. Zool. **172**, 283–302 (1969).

Umminger, B. L.: Osmoregulation by killifish in fresh water at temperatures near freezing. Nature (London) **210**, 421–422 (1970).

Weber, G., Singhal, R. L., Srivastava, S. K.: Action of glucocorticoid as inducer and insulin as suppressor of biosynthesis of hepatic gluconeogenic enzymes. Advan. Enzyme Regulation **4**, 43–75 (1965).

Wohlschlag, D. E.: Respiratory metabolism and ecological characteristics of some fishes in McMurdo Sound, Antarctica. In: Lee, Mo. O. (Ed.): Biology of the Antarctic Seas, Washington, D. C.: American Geophys. Union. **1**, 33–62 (1964).

Mechanisms of Insect Thermoregulation

Bernd Heinrich

With 6 Figures

In insects, as in other animals, the rates of most physiological processes and associated functions are related to body temperature. For the most part insects can be considered as ectotherms. The production of heat by endothermy and the elevation of body temperature have been known for a long time (see Newport, 1837; Bachmetjew, 1899; Dotterweich, 1928; Himmer, 1932; Oosthuizen, 1939; Krogh and Zeuthen, 1941; and others). More recent evidence suggests that flight activity in adults, as well as rate of growth of the immatures in social insects, are sometimes dependent on elaborate mechanisms of temperature regulation. The purpose of this essay is to point out some of the mechanisms whereby individual insects regulate their body temperature.

Heat Production and Body Temperature in Flight

During flight, as during other muscular activity, inefficiency at the chemical and mechanical levels of organization results in the conversion of about 80% of the energy expenditure to heat (Weis-Fogh, 1972). Insects during flight rely nearly exclusively on oxidative metabolism and the rate of heat production can be computed when the substrate for flight and the rate of oxygen consumption are known. Body temperature, however, is related not only to heat production but also to the rate of cooling. Insects cool primarily by convection (Church, 1960) and their cooling rate is thus to a great extent related to body size (Digby, 1955). Since the metabolic rate during flight in most insects ranges from about 30–100 ml O_2/g/hr, while body weight may vary over a thousand-fold, it is obvious that the maximum difference between ambient temperature (T_A) and thoracic temperature (T_{Th}) must vary greatly between differently sized insects. Small flies could scarcely have a T_{Th} of one C above T_A (Digby, 1955), while a large (1 to 3 g) sphinx month that is insulated with a dense layer of scales (Church, 1960) can have a T_{Th} of 25 C above T_A as a result of the flight metabolism (Heinrich, 1970).

Variations of Heat Production

If the insect is to become and remain air-borne it must maintain a certain power output and wing-beat frequency. These interrelated parameters depend on the speed of muscle shortening, which is a function of muscle temperature (Machin *et al.*, 1962; Neville and Weis-Fogh, 1963). Many insects would remain permanently grounded if they were not capable of warm-up prior to flight. "Warm-up" is a process analogous to shivering (see Figs. 1 and 2) during which the thoracic flight musculature is activated but the animal does not attempt to fly. During warm-

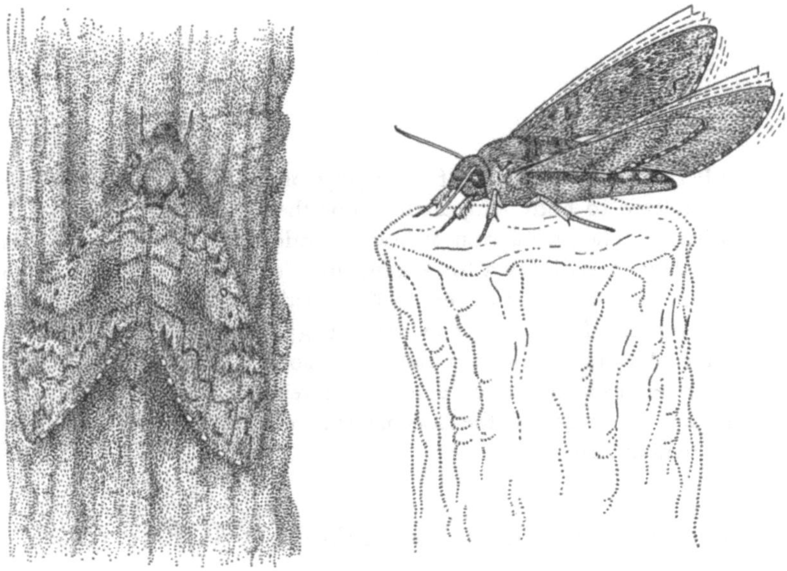

Fig. 1. The posture of a sphinx moth *(Manduca sexta)* at rest (left) and during pre-flight warm-up. Certain other moths (e.g. Saturniids) may hold their wings laterally during shivering, while in asynchronous flyers (Hemiptera, Hymenoptera, Coleoptera, Diptera) the wings remain folded dorsally and do not move during warm-up

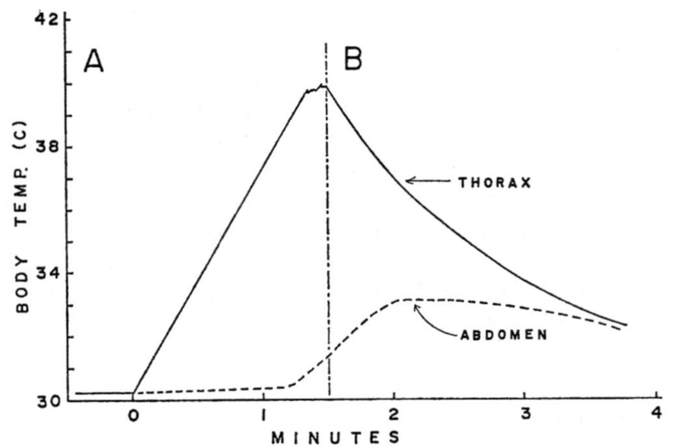

Fig. 2A and B. The approximate thoracic and abdominal temperatures of several sphinx and saturniid moths during pre-flight warm-up A and cooling B at an ambient temperature of 30 C. The animals were tethered by thermocouple leads during the measurements of thoracic temperature, and cool-down usually follows immediately after attempted flight (see also Fig. 4)

up in synchronous flyers, such as lepidopterans (Kammer, 1968), the orthopteran *Neoconocephalus robustus* (Heath and Josephson, 1970), and sometimes in asynchronous flyers such as bumblebees (see Fig. 3) the antagonistic flight muscles are activated synchronously. During warm-up the muscles contract primarily against each other rather than on the wings. In bees the production of heat during warm-up as well as during flight is a function of action potential frequency (Bastian and Esch, 1970; Heinrich and Kammer, 1973).

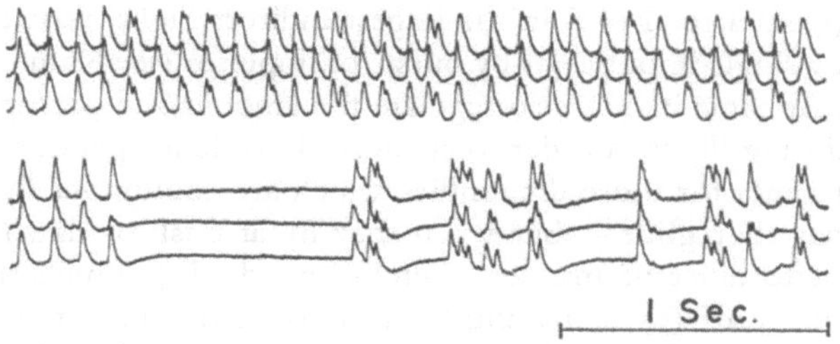

Fig. 3. Extracellular action potentials recorded from three different fibrillar muscles of a stationary bumblebee, *Bombus vosnesenskii*, during warm-up. The three traces are, from top to botton: Right dorso-ventral, left dorso-ventral and right dorso-longitudinal muscles. Thoracie temperature = 24 C. The action potentials can be "on", "off" or at a wide range of rates, particularly in stationary bees during the stabilization of T_{Th} (see Kammer and Heinrich, 1972; Heinrich and Kammer, 1973)

During any warm-up cycle the increase of T_{Th} is usually linear with time (Fig. 2). This can be expected on physical grounds if the rate of heat production is directly related to muscle temperature since the passive rate of heat loss is, according to Newton's law of cooling, directly related to the difference between T_{Th} and T_A.

Until recently it was thought that the rate of pre-flight warm-up is independent of T_A and muscle temperature in several species of moths (Heath and Adams, 1967; Hanegan and Heath, 1970a; McCrea and Heath, 1972). From this it was logically concluded that these moths regulate their rate of heat production during warm-up. However, high-speed recordings of T_{Th} during warm-up indicate, to the contrary, that the rate of heat production is *directly related* to thoracic temperature (Heinrich and Bartholomew, 1972; Bartholomew and Casey, 1973), and is thus not regulated with respect to thermoregulation. The moths warm up as rapidly as possible, thereby reducing the time that they are incapable of flight while being conspicuous because of the wing vibrations.

After the initiation of flight heat production is obligatory and continuous. Measurements of the metabolic rate of sphinx moths in free flight (Heinrich, 1971; Heinrich and Casey, 1973) and electrophysiological data of bumblebees in fixed flight (Heinrich and Kammer, 1973) indicate that the heat which is produced is a by-product of the flight effort alone under our conditions of measurement. The amount of heat which is produced does not vary with regard to T_A, and during tethered flight the insects do not vary the flight effort to offset heat loss (Heinrich, 1971; Heinrich and Kammer, 1973). To the contrary, the T_{Th} of the tethered animals usually *declined* rapidly during tethered flight, when the animals were not airborne by their own power. This is to be expected since, in various taxa of insects, the rate of heat production during flight varies with the fligth effort (see Weis-Fogh, 1964; Heinrich, 1971). During *free* flight, however, the animals would presumably receive information from exteroceptors which would inform them of their flight performance (as determined by the velocity of flight and the maintenance of altitude). Flight performance would deteriorate when thoracic temperature declines, but corrections of the flight effort during free flight in order to stay aloft could result in the can production of sufficient heat to elevate T_{Th}.

Flight effort can be regulated by activating varying numbers of motor units (from a total of 5 in the largest indirect flight muscles) and/or by activating them by single impulses or by bursts of impulses (see Kammer, 1968, 1970). It is probable that some insects also regulate the amount and perhaps the rate of heat production during flight in order to stabilize thoracic temperature. Corbet (1963), for example, reports for those dragonflies which are continuous flyers: "During the heat of the day they glide –, but when they fly at dusk or dawn they dart rapidly about in a quite different manner, thereby, no doubt, compensating for the lower air temperature" (by increasing the activity of the flight muscles and the rate of heat production). However, to my knowledge no data have been published that demonstrate the regulation of heat production during flight in any insect. Except in moths, few efforts have been made to investigate this problem.

The amount of heat which is produced by moths that are prevented from engaging in continuous flight (by being confined in a small respirometer or by being tethered by thermocouple leads) depends, in part, on the durations of warm-up, rest and attempted flight. For example, sphinx moths usually initiate continuous flight after warm-up, but when they are attached to a tether they usually only fly to the end of it, stop and immediately begin to cool. The tethered moths may attempt flight only once (Fig. 2) or a number of times (see Fig. 4) before cooling to air temperature. Such data of T_{Th} from moths during "uniform activity" have been used to support the theory that moths regulate their rate of heat production (Adams and Heath, 1964; Heath and Adams, 1965, 1967).

Other data used to support the theory that "Large moths increase metabolism during active periods to offset heat loss and thereby maintain a relatively constant internal temperature" (Heath and Adams, 1967) comes from oxygen consumption measurements of moths in a "microrespirometer". The metabolic rate of these animals (which were not required to expend the energy to support themselves in flight) extrapolate to zero metabolism at about 37 C (see Table 1). The conclusion that the moths regulate body temperature was considered to be "inescapable" (Heath and Adams, 1967). However, if the moths were maintaining a high T_{Th} in order to initiate flight (see Fig. 4), but were prevented from flying in the microrespirometer, they would necessarily have a greater energy expenditure at the lower than at the higher T_A. The above data have frequently been assumed to refer to flight. Similarly, Hanegan and Heath (1970a) summarize that "Directly measured rates of oxygen consumption in flying animals *(Hyalophora cecropia)* increase as air temperature decreases." However, their data indicating this trend (see Table 1) show that these animals, which are presumed to be "flying," (although wings were amputated, pers. comm. Hanegan), have a metabolic rate nearly identical to resting metabolism at T_A of 37 C. Such a metabolic rate is about 10 to 30 times less than that of other moths in flight (see Table 1). Data from *those* moths which were *not* in flight has been used repeatedly to discuss flight metabolism, and to found theories of temperature regulation during flight. It is obvious that such data are not applicable to answer physiological questions about flight when the methods of data gathering are taken into consideration. As a first approximation of flight (to answer questions of thermoregulation) wingbeats must be without a pause, at a minimum rate, and the animal must generate sufficient lift to remain airborne.

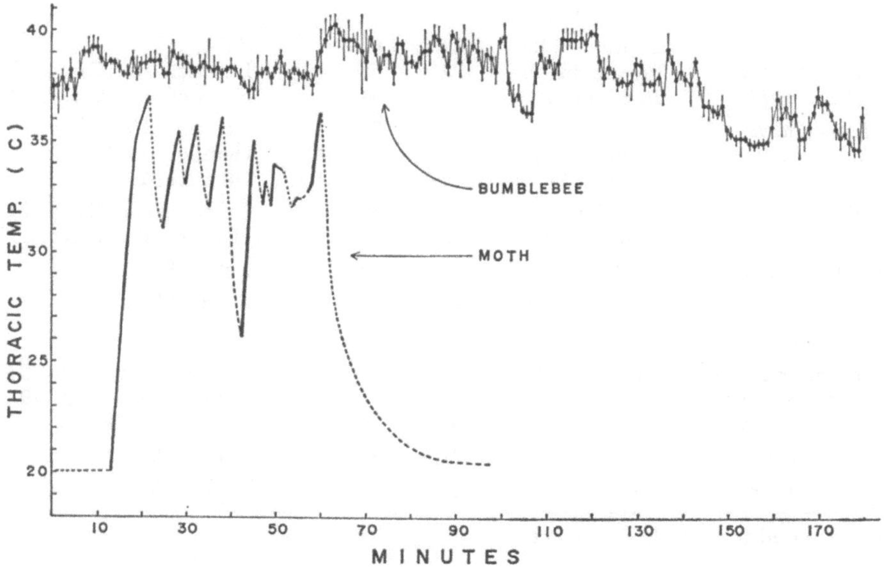

Fig. 4. Thoracic temperatures of a queen bumblebee, *B. vosnesenskii*, (Heinrich, unpubl. record) and of a sphinx moth, *Manduca sexta*, at 19–20 C. The relatively stationary bee was in the vicinity of her brood clump, where she maintained a high T_{Th} throughout the night without making attempts to fly. Her T_{Th} was recorded at 12 sec intervals, and the points represent the average for 1 min durations. The vertical lines represent the range of T_{Th} during this time. Thoracic temperature fluctuates rapidly in a small range, and "on" and "off" periods of heat production are usually indistinct or of short duration. The T_{Th} of the tethered moth (adapted from McCrea and Heath, 1971), on the other hand, fluctuates widely as the animal alternates between discrete periods of pre-flight warm-up (s. Fig.) usually followed by attempted flight after which the animal rests temporarily (-----) and T_{Th} declines

In bees, in contrast to moths, heat production is related not only to pre-flight warm-up and flight but also to nest temperature regulation. Bees, unlike moths, often maintain a high T_{Th} while they are stationary without making attempts to initiate flight (see Fig. 4). The mechanisms of heat production in "resting" bees have recently been investigated from the standpoints of biochemistry, electrophysiology and behavior.

Tissue extracts from bumblebee fibrillar muscles contain high catalytic activities of fructose diphosphatase and phosphofructokinase, which are thought to play a role in non-shivering thermogenesis of these muscles (Newsholme *et al.*, 1972). However,

Table 1. Rate of oxygen uptake[a] in sphinx and saturniid moths at various activities and air temperatures

Species	Activity	Temp. (C)	ml O_2/g/hr	Reference
Sphingidae				
Celerio (Hyles) lineata	"uniform activity"	31.5	15	Heath and Adams, 1967
Celerio (Hyles) lineata	(in a micro-respirometer)	26	26	Heath and Adams, 1967
Hyles euphorbia	Contin. flight	22.3	64	Zebe, 1954
Hyles euphorbia	Free flight (in	22	55	Heinrich and Casey, 1973
Hyles euphorbia	an 8.7 l jar)	15	54	Heinrich and Casey, 1973
Manduca sexta	Free flight (in	30	51	Heinrich, 1971
Manduca sexta	an 10.2 l jar)	15	44	Heinrich, 1971
Manduca sexta	Fixed flight	25	21	Heinrich, 1971
Deilephila elpenor	Free flight (in	22	59	Heinrich and Casey, 1973
Deilephila elpenor	an 8.7 l jar)	15	61	Heinrich and Casey, 1973
Metopsilus porcellus	Contin. flight	21–25	88	Zebe, 1954
Mimias tiliae	Contin. flight	22–24	56	Zebe, 1954
Saturniidae				
Hyalophora cecropia	"flying" (with	35	3	Hanegan and Heath, 1971 a, b
Halophora cecropia	wings amputated)	20	11	Hanegan and Heath, 1971 a, b
Hyalophora cecropia	Torpor	35	4	Hanegan and Heath, 1971 a, b
Saturnia pavonia	Contin. flight	21.5	81	Zebe, 1954
Antheraea pernyi	Contin. flight	22.6	29	Zebe, 1954

[a] Converted to mean body weights

the production of sufficient heat to elevate body temperature has so far not been demonstrated in resting insect muscle.

In asynchronous flyers like bees the indirect muscles appear to be functionally "unclutched" from the wings when the animal is at "rest", so that the muscles can be mechanically active while the wings remain motionless (Esch and Bastian, 1968). While stationary, bumblebees can activate their indirect muscles at great ranges of frequencies, and, like honeybees (Allen, 1959), thus regulate their "rate" of heat production (although heat production may also be "on" and "off" over durations of seconds or msec) from resting metabolism to metabolic rates such as those in flight. Thoracic temperature can be stabilized despite varying rates of heat loss from thorax to the environment or to the abdomen (Heinrich and Kammer, 1973).

Bumblebees also regulate their T_{Th} while foraging in the field (Fig. 5), where they alternate short flights and stops in rapid succession. The bees can forage at T_A near the freezing point while maintaining T_{Th} above 32 C (Heinrich, 1972a). Heat is produced by the thoracic muscles during flight between flowers, as well as (at low T_A) while perched on the flowers. In terms of energetics such endothermy is costly. However, it permits these insects to visit many flowers in rapid succession at low T_A where many other insects are unable to fly. Many flowers produce sufficient nectar so that the energetic cost of foraging at low T_A is justified (Heinrich, 1972a; 1972b).

Ectothermy

Insects of many orders increase their body temperature with negligible energy expenditure by basking and utilizing the solar energy of direct sunlight. Basking involves specific postural adjustments with respect to the incoming rays, and body temperature can increase as rapidly, and remain as high, as by endothermy. Basking permits locusts (which do not engage in endothermic warm-up) to initiate flight at low T_A (Fraenkel, 1930; Stower and Griffiths, 1966). Indeed, the ability to bask may permit some butterflies to be active at low temperatures at high elevations (Watt, 1969) and in the Arctic (Kevin and Shorthouse, 1970) where they might otherwise be excluded. However, basking is also observed in animals such as the desert cicada *Diceroprocta apache* (Heath and Wilkin, 1970) inhabiting "hot" environments.

Basking has been studied in butterflies more than in other insects, and several patterns of basking bevavior have been observed in this order. Some species such as *Danaus plexippus* bask as well as shiver (Kammer, 1970), others such as *Colias eurytheme* bask but do not shiver (see Watt, 1968) while still others (from near the equator) apparently do neither (Heinrich, 1972c). At least three disticnt postures have been observed in those butterflies that bask. The "dorsal baskers" usually point their head away from the sun and present their dorsal body and wing surface to the sun's rays (Vielmetter, 1958) The *Hesperidae* bask in a similar manner except that their fore-wings are in parallel with the sun's rays while their underwings are perpendicular to them. The "lateral baskers", in contrast, fold both sets of wings dorsally and orient with one of their lateral wing and body surfaces at right angles to the sunshine. In the lateral baskers the lower portions of the wing surfaces are in contact with the body. Heat transfer through the thin wings into the body is rapid, particularly when the wings are darkly pigmented (Watt, 1969). However, the claim that the wings of dorsal baskers act as radiometers to help capture heat energy from the sun (Clench, 1966) to increase body temperature could not be substantiated experimentally (Heinrich, 1972c; Kammer and Brachi, 1973).

Heat Loss

Most insects avoid high T_A or intense solar radiation. Small size allows them to utilize a diversity of microhabitats. In addition, structural features such as long legs and air space under the elytra may help to protect some beetles, for example, from a hot substrate and from intense solar radiation (Franz, 1930). Forced convection during flight can also prevent overheating from solar radiation in a small insect such as the butterfly *Precis villida* (Heinrich, 1972c).

Few insects can afford to lose water for evaporative cooling. However, the tsetse fly *Glossina morsitans* of equatorial Africa reduces its body temperature up to 1.6 C through evaporative cooling after opening its spiracles at high T_A (Edney and Barass, 1962). In the field this mechanism presumably permits the insects to take a blood meal while perched in direct sunshine on the hot hide of an animal, where there is a temporary superabundance of liquid.

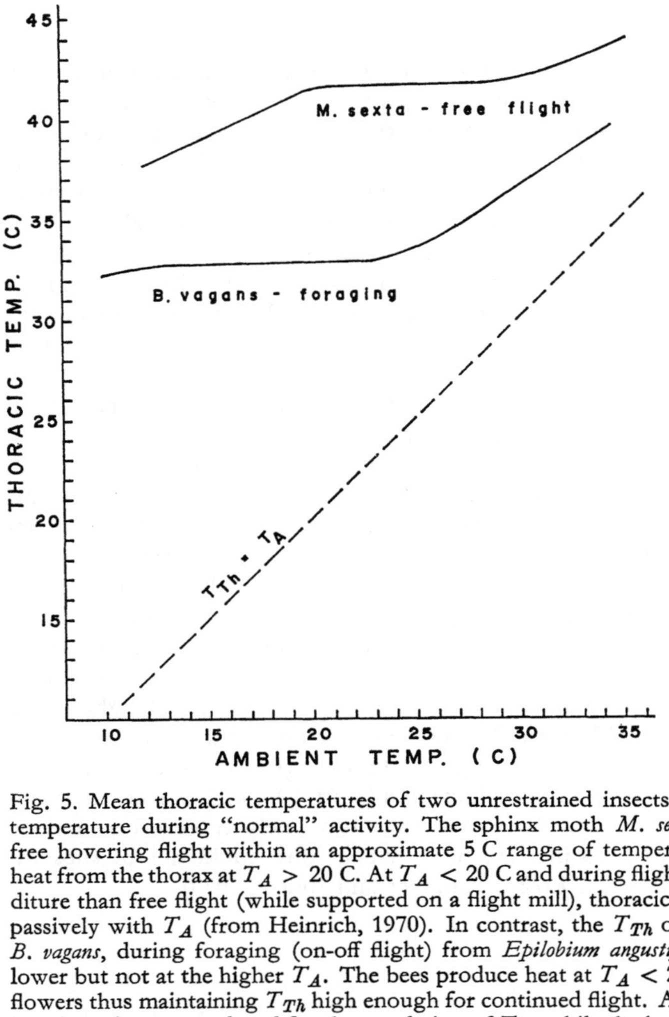

Fig. 5. Mean thoracic temperatures of two unrestrained insects as a function of ambient temperature during "normal" activity. The sphinx moth *M. sexta* regulates T_{Th} during free hovering flight within an approximate 5 C range of temperatures by shunting excess heat from the thorax at $T_A > 20$ C. At $T_A < 20$ C and during flight at a lower energy expenditure than free flight (while supported on a flight mill), thoracic temperature varies nearly passively with T_A (from Heinrich, 1970). In contrast, the T_{Th} of the bumblebee workers, *B. vagans*, during foraging (on-off flight) from *Epilobium angustifolium*, is regulated at the lower but not at the higher T_A. The bees produce heat at $T_A < 24$ C while stopped on the flowers thus maintaining T_{Th} high enough for continued flight. At the higher T_A additional heat is no longer produced for the regulation of T_{Th} while the bee is perched on the flowers and T_{Th} then varies passively with T_A. The magnitude of the difference between T_{Th} and T_A then presumably varies with the percentage of flight, and possible variations of flight effort

Overheating of the musculature during flight is not a general problem in the insects. Although their metabolic rates during free flight exceed those of all other organisms, the large surface/volume ratio of most insects necessarily results in the rapid dissipation of the endothermic heat to the environment. However, in the large heavily insulated sphinx moths, most of which remain in continuous flight throughout their activity period, the heat produced as a by-product of the flight metabolism can be sufficient to cause overheating of the thoracic muscles if the animals relied on passive heat loss from the thorax alone. At $T_A > 20$ C, overheating of the thoracic musculature in *Manduca sexta* is prevented by shunting excess heat in the haemolymph to the abdomen. Abdominal temperature remains relatively low as heat is lost from it to the environment by convection.

Bumblebees also regulate heat loss to the abdomen from the thorax, particularly while they are stationary and incubating their brood (Heinrich, 1972d). In the bumblebees, however, the heat flows from the abdomen by conduction directly into the brood clump containing the immatures (Fig. 6), rather than being dissipated to the environment by convection as a waste product of metabolism. The agent of this heat transfer remains to be elucidated.

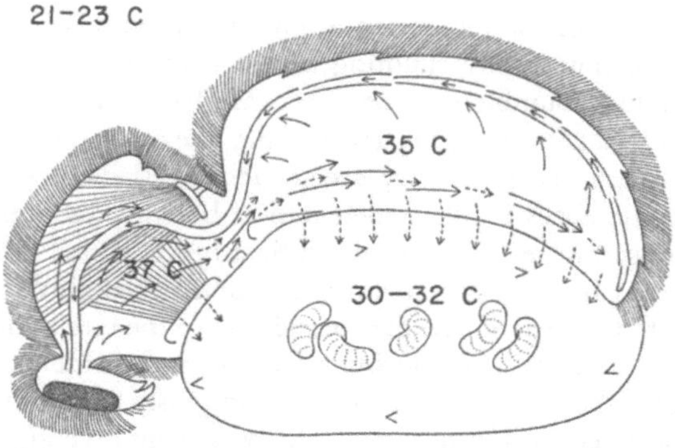

Fig. 6. Diagrammatic representation of the sagittal section of a queen bumblebee incubating her brood clump at room temperature (see Plath, 1943). The directions of heat flow (-----) from the thoracic muscles to the poikilothermic larvae (or eggs, pupae) which are being incubated, and the direction of blood flow (—) are indicated by arrows. (Adapted from Heinrich, 1972d)

The dorsal vessel of bees from some species is highly convoluted in the style (Wille, 1958). The anatomical arrangement of the vessel in the area between thorax and abdomen conforms to a counter-current heat exchanger, which could reduce heat transfer from thorax to abdomen during the circulation of blood between these two areas. However, in order to control heat retention in the thorax by counter-current mechanisms, and to affect heat loss to the abdomen by eliminating the counter-current, it may be necessary for the bees to regulate blood flow. At the present time it is not known if the flow of blood through the style is in both directions at the same time thus facilitating counter-current heat exchange, or if it is in pulses of fluid that alternate in direction of flow and thus facilitate heat transfer to the abdomen.

There is presumably a small amount of heat transfer from thorax to abdomen when the flow of air through the tracheal system is in that direction, but on the basis of the specific heat of air such heat transfer should be slight. Nevertheless, control of the direction of air movement through the tracheal system could be a significant factor in heat distribution in unsaturated air, when it could be combined with the evaporation and condensation of water. For example, heat in the thorax would be used to produce water vapor, and some of this heat could be released if the vapor condenses in the cooler abdomen. As with many other aspects of thermoregulation in insects there are few answers at the present time. Few questions have been formulated.

Summary

It is probable that the body temperature of most small insects varies nearly directly with ambient thermal conditions during rest as well as during activity. However, the metabolism of the flight musculature can be sufficient for some insects, particularly the larger ones, to generate a thoracic temperature of more than 30 C above air temperature. Some of these insects in which a high thoracic temperature is a prerequisite for flight, warm up prior to flight by activating their flight muscles, or by basking. Members of at least one group of insects, the sphinx moths, can stabilize their thoracic temperature during continuous free flight at a wide range of ambient temperatures by shunting excess heat from thorax to abdomen via the circulatory system. Bumblebees, on the other hand, can stabilize their thoracic temperature not only during foraging (when they are in on-off flight) but also while remaining stationary, as when they are incubating their eggs, larvae or pupae. The stationary bumblebees regulate their rate of heat production and heat loss to stabilize thoracic temperature. Abdominal temperature is varied independently from thoracic temperature by heat loss from the thorax. Heat flows by conduction from the abdomen into the brood. The data pertaining to the rate of heat production of moths reputed to be "in flight", and during warm-up, differ widely between laboratories. At the present time there are no data from continuously flying insects that indicate regulation of heat production during flight. The data from several species of sphinx moths indicate that the metabolic rate during continuous flight as well as during warm-up, is independent of T_A.

Acknowledgements

I thank Dr. R. H. Dadd for reading a draft of the manuscript. The author was supported by NSF grant GB-31542 during the writing, and some of the research, of this paper.

References

Adams, P. A., Heath, J. E.: Temperature regulation in the sphinx moth, *Celerio lineata.* Nature (Lond.) **201**, 20–22 (1964).

Allen, M. D.: Respiration rates of worker honeybees of different ages and at different temperatures. J. Exptl. Biol. **35**, 92–101 (1959).

Bachmetjew, P.: Über die Temperatur der Insekten nach Beobachtungen in Bulgarien. Z. Wiss. Zool. **66**, 521–604 (1899).

Bartholomew, G. A., Casey, T. M.: The effects of ambient temperature on warm-up in the moth, *Hyalophora cecropia.* J. Exptl. Biol. **58**, 503–510 (1973).

Bastian, J., Esch, H.: The nervous control of the indirect flight muscles of the honeybee. Z. Vergleich. Physiol. **67**, 307–324 (1970).

Church, N. S.: Heat loss and body temperature of flying insects. II. Heat conduction within the body and its loss by radiation and convection. J. Exptl. Biol. **37**, 186–212 (1960).

Clench, H. K.: Behavioral Thermoregulation in Butterflies. Ecology **47**, 1021–1034 (1966).

Corbet, P. S.: A biology of dragonflies, Chicago: Quadrangle Books, Inc. 1963.

Digby, P. S. B.: Factors affecting the temperature excess of insects in sunshine. J. Exptl. Biol. **32**, 279–298 (1955).

Dotterweich, H.: Beiträge zur Nervenphysiologie der Insekten. I. Das Schwirren der Schmetterlinge vor dem Fluge. Zool. Jahrb. Abt. Physiol. **44**, 399–450 (1928).

Edney, E. B., Barrass, R.: The Body Temperature of the Tsetse fly, *Glossina morsitans* Westwood (Diptera, Muscidae). J. Insect Physiol. **8**, 469–481 (1962).

Esch, H., Bastian, J.: Mechanical and electrical activity in the indirect flight muscles of the honeybee. Z. Vergleich. Physiol. **58**, 429–440 (1968).

Fraenkel, G.: Die Orientierung von *Schistocerca gregaria* zu strahlender Wärme. Z. Vergleich. Physiol. **13**, 300–313 (1930).

Franz, H.: Untersuchungen über den Wärmehaushalt der Poikilothermen. Biol. Zentr. **50**, 158–182 (1930).

Hanegan, J. L., Heath, J. E.: Mechanisms for the control of body temperature in the moth, *Hyalophora cecropia*. J. Exptl. Biol. **53**, 349–362 (1970a).

Hanegan, J. L., Heath, J. E.: Activity patterns and energetics of the moth, *Hyalophora cecropia*. J. Exptl. Biol. **53**, 611–627 (1970b).

Heath, J. E., Adams, P. A.: Temperature regulation in the sphinx moth during flight. Nature (Lond.) **205**, 309–310 (1965).

Heath, J. E., Adams, P. A.: Regulation of heat production by large moths. J. Exptl. Biol. **47**, 21–33 (1967).

Heath, J. E., Josephson, R. K.: Body temperature and singing in the Katydid, *Neoconocephalus robustus* (Orthoptera, Tettigoniidae). Biol. Bull. **138**, 272–285 (1970).

Heath, J. E., Wilkin, P. J.: Temperature responses of the desert cicada, *Diceroprocta apache* (Homoptera, Cicadidae). Physiol. Zool. **43**, 145–154 (1970).

Heinrich, B.: Thoracic temperature stabilization by blood circulation in a free-flying moth. Science **168**, 580–582 (1970).

Heinrich, B.: Temperature regulation of the sphinx moth, *Manduca sexta*. I. Flight energetics and body temperature during free and tethered flight. J. Exptl. Biol. **54**, 141–152 (1971).

Heinrich, B.: Temperature regulation in the bumblebee, *Bombus vagans*: a field study Science **175**, 185–187 (1972a)

Heinrich, B.: Energetics of temperature regulation and foraging in a bumblebee, *Bombus terricola* Kirby. J. Comp. Physiol. **77**, 49–64 (1972b).

Heinrich, B.: Thoracic temperatures of butterflies in the field near the equator. Comp. Biochem. Physiol. **43A**, 459–467 (1972c).

Heinrich, B.: Physiology of brood incubation in the bumblebee queen, *Bombus vosnesenskii*. Nature (Lond.) **239**, 223–225 (1972d).

Heinrich, B., Bartholomew, G. A.: An analysis of pre-flight warm-up in the sphinx moth, *Manduca sexta*. J. Exptl. Biol. **55**, 223–239 (1972).

Heinrich, B., Casey, T. M.: Metabolic rate and enothermy in sphinx moths. J. Comp. Physiol. **82**, 195–206 (1973).

Heinrich, B., Kammer, A. E.: Activation of the fibrillar muscles in bumblebees during warm-up, thermoregulation and flight. J. Exptl. Biol. (in press).

Himmer, A.: Die Temperaturverhältnisse bei den sozialen Hymenopteren. Biol. Rev. **7**, 224–253 (1932).

Kammer, A. E.: Motor patterns during flight and warm-up in Lepidoptera. J. Exptl. Biol. **48**, 89–109 (1968).

Kammer, A. E.: Thoracic temperature, shivering and flight in the monarch butterfly, *Danaus plexippus* (L.), Z. Vergleich. Physiol. **68**, 334–344 (1970).

Kammer, A. E., Brachi, J.: Role of the wings in the absorption of radiant energy by a butterfly. Comp. Biochem. Physiol. A. **45A**, 1057–1063 (1973).

Kammer, A. E., Heinrich, B.: Neural control of bumlebee fibrillar muscle during shivering. J. Comp. Physiol. **87**, 337–345 (1972).

Kevin, P. G., Shorthouse, J. D.: Behavioral thermoregulation by High Arctic butterflies. J. Arctic Inst. N. Amer. **23**, 269–279 (1970).

Krogh, A., Zeuthen, E.: The mechanism of flight preparation in some insects. J. Exptl. Biol. **18**, 1–10 (1941).

Machin, K. E., Pringle, J. W. S., Tamasige, M.: The physiology of insect fibrillar muscle IV. The effect of temperature on a flight muscle. Proc. Roy. Soc. B. **155**, 493–499 (1962).

McCrea, M. J., Heath, J. E.: Dependence of flight on temperature regulation in the moth, *Manduca sexta*. J. Exptl. Biol. **54**, 415–435 (1971).
Neville, A. C., Weis-Fogh, T.: The effect of temperature on locust flight muscle. J. Exptl. Biol. **40**, 111–121 (1963).
Newport, G.: On the temperature of insects, and its connexion with the functions of respiration and circulation in this class of Invertebrated animals. Phil. Trans. Roy. Soc., London **127** (2), 259–339 (1837).
Newsholme, E. A., Crabtree, B., Higgins, S. J., Thornton, S. D., Stuart, C.: The activities of fructose diphosphatase in flight muscles from the bumble-bee and the role of this enzyme in heat generation. Biochem. J. **128**, 89–97 (1972).
Oosthuizen, M. J.: The body temperature of *Samia cecropia* Linn. (Lepidoptera, Saturniidae) as influenced by muscular activity. J. Entomol. Soc. S. Afr. **2**, 63–73 (1939).
Plath, O. E.: *Bumblebees and their ways*. New York: MacMillan 1934.
Stower, W. J., Griffiths, J. F.: The body temperature of the desert locust *(Schistocerca gregaria)*. Entomol. Exptl. Appl. **9**, 127–178 (1966).
Vielmetter, W., Physiologie des Verhaltens zur Sonnenstrahlung bei dem Tagfalter *Argynnis paphia* L. I. Untersuchungen im Freiland. J. Insect Physiol. **2**, 13–37 (1958).
Watt, W. B.: Adaptive significance of pigment polymorphism in *Colias* butterflies. I. Variation of melanin pigment in relation to thermoregulation. Evol. **22**, 437–458 (1968).
Watt, W. B.: Adaptive significance of pigment polymorphism in *Colias* butterflies, II. Thermoregulation and photoperiodically controlled melanin variation in *Colias eurytheme*. Proc. Natl. Acad. Sci. **63**, 767–774 (1969).
Weis-Fogh, T.: Biology and physics of locust flight. VIII. Lift and metabolic rate of flying locusts. J. Exptl. Biol. **41**, 257–271 (1964).
Weis-Fogh, T.: Energetics of hovering flight in hummingbirds and in *Drosophila*, J. Exptl. Biol. **56**, 79–104 (1972).
Wille, A.: A comparative study of the dorsal vessel of bees, Ann. Entomol. Soc. Am. **51**, 538–546 (1958).
Zebe, E.: Über den Stoffwechsel der Lepidopteren. Z. Vergleich. Physiol. **36**, 290–317 (1954).

Environmental Factors Affecting the Acclimatory Responses of Ectotherms

R. C. Newell

With 9 Figures

Patterns of Acclimatory Responses in Ectotherms

Most studies on thermal acclimation of whole organisms have involved direct or indirect measurements of the effects of short-term experimental temperature change on easily measurable phenomena such as growth rates (Dehnel, 1955), heart rates (Segal, Rao and James, 1953; Pickens, 1965; Ahsanullah and Newell, 1971), cirral activity in barnacles (Crisp and Ritz, 1967a, b; Ritz and Foster, 1968), feeding rates (Rao, 1953a, b), radular activity (Newell, Pye and Ahsanullah, 1971a) or temperature tolerance (Fry, 1957; Vernberg, Schlieper and Schneider, 1963; Newell, Pye and Ahsanullah, 1971b). As is well-known, the rates of activity commonly vary directly with experimental temperature with a Q_{10} of approximately 2.0 over the normal environmental temperature range of the organism. Higher values than 2.0 for the Q_{10} are obtained at low experimental temperatures and lower values than this at high experimental temperatures. Compensation for long-term changes in temperature has been discussed in detail by Bullock (1955), Precht, Christophersen and Hensel (1955), Prosser (1955, 1958), Precht (1958) and Prosser and Brown (1961). Briefly, one may recognise (a) a relatively non-specific increase in enzyme concentration in response to cold acclimation, (b) *translation* or (c) *rotation* of the rate: temperature curve or a combination of these responses following thermal acclimation.

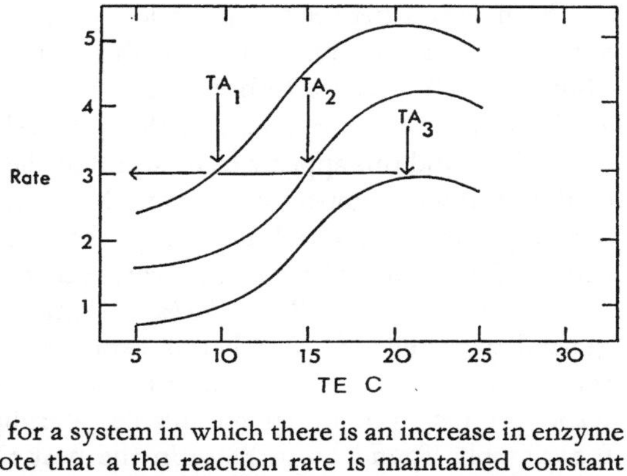

Fig. 1. Idealised rate: temperature curves for a system in which there is an increase in enzyme concentration with cold acclimation. Note that a the reaction rate is maintained constant at TA_1, TA_2 and TA_3. b The thermal tolerance of the system is the same at each acclimation temperature (TA)

An example of compensation by increase in enzyme concentration at low temperatures is shown in Fig. 1 whilst that of translation of the rate: temperature curve is shown in Fig. 2. It will be noticed that although each curve is markedly dependent upon experimental temperature, nevertheless the rate of reaction at acclimation temperature TA_3 equals the rate at TA_2 which in turn equals the rate at TA_1. Thus when the rate of reaction is plotted against acclimation temperature, perfect compensation may occur. Complete, or perfect, compensation of this kind has been reported in a wide variety of biological systems and occurs, for example, in the standard oxygen consumption of the sunfish *Lepomis gibbosus* over the temperature range 10–17.5 C (Roberts, 1964, 1967). This type of response is well-documented and represents one way in which an essentially temperature-dependent process may be compensated for long-term changes in temperature.

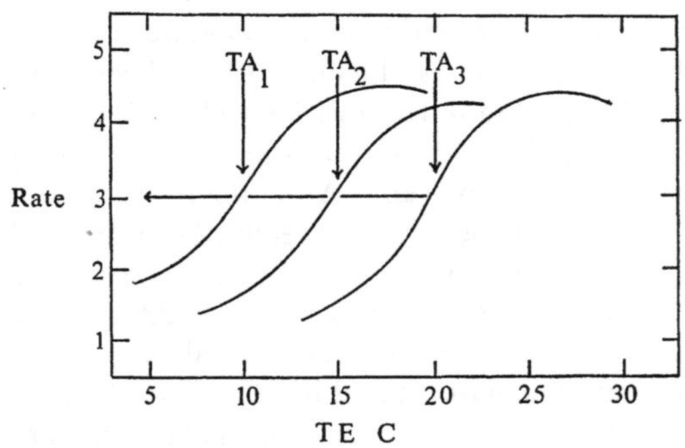

Fig. 2. Idealised graphs showing the translation of rate: temperature curves to the right following warm acclimation. Note that a this results in complete acclimation of the rate function which is the same at each acclimation temperature, b the thermal tolerance of the system is raised in response to warm acclimation

One of the earlier examples of thermal acclimation involved the changed heat tolerance of organisms with latitude or with season, and a general correspondence has been found between upper and lower lethal temperatures and the normal environmental temperature range to which the organisms are exposed (for review, see Gunter, 1957). Many ectotherms can raise their heat tolerance in response to warm acclimation (for reviews, see Fry, 1957; Newell, 1970; also Newell, Pye and Ahsanullah, 1971 b) and this appears to be accomplished essentially by a process of translation of the mortality: temperature curves. Much the same type of shift in the rate: temperature curve for cirral activity in the intertidal barnacles *Balanus balanoides* and *Chthamalus stellatus* in response to warm acclimation has been described by Southward (1964; also Crisp and Ritz, 1967 a, b; Ritz and Foster, 1968) and appears to be of general occurrence in ectotherms exposed to seasonal changes in temperature. In ectotherms which live in a thermostable environment, however, the rate: temperature responses may not be so modifiable. Vernberg, Schlieper and Schneider (1963), for example, found no increase in the heat tolerance of isolated gill tissue excised from the sublittoral *Aequipecten* following warm-acclimation, although tissue from the intertidal bivalves *Crassostrea* and *Modiolus* did show a raised heat-tolerance.

Again, there is little evidence of an alteration in the rate: temperature curves for the heart rate of the crab *Carcinus maenas* following warm acclimation (Ahsanullah and Newell, 1971) and this is an ectotherm which can evade thermal stress by behavioural thermoregulation.

Such translation of rate: temperature curves to the left following cold acclimation or right following warm acclimation is by no means confined to whole organisms. There are many examples where enzymes from cold acclimated organisms show higher levels of activity compared with those extracted from warm acclimated ones (for review, see Prosser, 1962). Complete compensation has been reported for the cholinesterase activity of brain tissue of the fish *Fundulus heteroclitus* at 13 and 30 C (Baslow and Nigrelli, 1964). Rao (1967) has cited data showing compensatory increases in the activity of succinic-, malic-, lactic-, and pyruvic-dehydrogenase from the hepatopancreas of cold-acclimated scorpions in response to cold acclimation. Similar changes in the activity of cytochrome *c*, cytochrome oxidase and lipase from the earthworm *Lampito mauritii* also occur in response to cold or warm acclimation (Rao, 1967). Compensatory increases in enzyme activity in response to cold acclimation have even been reported in succinic- and malic-dehydrogenase and in cytochrome *c* oxidase from rat tissue (Hannon, 1960; Hannon and Vaughan, 1960). Translation of a temperature-dependent curve to the right following warm acclimation or to the left following cold-acclimation may thus be regarded as a common pattern of response which has been demonstrated in whole ectotherms, their isolated tissues as well as in specific enzyme systems.

Although such translation may compensate for seasonal or latitudinal changes in environmental temperature, nevertheless it remains true that short-term temperature fluctuations such as would occur diurnally, or with the ebb and flow of the tide, result in marked alteration of the rate function. Where compensation for short-term temperature change is required, a *rotation* of the rate: temperature curve in response to thermal acclimation occurs. Many examples of such rotation of rate: temperature curves have been described, particularly in the metabolism of ectotherms. However, other studies have yielded more diverse results including combinations of both rotation and translation of the metabolism: temperature curves in response to thermal acclimation. It is my main purpose to show how these data for both enzymic studies and whole animal metabolism can be integrated to form a model for thermal acclimation in ectotherms.

Effects of Temperature on Ectotherm Metabolism

One of the complicating factors which affects measurements of the oxygen consumption of animals is the level of activity which they show during the course of an experiment. Some animals remain inactive or active in a respirometer for long enough periods to record their oxygen consumption. In those animals which always show some level of activity in the respirometer, however, it may be possible to simultaneously measure oxygen consumption and activity level. Then oxygen consumption can be plotted as a function of locomotory activity and by extrapolation the respiration rate of the organism at zero activity can be recorded. Adopting the nomenclature of Fry (1947, 1957; see also Halcrow and Boyd, 1967), we may then recognise a rate of oxygen consumption during quiescence "the standard rate" and a rate corresponding with maximal activity "the active rate". Between these two rates is a variety of "routine rates" corresponding with spontaneous movements of the organism concerned. It is of importance to distinguish between the active and standard rates because, as will be shown below, experimental temperature may affect the two rates differently.

It has been shown above that many processes such as ciliary activity of bivalves (Gray, 1923), cirral activity of barnacles (Southward, 1964), radular activity of gastropods (Newell, Pye and Ahsanullah, 1971) and the rates of movement of a wide variety of ectotherms are temperature-dependent and have a Q_{10} of approximately 2.0 over the normal environmental temperature range. Not surprisingly, the oxygen consumption of such active animals shows a similar curve in response to experimental temperature, since activity itself is an energy-consuming process and would be expected to dominate the metabolic requirements of the organism.

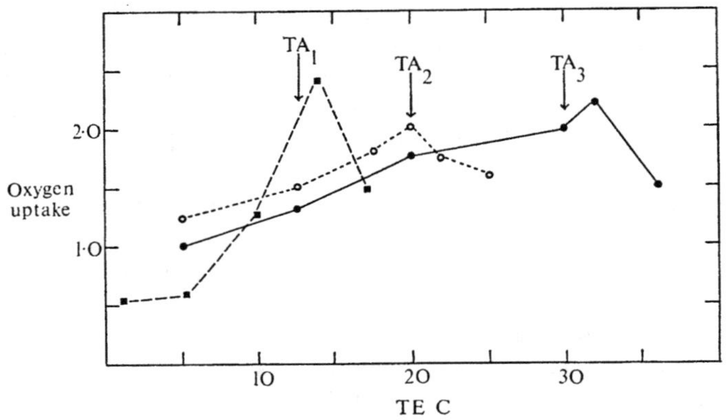

Fig. 3. Graphs showing the effect of experimental temperature on the rate of oxygen consumption (μl O_2/mg dry protein/hr) of active *Littorina littorea*. Data for a specimen of 30 mg dry protein weight. (After Newell and Pye, 1971a)

Fig. 3 shows the effect of experimental temperature on the active rate of oxygen consumption of 30 mg specimens of the intertidal gastropod *Littorina littorea* collected when the environmental temperature was A) 12.5 C, B) 20 C and C) 30 C. It is evident that the oxygen consumption is not only temperature-dependent, but that warm acclimation results in both translation and rotation of the rate: temperature curve, much as would be expected from the data cited for direct observations on activity in ectotherms. Since most organisms are to a greater or lesser extent active in a respirometer, it is not surprising that temperature-dependent rates of oxygen consumption are commonly reported in the literature.

In contrast, the metabolism of quiescent *Littorina* is at a low level and has been found to be markedly independent of experimental temperature (Newell, 1969; Newell and Pye, 1970a, b; 1971a, b). The same is true of a variety of other ectotherms (Newell and Northcroft, 1965, 1967; Halcrow and Boyd, 1967; Collardeau-

Roux, 1966; Roux and Roux, 1967; McFarland and Pickens, 1965; Siefken and Armitage, 1968; Barnes and Barnes, 1969). Curiously enough, similar data for other ectotherms including the earthworm *Lumbricus terrestris*, the frog *Rana temporaria* and the newt *Triturus vulgaris* were recorded by Vernon (1897) but they received little attention until they were replotted by Bullock (1955). It follows, therefore, that the metabolism of such organisms is endowed with considerable flexibility and may vary from a low rate which is virtually independent of experimental temperature to a much faster active rate which is dependent upon experimental temperature. This range may be regarded as "the scope for activity" (Fry, 1957) and is obtained by subtraction of the standard from the active rate.

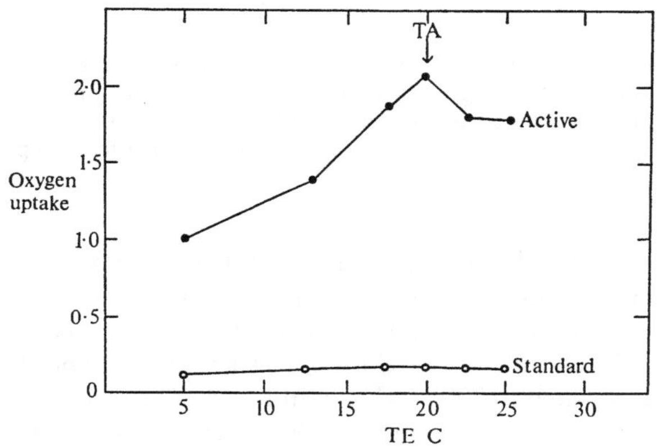

Fig. 4. Graphs showing the effect of experimental temperature on the active (solid symbols) and standard (open symbols) oxygen consumption of *Littorina littorea* (μl O_2/mg dry protein/hr). Data for an animal of 30 mg dry protein weight acclimated to 20 C. (After Newell and Pye, 1971 a)

Both the active rate and the standard rate of metabolism of a specimen of *Littorina littorea* which had been collected when the environmental temperature was 20 C are shown in Fig. 4. It is evident that the two rates reach their maximum divergence at the acclimation temperature of the animal and thereafter they converge towards heat-lethal temperatures. This implies that the organism is endowed with its maximum scope for activity near the acclimation temperature. Further, the convergence of the active and standard rates results, at high temperatures, in zero scope for activity. This corresponds with the onset of heat-coma which occurs in many ectotherms prior to heat death. Thermal acclimation, involving a translation to the right of the curves for the active and standard rate, then allows the maximum scope for activity to correspond once again with environmental conditions. A series of graphs showing the scope for activity of *Littorina* acclimated to 12.5, 20, and 30 C is shown in Fig. 5.

One of the interesting problems which poses itself is how one can envisage a cellular mechanism which on the one hand is temperature-dependent in the active animal and yet can also be independent of temperature in the inactive animal. Our first question was whether temperature-independent metabolism could be de-

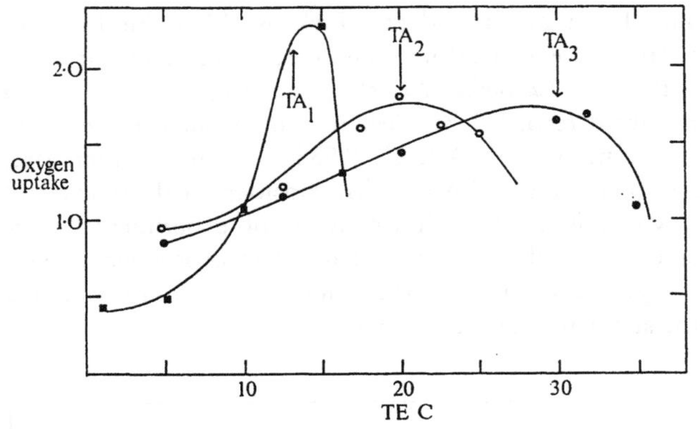

Fig. 5. Figure showing the effect of experimental temperature on the scope for activity (O_2 uptake, μl O_2/mg dry protein/hr) in *Littorina* acclimated to 12.5, 20 and 30 C. Data for a specimen of 30 mg dry protein weight. Note that thermal acclimation results in the maximum scope for activity coinciding with the acclimation temperature. (Based on Newell and Pye, 1971 a)

monstrated at a subcellular level. If one homogenises the tissues, or makes a crude mitochondrial preparation of the tissues of *Littorina* whose active and standard rates of oxygen consumption are known, one can compare the rates obtained quantitatively with those of the intact animal. A typical result obtained with an animal acclimated to 20 C is shown in Fig. 6. The main point of interest is that in the homogenate there is a region of the curve which is virtually independent of temperature which thus, like the standard rate of the intact animal, shows "metabolic homeostasis". But the rate is higher than the standard rate and the form of the curve is different from either of the rates recorded in the intact organism.

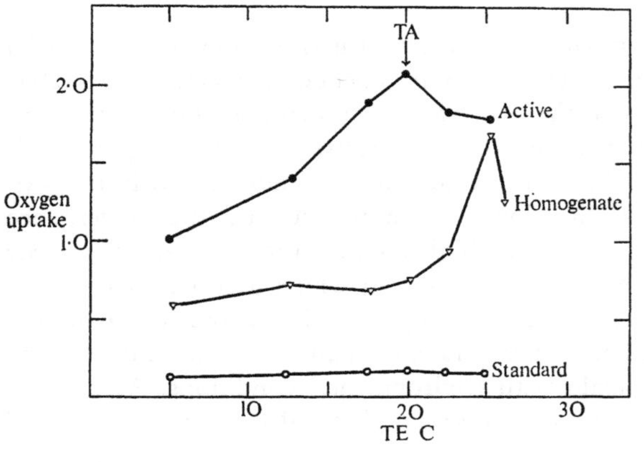

Fig. 6. Graphs showing the relationship between the active and standard metabolism of intact *Littorina* and that of crude homogenates in relation to experimental temperature. Data for intact animals of 30 mg dried protein weight and a homogenate containing the same weight of protein suspended in pasteurised seawater (based on Newell and Pye, 1971 a). Rates expressed as μl O_2/mg dry protein/hr

The results were of sufficient interest to prompt a further investigation of these results using a mitochondrial preparation rather than a crude homogenate. It should be emphasised, however, that in so doing, many of the possibilities for metabolic compensation may also be removed, so that in one sense the system might be expected to show such compensation only within rather a limited set of environmental conditions. Fig. 7 shows the effect of temperature on the rate of oxygen consumption of mitochondrial preparations extracted from *Littorina*. The preparations were suspended in pasteurised seawater to which various concentrations of pyruvate had been added. The rates of oxygen consumption are expressed per unit of protein in the original homogenate from which the mitochondria were extracted so that the rates are directly comparable with those of the intact animal shown in Fig. 6. It is evident that one can induce rates identical with the active rate by supplying high substrate levels (above 1.0 mM pyruvate). One obtains low and temperature-independent rates identical in level with the standard rate of inactive animals at low substrate concentrations of 0.01 mM. Moreover, rates similar to those of homogenates are obtained with intermediate substrate levels. In this connection it is of interest that Somero and Hochachka (1969) quote values for the pyruvate concentration of King Crab muscle which are very similar to those associated with the temperature-independent rates shown in Fig. 6 for *Littorina littorea*.

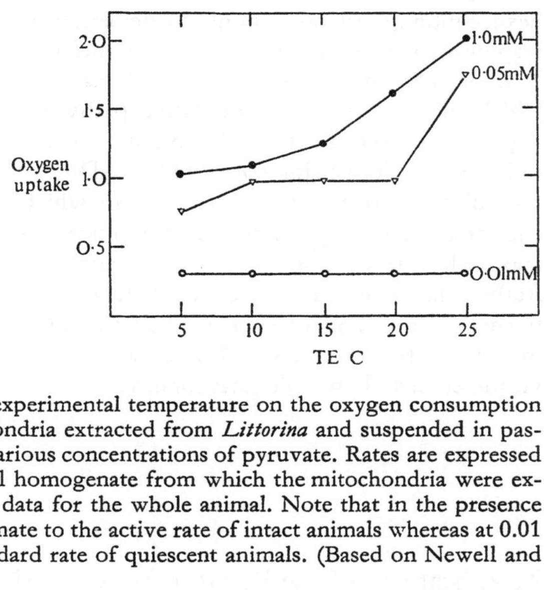

Fig. 7. Graphs showing the effect of experimental temperature on the oxygen consumption ($\mu l\ O_2$/mg dry protein/hr) of mitochondria extracted from *Littorina* and suspended in pasteurised seawater in the presence of various concentrations of pyruvate. Rates are expressed in terms of the protein in the original homogenate from which the mitochondria were extracted, to facilitate comparison with data for the whole animal. Note that in the presence of 1.0 mM pyruvate the rates approximate to the active rate of intact animals whereas at 0.01 mM their rates are similar to the standard rate of quiescent animals. (Based on Newell and Pye, 1971a)

We may thus infer that the variety of rates recorded may reflect variations in the availability of substrate to the mitochondria. During active metabolism, there may be a mobilisation of metabolic reserves whilst the onset of quiescence, which occurs in many ectotherms during adverse environmental conditions, there is a fall in available substrate. It follows from this, that after depletion of the metabolic reserves by starvation of the intact animal, one would expect the induction of a low and temperature-independent rate of oxygen consumption. We have not yet established whether this is so in *Littorina* but Marsden and I (in press) have found that this occurs in the shore crab *Carcinus maenas*, and Bayne (this symposium) has demonstrated this also in the mussel *Mytilus edulis*. In the barnacle, *Balanus balanoides*, the Q_{10} also becomes 1.0 following a period of starvation (Barnes, Barnes and Finlayson, 1963).

Somero (1969) and Hochachka and Somero (1971) have recently reviewed the evidence which suggests that positive or negative modulation of enzyme-substrate affinity with temperature may maintain reaction rates of a variety of crude enzyme preparations relatively independent of environmental temperature at low substrate concentrations in several ectotherms. Such modulation does not occur at substrate concentrations above K_m for the system and the rate of reaction is temperature-dependent at saturating substrate concentrations. Clearly, such a system agrees well with our data for *Littorina* and with the effects of starvation on intact crabs, barnacles and mussels.

Although this scheme is a relatively simple one which suggests that substrate availability is one factor which controls both the level and the temperature dependence of ectotherm metabolism, it is obvious that there are many other mechanisms which may be implicated in the control of metabolism. Hochachka (1967), for example, has cited the possibility of gluconeogenesis, glycogen synthesis, lipogenesis and hexose-monophosphate shunt participation in the regulation of metabolism in response to temperature. Such shunt mechanisms would be capable of operation at high substrate concentrations and may function in subcellular preparations in which the possibility of alternative pathways in metabolism have been retained. Again, the onset of activity in an animal is associated with a fall in the ATP/ADP ratio since ATP is being converted to ADP. It would be easy to envisage a terminal control of mitochondrial respiration in which the ATP suppressed further mitochondrial activity by endproduct inhibition. However tempting to a "reductionist" approach, it seems unlikely that one principal controlling mechanism will be found. Rather, the great flexibility of the metabolism of ectotherms within the limits set by the active and standard rates, is likely to be regulated at a variety of different points in the metabolic process and be capable of functioning over a wide variety of environmental and physiological conditions.

Modifications of the Relation between Metabolism and Temperature

So far, we have been mainly concerned with a consideration of the flexibility of ectotherm metabolism, and how relatively simple biochemical mechanisms could account for both the temperature dependence of the active rate as well as the independence of the standard rate. Many organisms, as has been shown on p. 152, can also alter their whole range of thermal tolerance. The patterns I have discussed, therefore, can be shifted up or down the temperature range to conform with local seasonal or geographic conditions. In much the same way as the active rate of *Littorina* is translated to the right following warm acclimation, so too are the rate: temperature curves for homogenates of these animals. Fig. 8 shows the curves obtained with homogenates of *Littorina* acclimated to a variety of temperatures for two weeks. One of the interesting features of some of our data is that they suggest that translation to the right following warm acclimation is achieved in a stepwise fashion, and that one particular curve may be characteristic of several different acclimation temperatures.

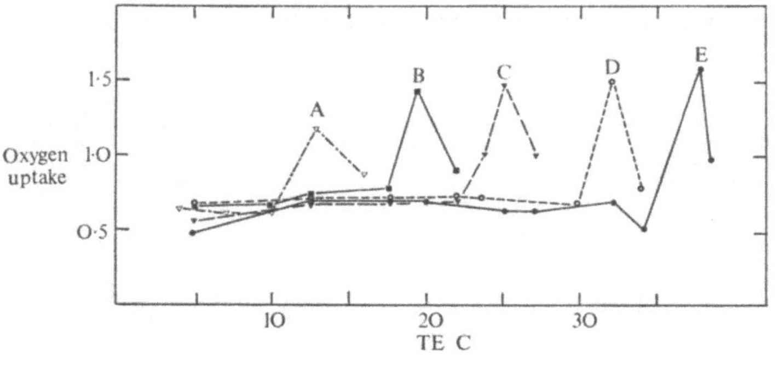

Fig. 8. Rate: temperature curves for homogenates of *Littorina* acclimated to various temperatures. Each curve was obtained by pooling the data obtained for homogenates suspended in pasteurised seawater following acclimation for two weeks. *A* Curve obtained with animals acclimated to 5, 7 and 10 C. *B* Curve obtained with animals acclimated to 13 and 15 C. *C* Curve obtained with animals acclimated to 18 and 22 C. *D* Curve obtained with animals acclimated to 25 and 28 C. *E* Curve obtained with animals acclimated to 30 and 32 C. Curves *D* and *E* were obtained from homogenates of animals collected from the shore rather than acclimated in the laboratory. Rates expressed as μl O_2/mg dry protein/hr from unpublished data of Newell and Marsden

It is clear then that the thermal tolerance of the respiratory system can be raised in response to warm acclimation and it is tempting to think of this in terms of changes in some important regulatory enzyme system. Several key enzymes in metabolism are now known to occur as isozymes, each of which may function optimally at a different pH or temperature regime. For example, in salmonids LDH isozymes are induced in response to cold acclimation and their rate of activity is sensitive to metabolic control (Hochachka, 1965, 1967). Again, in the Alaskan King Crab there is a temperature-induced interconversion between two forms of pyruvate kinase (Somero, 1969). These and other isozymes, moreover, have kinetic properties which make them ideal candidates for the control of the type of metabolic homeostasis we have been considering. At low substrate concentrations, for example, LDH isozymes show a decrease in enzyme-substrate affinity with an increase in temperature. This offsets the tendency for the reaction rate to increase which would be expected on thermodynamic grounds. Equally, the stepwise alteration in the rate: temperature curves which are shown in Fig. 8 may be attributable to the presence of different regulatory isozymes appropriate to the particular acclimation regime.

Conclusion

I wish to conclude by emphasising that we do not yet know whether changes in enzyme-substrate affinity of LDH, pyruvate kinase, or some other enzyme is of prime importance in the control of metabolism in ectotherms. Many other enzymes, the terminal steps of oxidative phosphorylation, or alternative pathways may be of importance. We could imagine, however, that, as is illustrated in Fig. 9, one factor which allows an organism to live on land compared with life in a thermostable environment such as in deposits or on the sea floor, is its ability to synthesise multiple forms of key enzymes. Such isozymes may be induced in response to ionic, or pH changes in the cellular environment, by hormonal or nervous stimuli induced by

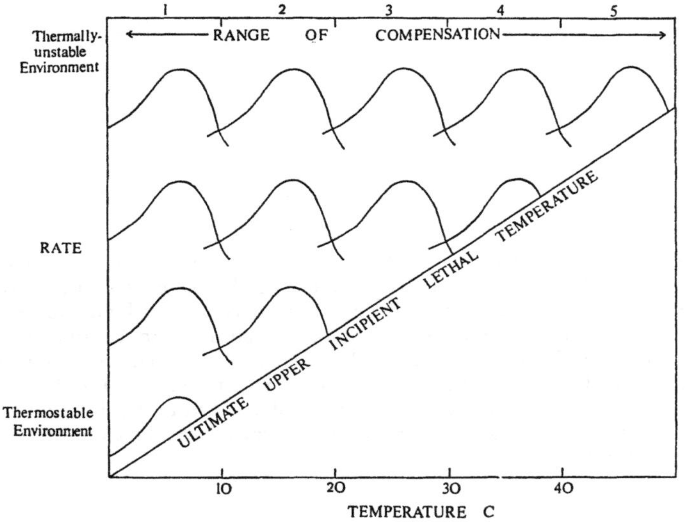

Fig. 9. Model showing the ecological pattern of variation in the thermal tolerance of ectotherms in relation to acclimation temperature. Note that animals from the thermostable environment cannot alter the temperature tolerance either of the whole organism or its isolated tissues in response to thermal acclimation. The rate: temperature curves depicted illustrate the type of response obtained with measurements of the activity of animals and could also depict the percentage mortality versus temperature after exposure to acclimation regimes 1–5

seasonal changes in photoperiod, or even by temperature itself. The presence of such isozymes, however they have been induced, would give great flexibility to the regulatory processes in metabolism and allow new environments to be colonised whilst retaining the same essential mechanism controlling the level of the metabolic rate. It is possible, then, that conformational changes in protein tertiary and quaternary structure may represent a basic mechanism of temperature compensation, but there are many ways of inducing such changes in enzyme-substrate affinitiy and these require a great deal of further study.

Summary

1. The patterns of acclimatory response of ectotherms are described, and examples given of the occurrence of translation and rotation of the rate: temperature curves for a variety of processes.

2. The effects of temperature on ectotherm metabolism are described. It is shown that acute temperature change affects the active and standard rates of some ectotherms differently, the active rate being more dependent upon temperature than the standard rate.

3. The metabolism of such ectotherms is thus variable in level within the limits set by the active and standard rates and moreover shows variable effects of experimental temperature depending upon activity. This range of variability may be regarded as the "scope for activity" and is obtained by substraction of the standard from the active rate.

4. Warm acclimation results in a translation to the right of the curves for the standard and active rate and allows the maximum scope for activity to correspond once again with environmental conditions.

5. Rates of oxygen consumption identical with the active rate of intact *Littorina* can be induced in mitochondrial preparations by supplying with high substrate levels (more than 1.0 mM pyruvate). Low and temperature-independent rates identical in level with the standard rate of inactive animals are obtained at low substrate concentrations (0.01 mM pyruvate). Rates similar to those of crude homogenates are obtained with intermediate substrate levels.

6. It is suggested that variations in the level and temperature-dependence of metabolism in *Littorina*, and perhaps also in other ectotherms, is associated with mobilisation or reduction in substrate availability to the mitochondria. This suggestion is supported by the fact that starvation of a variety of invertebrates results in a low and temperature-independent rate of oxygen consumption.

7. The temperature-independence of respiration at low substrate concentrations may be attributable to variations in enzyme-substrate affinity of some important regulatory enzyme with temperature. The temperature-dependence of the active rate of intact animals and the respiration of mitochondria in the presence of high substrate concentrations, however, is attributable to the fact that at saturating levels of substrate, enzyme-substrate affinity is not rate limiting and hence many enzyme reaction rates are temperature-dependent.

8. Translation of such curves to the right following warm acclimation may be attributable to the mobilisation of isozymes capable of functioning at different acclimation regimes.

9. It is suggested that one factor which allows ectotherms to live in a thermally unstable environment may be an ability to synthesise multiple forms of regulatory enzymes. Such isozymes would give great flexibility to the processes of metabolism and allow new environments to be colonised whilst retaining the same essential mechanisms controlling the level of the metabolic rate.

References

Ahsanullah, M., Newell, R. C.: Factors affecting the heart rate of the shore crab *Carcinus maenas* (L.). Comp. Biochem. Physiol. **39 A**, 277–287 (1971).

Barnes, H., Barnes, M.: Seasonal changes in the acutely determined oxygen consumption and effect of temperature for three common cirripedes, *Balanus balanoides* (L), *B. balanus* (L) and *Chthamalus stellatus* (Poli). J. Exp. Mar. Biol. Ecol. **4**, 35–50 (1969).

Barnes, H., Barnes, M., Finlayson, D. M.: The metabolism during starvation of *Balanus balanoides*. J. Marine Biol. Assoc. U.K. **43**, 213–233 (1963).

Baslow, M. M., Nigrelli, R. F.: Effects of thermal acclimation on brain cholinesterase activity of the killifish, *Fundulus heteroclitus*. Zoologica **49**, 41–51 (1964).

Bullock, T. H.: Compensation for temperature in the metabolism and activity of poikilotherms. Biol. Rev. **30**, 311-342 (1955).

Collardeau-Roux, C.: Influence de la température sur la consommation d'oxygène de quelques larves de Trichoptères. Hydrobiologica **18**, 252–264 (1966).

Crisp, D. J., Ritz, D. A.: Changes in temperature tolerance of *Balanus balanoides* during its life-cycle. Helgoländer Meeresuntersuch. **15**, 98–115 (1967a).

Crisp, D. J., Ritz, D. A.: Temperature acclimation in barnacles. J. Exp. Mar. Biol. Ecol. **1**, 236–256 (1967b).

Dehnel, P. A.: Rate of growth of gastropods as a function of latitude. Physiol. Zool. **28**, 115–144 (1955).

Fry, F. E. J.: Effects of the environment on animal activity. Univ. Toronto Stud. Biol. 55 (Publ. Ontario Fish. Res. Lab.) **68**, 1–62 (1947).

Fry, F. E. J.: The lethal temperature as a tool in taxonomy. Ann. Biol. **33**, 205–219 (1957a).

Fry, F. E. J.: The aquatic respiration of fish. In: Brown, M. E. (Ed.): The Physiology of fishes 1, p. 1–63. New York: Academic Press (1957b).

Gray, J.: The mechanism of ciliary movement III. The effect of temperature. Proc. Roy. Soc. (Lond.) (B) **95**, 6–15 (1923).

Gunter, G.: Temperature. In: Hedgpeth, J. W. (Ed.): Treatise on Marine Ecology and Palaeoecology. Mem. (67) Geol. Soc. Am. **1**, 159–184 (1957).

Halcrow, K., Boyd, C. M.: The oxygen consumption and swimming activity of the amphipod *Gammarus oceanicus* at different temperatures. Comp. Biochem. Physiol. **23**, 233–242 (1967).

Hannon, J. P.: Effect of prolonged cold exposure on components of the electron transport system. Am. J. Physiol. **198**, 740–744 (1960).

Hannon, J. P., Vaughan, D. A.: Effect of prolonged cold exposure on the glycolytic enzymes of liver and muscle. Am. J. Physiol. **198**, 375–380 (1960).

Hochachka, P. W.: Isoenzymes in metabolic adaptation of a poikilotherm and subunit relationships in lactic dehydrogenases of goldfish. Arch. Biochem. Biophys. **111**, 96–103 (1965).

Hochachka, P. W.: Organisation of metabolism during temperature compensation. In: Prosser, C. L. (Ed.): Molecular Mechanisms of Temperature Adaptation, p. 177–203. Am. Assoc. Advan. Sci. (84), Washington D. C. (1967).

Hochachka, P. W., Somero, G. N.: Biochemical adaptation to the environment. In: Hoar, W. S., Randall, D. J. (Eds.): Fish Physiology 6, p. 99–156. New York: Academic Press (1971).

McFarland, W. N., Pickens, P. E.: The effects of season, temperature and salinity on standard and active oxygen consumption of the grass shrimp *Palaemonetes vulgaris* (Say). Can. J. Zool. **43**, 571–585 (1965).

Newell, R. C.: The effect of temperature fluctuation on the metabolism of intertidal invertebrates. Am Zool. **9**, 293–307 (1969).

Newell, R. C.: Biology of Intertidal Animals, p. 555. London: Logos Press (1970).

Newell, R. C., Northcroft, H. R.: The relationship between cirral activity and oxygen uptake in *Balanus balanoides*. J. Marine Biol. Assoc. U.K. **45**, 387–403 (1965).

Newell, R. C., Northcroft, H. R.: A re-interpretation of the effect of temperature on the metabolism of certain marine invertebrates. J. Zool. (Lond.) **151**, 277–298 (1967).

Newell, R. C., Pye, V. I.: Seasonal changes in the effect of temperature on the oxygen consumption of the winkle *Littorina littorea* (L.) and the mussel *Mytilus edulis* L. Comp. Biochem. Physiol. **34**, 367–383 (1970a).

Newell, R. C.,Pye, V. I.: The influence of thermal acclimation on the relation between oxygen consumption and temperature in *Littorina littorea* (L). and *Mytilus edulis* L. Comp. Biochem. Physiol. **34**, 385–397 (1970b).

Newell, R. C., Pye, V. I.: Quantitative aspects of the relationship between metabolism and temperature in the winkle *Littorina littorea* (L.). Comp. Biochem. Physiol. **38 B**, 635–650 (1971 a).

Newell, R. C., Pye, V. I.: Temperature-induced variations in the respiration of mitochondria from the winkle *Littorina littorea* (L.). Comp. Biochem. Physiol. **40 B**, 249–261 (1971 b).

Newell, R. C., Pye, V. I., Ahsanullah, M.: Factors affecting the feeding rate of the winkle *Littorina littorea*. Marine Biol. **9**, (2) 138–144 (1971 a).

Newell, R. C., Pye, V. I., Ahsanullah, M.: The effect of thermal acclimation on the heat tolerance of the intertidal prosobranchs *Littorina littorea* (L.) and *Monodonta lineata* (Da Costa). J. Exptl. Biol. **54**, 525–533 (1971 b).

Pickens, P. E.: Heart rates of mussels as a function of latitude, intertidal height and acclimation temperature. Physiol. Zool. **38**, 390–405 (1965).

Precht, H.: Concepts of temperature adaptation of unchanging reaction systems of cold-blooded animals. In: Prosser, C. L. (Ed.): Physiological Adaptation, p. 50–78. Am. Physiol. Soc., Washington, D. C. (1958).

Precht, H., Christophersen, J., Hensel, H.: Temperatur und Leben, p. 514. Berlin–Göttingen–Heidelberg: Springer 1955.

Prosser, C. L.: Physiological variation in animals. Biol. Rev. **30**, 229–262 (1955).

Prosser, C. L.: The nature of physiological adaptation. In: Prosser, C. L. (Ed.): Physiological Adaptation, p. 167–180. Am. Physiol. Soc., Washington, D. C. (1958).

Prosser, C. L.: Acclimation of poikilothermic vertebrates to low temperatures. In: Hannon, J. P., Viereck, E. (Eds.): Comparative physiology of temperature regulation, III, p. 1–44. Arctic Aeromedical Lab., Fort Wainwright, Alaska (1962).

Prosser, C. L., Brown, F. A.: Comparative Animal Physiology, 2nd ed., p. 688. Philadelphia: Saunders (1961).

Rao, K. P.: Rate of water propulsion in *Mytilus californianus* as a function of latitude. Biol. Bull. (Woods Hole) **104**, 171–181 (1953 a).

Rao, K. P.: Shell weight as a function of intertidal height in a littoral population of pelecypods. Experimentia **9**, 465 (1953 b).

Rao, K. P.: Biochemical correlates of temperature acclimation. In: Prosser, C. L. (Ed.): Molecular Mechanisms of Temperature Adaptation, p. 227–244. Am. Assoc. Advan. Sci., Symp. (84), Washington, D. C. (1967).

Ritz, D. A., Foster, B. A.: Comparison of the temperature responses of barnacles from Britain, South Afrika and New Zealand, with special reference to temperature acclimation in *Elminius modestus*. J. Marine Biol. Assoc. U. K. **48**, 545–559 (1968).

Roberts, J. L.: Metabolic responses of freshwater sunfish to seasonal photoperiods and temperatures. Helgoländer Wiss. Meeresuntersuch. **9**, 459–473 (1964).

Roberts, J. L.: Metabolic compensations for temperature in sunfish. In: Prosser, C. L. (Ed.): Molecular Mechanisms of Temperature Adaptation, p. 245–262. Amer. Assoc. Advan. Sci., Symp. (84), Washington, C. D. 1967.

Roux, C., Roux, A. L.: Température et métabolisme respiratoire d'espèces sympatriques de gammares du groupe *pulex* (Crustacés, Amphipodes). Ann. de Limnol. **3**, 3–16 (1967).

Segal, E., Rao, K. P., James, T. W.: Rate of activity as a function of intertidal height within populations of littoral molluscs. Nature (Lond.) **172**, 1108–1109 (1953).

Siefken, M., Armitage, K. B.: Seasonal variations in metabolism and organic nutrients in three *Diaptomus* (Crustacea: Copepoda). Comp. Biochem. Physiol. **24**, 591–609 (1968).

Somero, G. N.: Enzymic mechanisms of temperature compensations: Immediate and evolutionary effects of temperature on enzymes of aquatic poikilotherms. Am. Naturalist **103**, 517–530 (1969).

Somero, G. N., Hochachka, P. W.: Isoenzymes and short-term compensation in poikilotherms: Activation of lactate dehydrogenase isoenzymes by temperature decreases. Nature (Lond.) **233**, 194–195 (1969).

Southward, A. J.: The relationship between temperature and rhythmic cirral activity in some cirripedia considered in connection with their geographical distribution. Helgoländer Wiss. Meeresuntersuch. **10**, 391–403 (1964).

Vernberg, F. J., Schlieper, C., Schneider, D. E.: The influence of temperature and salinity on ciliary activity of excised gill tissue of molluscs from North Carolina. Comp. Biochem. Physiol. **8**, 271–285 (1963).
Vernon, H. M.: The relation of the respiratory exchange of cold-blooded animals to temperature. J. Physiol. (Lond.) **21**, 443–496 (1897).

The Influence of Starvation on Temperature Relationships of Metabolism in Ectotherms

Islay D. Marsden

With 5 Figures

Introduction

Although nutritional factors are well known to influence the level of metabolism in ectotherms, comparatively few studies have taken this factor into account during thermal acclimation. Dehnel (1960) for example studied the effect of temperature and salinity on the oxygen consumption of *Hemigrapsus oregonensis* and *H. nudus* which had been starved during the acclimation period. A similar practice has been adopted in most other studies on thermal acclimation, especially in Crustacea, where it is known that starvation prevents the onset of moulting. Roberts (1957) studied the influence of starvation and moulting in the metabolic acclimatory response of *Pachygrapsus crassipes* and reported that the rate of oxygen consumption of starved crabs was approximately 60 % of that in animals which had been fed on liver. Barnes, Barnes and Finlayson (1963a) have demonstrated that respiration in the barnacle *Balanus balanoides* varies according to the nutritional state of the animal. Davies (1966, 1967) showed that the metabolic rate could be correlated with the availability of food because the oxygen consumption of *Patella vulgata* collected from bare rock surfaces on the upper shore was low compared with that of limpets collected from the mud shore near algae. In the pulmonate *Arianta arbustorum*, starved specimens had a lower oxygen consumption than well-fed ones and if starvation was continued for 4 days then the rate fell to the aestivation level (Nopp, 1965). This effect has also been observed in the isolated tissue of hepatopancreas and gonad following several weeks starvation (Nopp and Farahat, 1967). More recently Hagerman (1970) has found that the active rate of oxygen consumption in the shrimp *Crangon vulgaris* declined following starvation, although no such effect was observed for the standard rate.

Studies on various invertebrates suggest that nutritional factors influence not only the level of oxygen consumption, but also the temperature relationships. In *Balanus balanoides* for example, starved animals are reported to have a Q_{10} value approaching 1.0 whereas the metabolism of fed specimens is more dependent on temperature. In *Littorina littorea* Newell and Pye (1971a, b) found that in subcellular preparations the temperature dependence and level of oxygen consumption varied according to the availability of metabolic substrates.

The following study was undertaken primarily to determine the influence of starvation on temperature relationships of metabolism in the shore crab *Carcinus maenas* and whether such effects could be demonstrated at a subcellular level.

Material and Methods

Specimens of *Carcinus maenas* of a size range of 18–46 mm carapace length were divided into two groups and kept at 15 C in aerated seawater for periods up to 23 days. Groups were either starved or fed daily on mussels and fish and the water was changed on alternate days. The oxygen consumption for all experiments was carried out using a Gilson respirometer with special flasks. The same group of animals was used to measure respiration over the range 5–30 C with an equilibration period of 30 minutes after the experimental temperature had been reached. Readings were done in the sequence 20 C, 5 C, 15 C, 25 C, 10 C, 30 C to minimise possible short term acclimatory responses.

Gill and hepatopancreas tissus was removed from 5 crabs at the top of the size range and this was used to measure the oxygen consumption of isolated tissue and tissue homogenates. Fresh tissue was used for each experimental temperature and an equilibration time of 15 minutes was allowed.

After each experiment the crabs were measured, killed and then dried to constant weight. The tissues and samples of the homogenates were dried at 60 C and the rates of oxygen consumption were expressed as ml O_2/g/hr.

Results

The Effect of Starvation on Intact Animals

The oxygen consumption is usually represented as units of oxygen/dry weight of tissue/hr. If starvation involved a large decrease in the body weight, then this would complicate weight-specific values of oxygen consumption. The carapace length could have been used but this would not allow comparison with other metabolic studies. However, an analysis showed that the ratio $\frac{\text{dry body weight}}{\text{carapace length}}$ did not vary significantly after feedings or starving the animals from 7 to 20 days.

Because of the small size of the respiration flasks, the crabs had limited powers of movement apart from irrigation. It is suggested, therefore, that the respiration rates recorded approach the standard rather than the active rate of metabolism. Measurements of oxygen consumption were recorded after starvation or feeding for the periods indicated above and the rates were expressed as ml O_2/hr. The results for the weight specific respiration at various experimental temperatures for different periods of starvation or feeding are illustrated in Fig. 1. All the results were tested by analysis of covariance for slope and elevation. After 1 week the rates of oxygen consumption for starved and fed animals are similar at all experimental temperature. The slopes of the regression lines are insignificantly different and the mean value for slope *b* for the experiments is 0.79. This value is similar to those quoted for other organisms (Zeuthen, 1947, 1953; Hemmingsen, 1950, 1960; Tashian, 1956; Roberts, 1957) and for *Carcinus maenas* (Wallace, 1972). After 2 weeks, however, the graphs show that oxygen consumption for the two groups is significantly different for the experimental temperatures 10 C, 15 C and 20 C. If data for animals weighing one gram and four grams respectively are extracted from Fig. 1 it becomes evident that

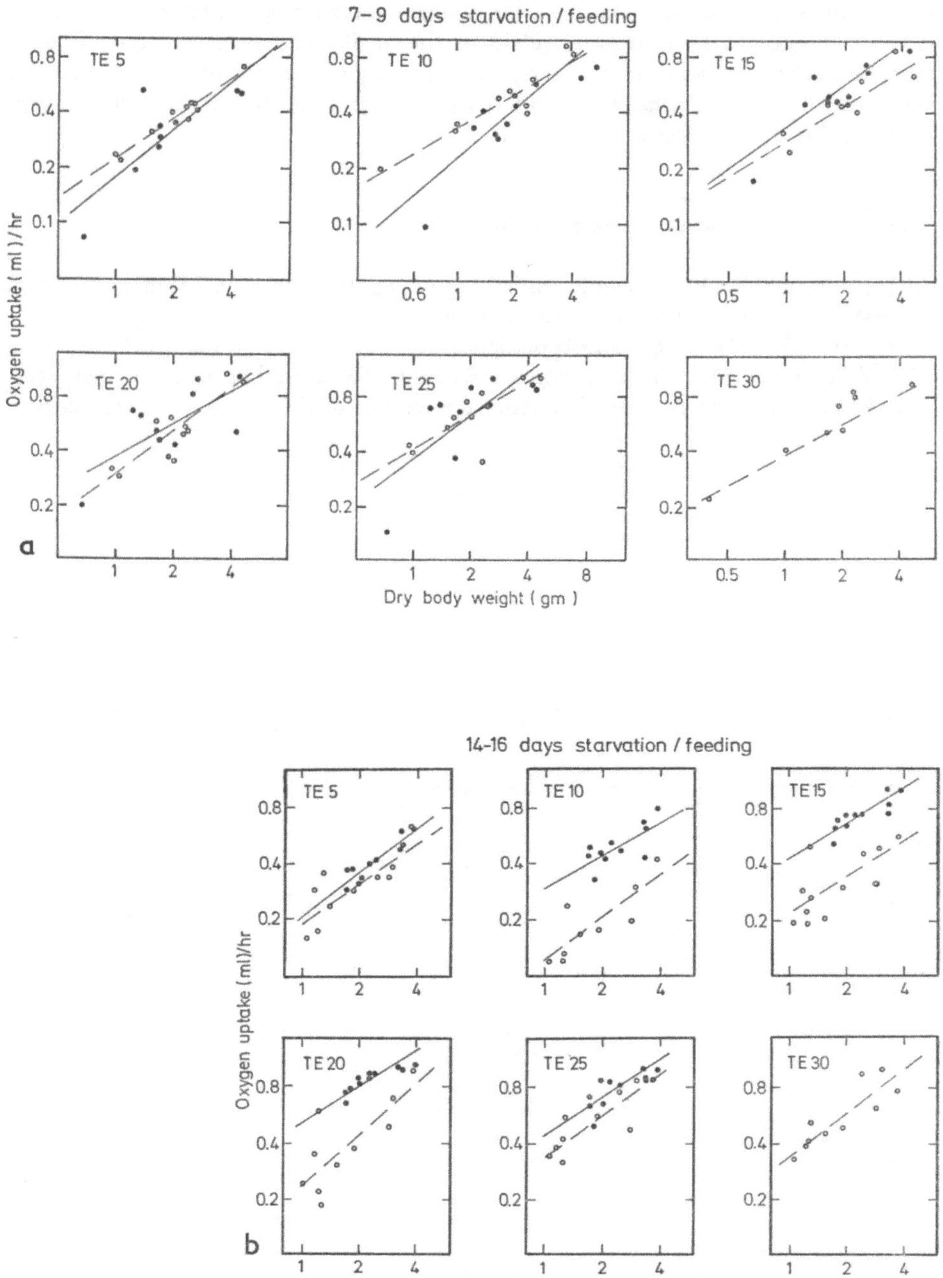

Fig. 1. a and b Graphs showing the relationship between oxygen consumption (ml) at various experimental temperatures (TE) and body size (dry weight in gms) following starvation or feeding for 7–9 days and 14–16 days. Open symbols indicate animals which have been starved and closed symbols animals which have been fed

starvation leads to a suppression of metabolism and to reduced temperature sensitivity. This appears to occur more quickly in the smaller organisms, and after 16 days starvation their metabolism is nearly independent of temperature. These results show that both nutritional state and body size are important in the temperature relationships of this species.

The Effect of Starvation on Isolated Tissues

As there is an effect in the whole animal following starvation for 2 weeks, it was of interest to see if this could be detected at either the tissue or the homogenate level. *R*:*T* curves for isolated gill and hepatopancreas tissue are illustrated in Fig. 2 and 3 after different periods of starvation or feeding. These graphs indicate that even after 23 days of starvation there is little suppression of the rate temperature curves

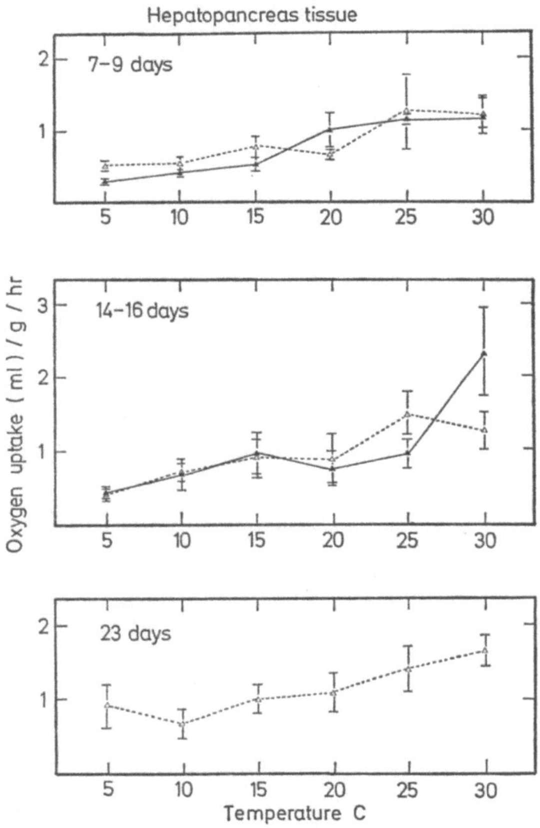

Fig. 2. *R*:*T* curves of isolated hepatopancreas after starvation or feeding for 7–9 days, 14–16 days and 23 days. Open symbols indicate tissue from starved crabs and closed symbols indicate tissue from fed crabs. The standard deviation is shown by a line

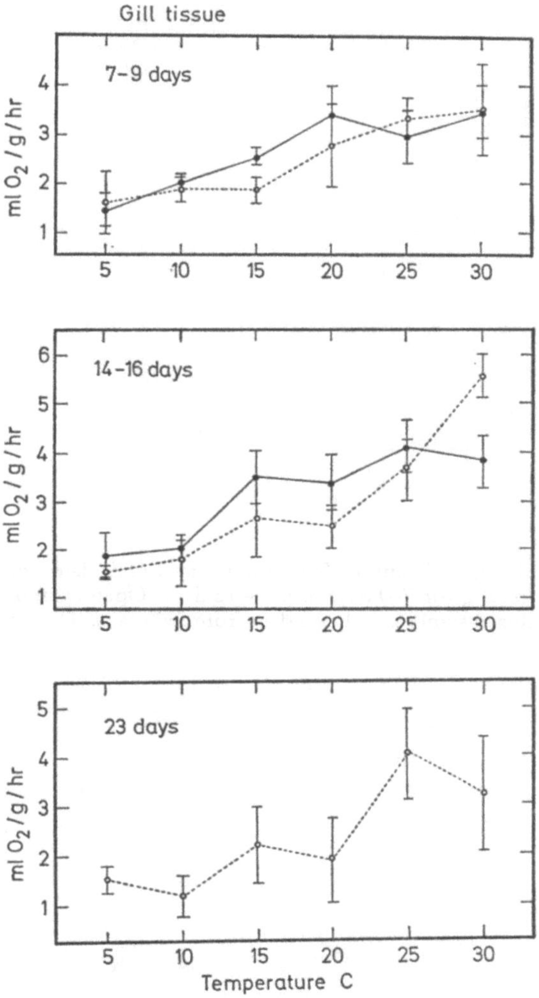

Fig. 3. *R*:*T* curves of isolated gill tissue after starvation or feeding for 7–9 days, 14–16 days and 23 days. Open symbols indicate tissue from starved crabs and closed symbols indicate tissue from fed crabs. The standard deviation is shown by a line

of either tissue. A similar result has been obtained with gill tissue taken from larger crabs, although the level of metabolism was lower. The data for tissue homogenates are shown in Fig. 4 and 5 with the main differences occurring in the gill preparations. After 7–9 days starvation there is little difference between the starved and the fed animals, but after 15 days starvation the metabolism is less temperature-dependent than in the fed animals. In the case of the gill preparations this represented a considerable suppression of the metabolic rate.

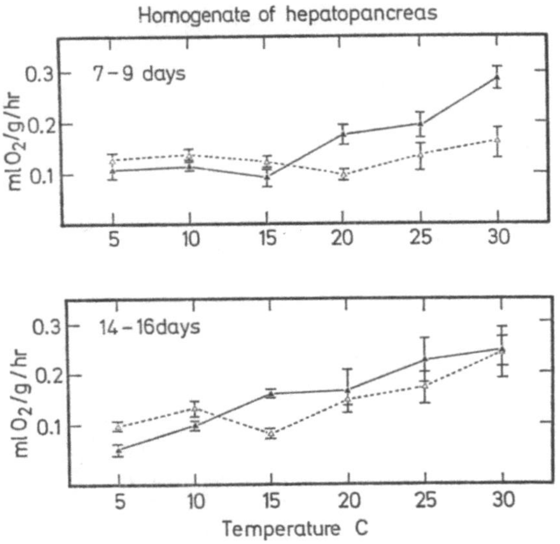

Fig. 4. R:T curves for homogenates of isolated hepatopancreas tissue after starvation or feeding for 7–9 days and 14–16 days. Open symbols indicate data from starved crabs and closed symbols indicate data from fed crabs. The standard deviation is shown by a line

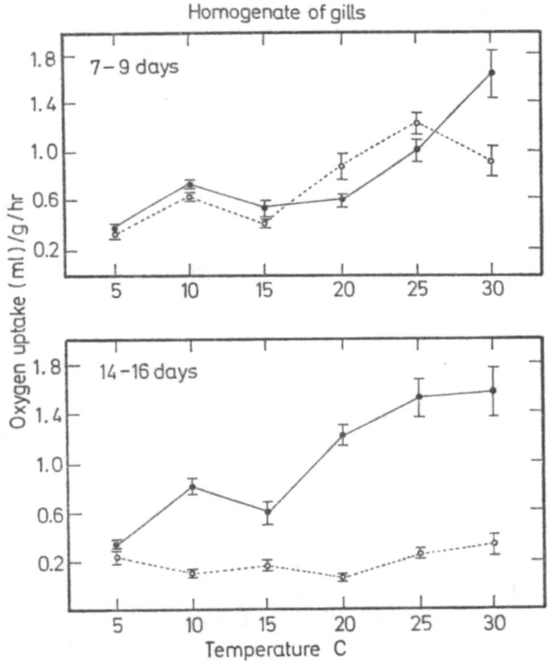

Fig. 5. R:T curves for homognates if isolated gill tissue after starvation or feeding for 7–9 days and 14–16 days. Open symbols indicate data from starved crabs and closed symbols indicate data from fed crabs. The standard deviation is shown by a line

Discussion

The results of this study demonstrate that starvation of *Carcinus maenas* in the laboratory at 15 C results in a progressive suppression of the metabolism compared with fully fed crabs. This effect can be observed after 2 weeks but may occur sooner in smaller specimens, or at higher acclimation temperatures.

These results are similar to those reported for a variety of other ectotherms. Saunders (1963) for example, has found that at an acclimation temperature of 15 C the rate of oxygen consumption of small specimens of *Gadus morhua* took 4–5 days to fall to starvation levels, whereas larger fish took 8 days. In much the same way, Bayne (this symposium) has shown that following 40 days of starvation the oxygen consumption of the mussel *Mytilus edulis* falls from its routine level of 0.4 ml O_2/g dry weight/hr to a basal level of approximately 0.16 ml O_2/g/hr. Following presentation of food, the rate increases to approximately 0.6 ml and this decreases over a period of 10 days to the routine rate again.

Apart from work on barnacles (Barnes, Barnes and Finlayson 1963b) there have been relatively few studies made on crustaceans. Hagerman (1971) has, however, demonstrated the effect of starvation during acclimation in the shrimp *Crangon vulgaris* to salinity and temperature. He found that the active rate of oxygen consumption fell steadily over a period of 8 days starvation at 10 C. The resting rate, however, was unaffected by starvation so that metabolism of starved crabs is similar to that of inactive ones much as has been shown in *Mytilus edulis* (Bayne, this symposium). The data which has been presented above for *Carcinus* refers to a routine rate of oxygen consumption and the suppression towards the standard rate following starvation may reflect a similar reduction in activity as occurs in *Crangon* and *Mytilus*.

There is very little information on the influence of starvation on the temperature relationships of metabolism in either intact organisms or in subcellular preparations. The data for *Carcinus*, however, demonstrates that in the intact crabs the rate of oxygen consumption is less dependent on temperature following starvation than in fully-fed crabs. These differences in the level and slope of the metabolism: temperature curves are also apparent in cell-free preparations of gill tissue when no exogenous substrate is supplied.

Starvation is undoubtedly an important ecological factor which may also be implicated in seasonal variations in metabolism. A suppressed rate of oxygen consumption can be induced within 8–14 days in most of the organisms cited above. This suggests that even in short term acclimation experiments the nutritional state of the animal must be taken into consideration. The effect of starvation on temperature relationships of ectotherm metabolism may thus be important in the interpretation of previous studies on thermal acclimatory responses in such organisms.

References

Barnes, H., Barnes, M., Finlayson, D. M.: The seasonal changes in body weight, biochemical composition, and oxygen uptake of two common boreo-arctic cirripedes, *Balanus balanoides* and *B. balanus*. J. Marine Biol. Assoc. U. K. **43**, 185–211 (1963a).

Barnes, H., Barnes, M., Finlayson, D. M.: The metabolism during starvation of *Balanus balanoides*. J. Marine Biol. Assoc. U. K. **43**, 213–223 (1963b).

Davies, P. S.: Physiological ecology of *Patella*. I. The Effect of Body Size and Temperature on Metabolic rate. J. Marine Biol. Assoc. U. K. **46**, 647–658 (1966).
Davies, P. S.: Physiological Ecology of *Patella*. II. Effect of environmental acclimation on the metabolic rate. J. Marine Biol. Assoc. U. K. **47**, 61–74 (1967).
Dehnel, P. A.: Effect of temperature and salinity on the oxygen consumption of two intertidal crabs. Biol. Bull. **118**, 215–249 (1960).
Hagerman, L.: The oxygen consumption of *Crangon vulgaris* (Fabricus) (Crustacea, Natantia) in relation to salinity. Ophelia **7**, 283–292 (1970).
Hemmingsen, A. M.: The relation of standard (basal) energy metabolism to total fresh weight of living organisms. Rept. Steno. Mem. Hosp. **4**, 7–58 (1950).
Hemmingsen, A. M.: Energy metabolism as related to body size and respiratory surfaces and its evolution. Rept. Steno. Mem. Hosp. **9**, 7–110 (1960).
Marsden, I. D., Newell, R. C., Ahsanullah, M.: The effect of starvation on the metabolism of the shore crab *Carcinus maenas*. Comp. Biochem. Physiol. **45A**, 195–213 (1973).
Newell, R. C., Pye, V. I.: Quantitative aspects of the relationship between metabolism and temperature in the winkle *Littorina littorea* (L.). Comp. Biochem. Physiol. **38B**, 635–650 (1971a).
Newell, R. C., Pye, V. I.: Temperature-induced variations in the respiration of mitochondria from the winkle *Littorina littorea* (L.). Comp. Biochem. Physiol. **40B**, 249–261 (1971b).
Nopp, H.: Temperaturbezogene Regulationen des Sauerstoffverbrauchs und der Herzschlagrate bei einigen Landpulmonaten. Z. Vergleich. Physiol. **50**, 641–659 (1965).
Nopp, H., Farahat: Temperatur und Zellstoffwechsel bei Heliciden. Z. Vergleich. Physiol. **55**, 103–118 (1967).
Roberts, J. L.: Thermal acclimation of Metabolism in the crab *Pachygrapsus crassipes* Randall. I. The influence of body size, starvation, and moulting. Physiol. Zool. **30**, 232–242 (1957).
Saunders, R. L.: Respiration of the Atlantic cod. J. Fisheries Res. Board Can. **20**, 373–386 (1963).
Tashian, R. E.: Geographic variation in the respiratory metabolism and temperature coefficient in tropical and temperate forms of the fiddler crab, *Uca pugnax*. Zoologica **41**, 39–47 (1956).
Wallace, J. C.: Activity and metabolic rate in the shore crab *Carcinus maenas*. Comp. Biochem. Physiol. **41A**, 523–533 (1972).
Zeuthen, E.: Body size and metabolic rate in the animal kingdom. C. R. Trav. Lab. Carlsb., Ser. chim. **26**, 15–161 (1947).
Zeuthen, E.: Oxygen uptake as related to body size in organisms. Am. Rev. Biol. **28**, 1–12 (1953).

Effect of Temperature on Active and Resting Metabolism in Polychaetes *

P. E. Coyer and C. P. Mangum

With 4 Figures

Introduction

Newell and Northcroft (1967) concluded that the minimal rate of oxygen consumption corresponding to a resting phase in four intertidal species, the lamellibranch *Cardium edule*, the actinian *Actinia equina*, the polychaete *Nepthys hombergi*, and the gastropod *Littorina littorea*, is thermally insensitive over the temperature range 5–22. C. Newell (1967) also presented evidence for varying degrees of temperature independence of mitochondrial oxidative activity in subtidal, intertidal, and terrestrial poikilotherms. Since this hypothesis was made, only Mangum and Sassaman (1969) have related independently measured activity patterns to oxygen consumption data to identify active and resting metabolic phases. Nimura and Inoue (1969), Tribe and Bowler (1968), and Halcrow and Boyd (1967) described methods of determining the metabolic rate in various marine and terrestrial invertebrates at low muscluar activity levels. However, these methods either involve anaesthetization, extrapolation from higher activity levels or presumably intermittent visual observation of the activity level. Recently, conflicting conclusions on the temperature coefficients for inactive metabolic rates were reached by Wallace (1972) and Lowe and Trueman (1972).

Characteristic patterns of alternating activity and rest comprise the ventilation rhythms of many polychaetous annelids, including the onuphid *Diopatra cuprea* (Bosc) and the glycerid *Glycera dibranchiata* (Ehlers), whose behavior results in rapid and slow rates of oxygen consumption (Mangum and Sassaman, 1969; Hoffman and Mangum, 1970). Since Mangum and Sassaman (1969) described thermal sensitivity of active and resting metabolism in *D. cuprea* only over a 10 C interval, this species has been further investigated over a wider range. In addition, *G. dibranchiata* and the terebellid *Amphitrite ornata* (Leidy) were chosen for study of the effects of temperature on what is unequivocally active and resting metabolism in these intertidal organisms.

Materials and Methods

Oxygen Consumption

Although Newell and Northcroft (1967) based their conclusions on acute measurements of metabolism, we have used acclimated measurements for two reasons:

* Supported by National Science Foundation Research Grant GB-20335.

1. Temperature shock disrupts the pattern of spontaneous rhythmic activity resulting in loss of identifiable phases of metabolism. At least in *D. cuprea*, the acclimation which ensues involves the behavior pattern but not aerobic metabolism (Mangum, 1972). 2. We believe that temperature changes occur in nature at a rate that is small enough so that the animal can simultaneously acclimate to them (Johnson, 1965; Mangum and Sassaman, 1969). Therefore, acclimated measurements have more pertinence to the responses of infaunal, tubicolous annelids, even those in intertidal habitats.

Active and resting oxygen consumption rates for three species of polychaetous annelids *Diopatra cuprea*, *Glycera dibranchiata*, and *Amphitrite ornata* were computed using the technique described by Mangum and Sassaman (1969). The animals were pre-exposed in the laboratory to each experimental temperature for a minimum of five days and a maximum of two weeks, depending on the departure from habitat temperature. During the period of exposure, the animal was placed in a glass tube whose length just barely exceeded that of the worm. Continuous recordings of oxygen depletion by *D. cuprea* were made in lucite chambers with the animal lying horizontally. *G. dibranchiata* was placed in cylindrical chambers with the animal oriented vertically. And finally, *A. ornata* was put in rectangular chambers housing a U-shaped tube. Measurements of oxygen consumption at oxygen concentrations above 75% air saturation were made in Millipore-filtered sea water (salinity 15.2 to 17.4‰), to which the animal was adapted, and at constant temperatures (± 0.5 C). As a control, the oxygen consumption rate of the probe itself was measured and subtracted from those of the animals; the observed amount was barely detectable, viz., less than 1% of the existing oxygen concentration. Recordings of oxygen consumption were made with a Yellow Springs Model 5420A probe and Model 54 meter. Both active and resting metabolic rates for the three species were corrected to common weights using the covariance method of adjusting treatment means (Steel and Torrie, 1960).

Diopatra cuprea

Collections of ragworms were made from August through October 1971 in the intertidal zone of the York River estuary at Sandy Point, Virginia. Five to twelve animals were measured at each of three temperatures: 12.5, 22.5, and 32.5 C, the highest reached by stepwise acclimation of 1 C per day from lower collecting temperatures. Values for the slope of the logarithmic regression line describing metabolism as a function of body weight were extracted from the results of Mangum and Sassaman (1969) at 17.5 and 27.5 C to correct oxygen consumption rates to a common weight of 0.88 gm, which is the value that they used.

Glycera dibranchiata

Bloodworms were obtained during the winter months from the Maine Bait Company. Upon arrival they were kept in darkened aerated aquaria under salinity conditions which the species encounters in nature (York River water, salinity 17.2‰). Five to eight animals were measured at 5 C intervals over the range 2.5–27.5 C. The *b* values for the relationship between oxygen consumption and body weight found by Hoffman and Mangum (1970) at 10 and 20 C were used to correct the data to a common weight of 4.0 gm, chosen because of its proximity to the sample mean. At 10 C the calculated *b* value = —0.470 (the correlation coefficient $r = 0.527$) and at 20 C $b = —0.542$ ($r = 0.732$); the difference between the two is not significant ($P > 0.05$), indicated by the overlap of 95% confidence intervals. To minimize temperature shock during the exposure period, the 2.5 and 7.5 C measurements were made during the coldest months, January and February, measurements at 17.5 and 12.5 C were made in late November and December, and the higher ones in March and April.

Amphitrite ornata

Worms were collected from the shores of the causeway at Beaufort, North Carolina, in February when the mean monthly water temperature is 9 C (Kirby-Smith and Gray, 1971). Animals were also obtained in the spring from the Supply Department, Woods Hole, and exposed to 5, 10, 15, 20, and 25 C during the months March through May. Oxygen consumption rates were plotted as a function of oxygen concentrations to characterize the effect of oxygen concentration on the species' aerobic metabolism. The slopes of the linear regression lines describing the linear relationship of oxygen consumption and oxygen concentrations were used to compare rates at 100% saturation according to the covariance method (Steel and Torrie, 1960). A *b* value (—0.955; $r = 0.898$) was calculated at 10 C for the relationship of oxygen consumption rates (N = 8) at 100% air saturation and body weights of Massachusetts and North Carolina animals. To test the validity of the calculated *b* values at 10 C over a wide temperature range, a second logarithmic regression analysis of data from Massachusetts animals was made at 20 C (N = 8). At 20 C $b = —0.869$ ($r = 0.863$). There is no significant difference ($P > 0.05$) between these values, indicating the homogeneous nature of the two slopes. At each temperature, active and resting metabolic rates were derived from 5–8 animals from Massachusetts only at the common weight of 3.50 gm.

Ventilatory Behavior of Amphitrite ornata

Worms were placed in U-shaped tubes whose diameters exceeded those of the animals by only several millimeters. Spontaneous irrigation cycles were recorded on a kymograph drum in an apparatus similar to that described by Mangum (1970). The tubes were oriented vertically to simulate the worm's natural position, and recordings were made from either the anterior or posterior end of the animal. A large darkened lucite aquarium maintained at a constant water level by running sea water housed the recording chamber.

A total of 122 hours of recordings from 11 animals was analyzed for the mean frequency, duration, and amplitude of bursts of pumping activity. By simulating activities of the worms with the addition of known volumes of water from below the recording chamber, the records were empirically calibrated and flow rates calculated (Dales *et al.*, 1970).

Statistical Methods

Differences between means for active and resting rates at each temperature were compared according to Student's *t* test.

For each species Q_{10} values over 5 C temperature intervals were tested for their significance against the null hypothesis that $Q_{10} = 1.0$. This test, devised by W. Van Winkle (personal communication), is assumed to follow the distribution for the *t* statistic. The test is:

$$t = \frac{\Delta T (\log Q_{10})}{10 \sqrt{2} (V \log R)}$$

where $V \log R$ ist the pooled variance (V) for log oxygen consumption rates (R), and T is the temperature.

A similar test, also devised by Van Winkle (personal communication), was used to test the significance between active and resting Q_{10} values over the same temperature interval. This test is:

$$t = \frac{\Delta T (\log Q_{10A} - \log Q_{10B})}{20 (V \log R)} .$$

Results

Oxygen Consumption

Continuous recordings of oxygen depletion by *Diopatra cuprea* and *Glycera dibranchiata* in airtight chambers reflect the patterns of rest and activity described by Mangum and Sassaman (1969) and Hoffman and Mangum (1970) for the spontaneous behavior of these animals. Recordings of the ventilatory cycles of *D. cuprea* in glass tubes are similar to those found in sand (Mangum *et al.*, 1968). The frequency of rapid and slow rates of metabolism observed in *G. dibranchiata* at 22.5 C closely agrees with that for bursts of activity at 23 C (0.70 ± 0.04/hr; Hoffman and Mangum, 1970). Because most oxygen consumption measurements for *G. dibranchiata* were made between 09.00 and 19.00 hr., active and resting rates determined in this study probably correspond to the first and second of the four discrete metabolic levels detected by Mangum and Miyamoto (1970) at 21.8–23.8 C. It is also interesting to note the agreement between Q_{10} values for bursts of activity in *D. cuprea* reported by Dales *et al.* (1970) and those we found for active metabolism in the same species over the temperature range 12.5–22.5 C. Beyond that interval, Q_{10} values for active metabolism are somewhat higher than those for ventilation frequencies, although activity measurements were not made above 25 C. Resting metabolism for the three species generally varies from 20 to 25 % of the active metabolic rate.

Over 5 C temperature intervals, all of the Q_{10} values for *D. cuprea*, eight of the ten values for *G. dibranchiata*, and five of the eight values for *A. ornata* are significantly greater than 1.0 ($P < 0.05$). With a few exceptions temperature change affects both active and resting metabolism to a similar degree. Significantly different ($P < 0.05$) Q_{10} values for both metabolic phases were observed only over the temperature interval 27.5–32.5 C in *D. cuprea*, which may indicate different responses of active and resting metabolism as the upper lethal limit above 32.5 C (Mangum and Sassaman, 1969) is approached. Q_{10} values for active and resting metabolism at the highest temperature interval in *G. dibranchiata* and at the lowest interval in *A. ornata* are not significantly different ($P > 0.05$). Finally, the magnitude of most of the Q_{10} values resembles those found in maldanid polychaetes where no discrimination between active and resting metabolism was made (Mangum, 1963).

Ventilatory Behavior of Amphitrite ornata

Between 18.0 and 19.8 C the mean frequency ($\pm$ S. E.) of rhythmic activity bursts is 0.58 ± 0.08/hr., the duration is 64.74 ± 7.36 min., and the amplitude is 4.76 $\pm$ 0.70 cm. deflection recorded on the kymograph drum. The rate of flow calculated from empirically calibrated records is 74.6 ml/hr. Under existing temperature and salinity (31–32$^0/_{00}$) conditions, the seawater contains around 5.5 ml O_2/l. making 0.418 ml O_2/hr. available to the animal. Based on the mean of active and resting metabolism at 20 C, a 3.5 gm. animal consumes 0.210 ml/worm/hr. Therefore, the oxygen utilization rate is about 50% which agrees closely with that (50–60%) reported by Dales (1961) for three other terebellid species. Fig. 4 is a kymograph recording of the spontaneous behavior patterns of *A. ornata*.

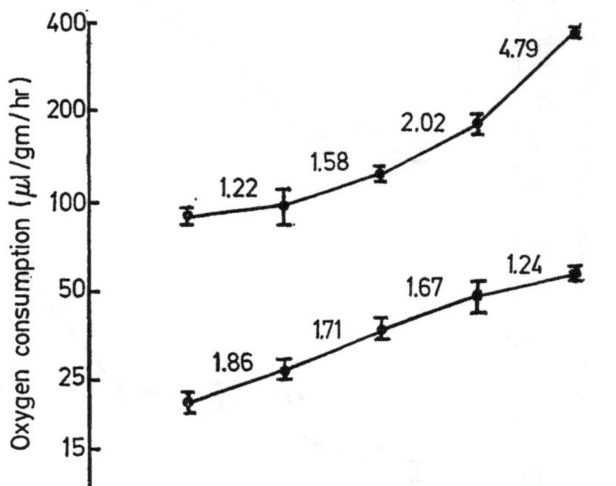

Fig. 1. *R*:*T* curves for active and resting metabolism in *Dipatra cuprea*. Q_{10} values given for each interval; vertical bars represent standard error

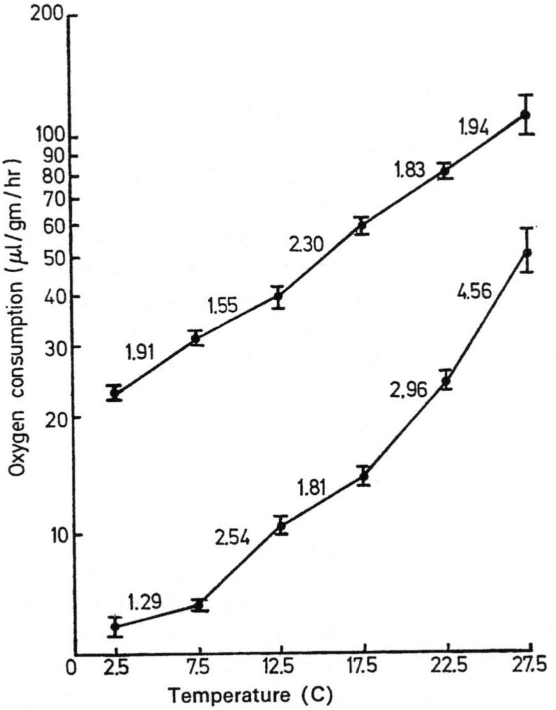

Fig. 2. $R:T$ curves for active and resting metabolism in *Glycera dibranchiata*. Q_{10} values given for each interval; vertical bars represent standard error

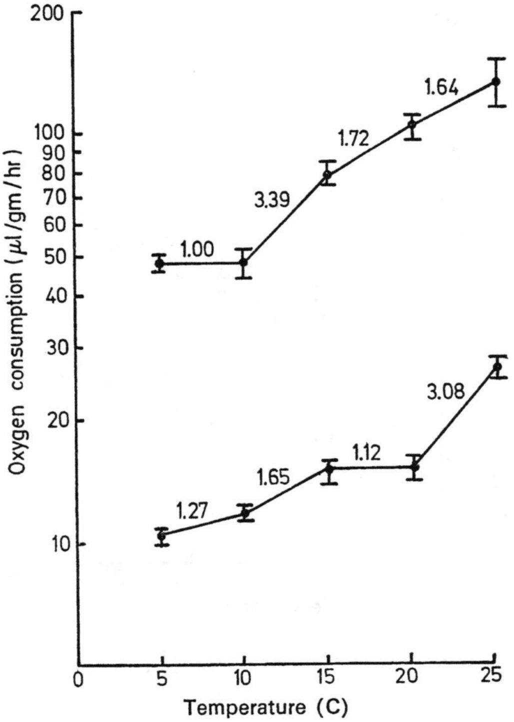

Fig. 3. $R:T$ curves for active and resting metabolism in *Amphitrite ornata*. Q_{10} values given for each interval; vertical bars represent standard error

Discussion

In our studies, neither active nor resting metabolism remained thermally insensitive over broad temperature ranges which these intertidal species encounter in nature. There is, in fact, no evidence from our data to support the existence of a mechanism operating at the whole animal level that reduces the effects of temperature. While low thermal sensitivity of metabolism over a small temperature range might be explained by the mechanisms proposed by Hochachka (1968) and Somero and Hochachka (1971), one would not expect these mechanisms be operative over temperature ranges as large as 20 C.

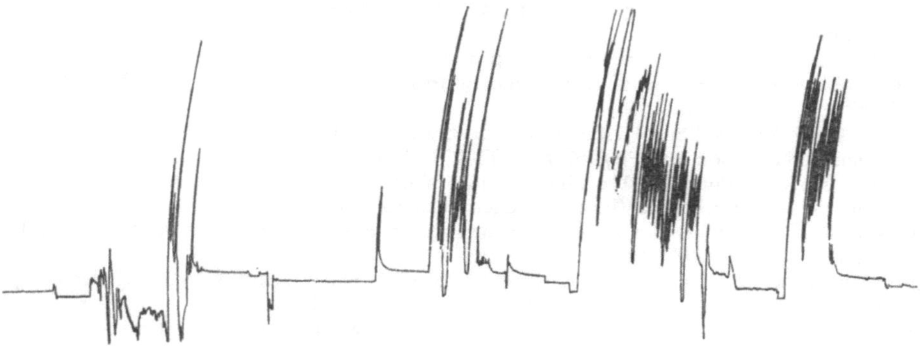

Fig. 4. Kymograph record of ventilation frequencies in *Amphitrite ornata*. Duration, 8.64 hrs; temperature, 18.5 C

An intrinsic weakness in many investigations of active and resting metabolism is the untested assumption that high and low measured levels of metabolism correspond to high and low levels of muscular activity. Fry (1957) and Tribe and Bowler (1967) discussed the necessity of measuring metabolic rates associated with quiescent behavior directly without unnatural extrapolation. Many different approaches to the problem of measurement have been employed, including two recent ones: Lowe and Trueman (1972) arrived at a low activity level of metabolism for *Mya arenaria* defined by reduced gill flow and heart rates, but they were forced to employ oxygen concentrations less than 20% air saturation in order to induce it. Thus the participation of anaerobic pathways may have been appreciable. The resting metabolism described by Wallace (1972) is that of the shore crab measured in air. Q_{10} values for this rate are temperature sensitive, but they may reflect only the response of aerial and not aquatic respiration.

Aquatic annelids are especially suitable for direct measurement because active and resting metabolic rates are normal correlates of spontaneous rhythmic activity. Unlike actinians in which respiratory rates do not reflect different phases of activity, probably because the energy expenditure in maintaining the retracted phase is so small (Sassaman and Mangum, 1970), parameters of rhythmic activity cycles in annelids can be correlated directly with those of aerobic respiration. Periods of activity and rest can be detected in recordings of both respiration and spontaneous behavior of *Amphitrite ornata*. Frequencies for bursts of activity independently observed in ventilation and oxygen depletion records agree closely in magnitude. And both active and resting metabolic rates for *A. ornata* and the two other polychaete species are clearly sensitive to temperature change.

References

Dales, R. P.: Oxygen uptake and irrigation of the burrow by three terebellid polychaetes: *Eupolymia*, *Thelepus*, and *Neoamphitrite*. Physiol. Zool. **34**, 306–311 (1961).

Dales, R. P., Mangum, C. P., Tichy, J. C.: Effects of changes in oxygen and carbon dioxide concentrations on ventilation rhythms in onuphid polychaetes. J. Marine Biol. Assoc. U. K. **50**, 365–380 (1970).

Fry, F. E. J.: The aquatic respiration of fish. Physiology of Fishes, p. 1–63. New York: Academic Press 1957.

Halcrow, K., Boyd, C. M : The oxygen consumption and swimming activity of the amphipod *Gammarus oceanicus* at different temperatures. Comp. Biochem. Physiol. **23**, 233–242 (1967).

Hochachka, P. W.: Action of temperature on branch points in glucose and acetate metabolism. Comp. Biochem. Physiol. **25**, 107–118 (1968).

Hoffman, R. J., Mangum, C. P.: The function of coelomic cell hemoglobin in the polychaete *Glycera dibranchiata*. Comp. Biochem. Physiol. **36**, 211–228 (1970).

Johnson, R. G.: Temperature variation in the infaunal environment of a sand flat. Limnol. Oceanog. **10**, 114–120 (1965).

Kirby-Smith, W. W., Gray, I. E.: A Checklist of Common Marine Animals of Beaufort, North Carolina. Duke University Marine Laboratory 1971.

Lowe, G. A., Trueman, E. R.: The heart and water flow rates of *Mya arenaria* (Bivalvia: Mollusca) at different metabolic levels. Comp. Biochem. Physiol. **41**, 487–494 (1972).

Mangum, C. P.: Studies on speciation in maldanid polychaetes of the North American coast-III. Intra- and inter-specific divergence in oxygen consumption. Comp. Biochem. Physiol. **10**, 335–344 (1963).

Mangum, C. P.: Respiratory physiology in annelids. Am. Scientist **58**, 641–647 (1970).

Mangum, C. P.: Temperature sensitivity of metabolism in offshore and intertidal onuphid polychaetes. Marine Biol. (in press).

Mangum, C. P., Miyamoto, D. M.: The relation between spontaneous activity cycles and diurnal rhythms of metabolism in the polychaetous annelid *Glycera dibranchiata*. Marine Biol. **7**, 7–10 (1970).

Mangum, C. P., Santos, S. L., Rhodes, W. R.: Distribution and feeding in the onuphid polychaete *Diopatra cuprea* (Bosc). Marine Biol. **2**, 33–40 (1968).

Mangum, C. P., Sassaman, C.: Temperature sensitivity of active and resting metabolism in a polychaetous annelid. Comp. Biochem. Physiol. **30**, 111–116 (1969).

Newell, R. C.: Oxidative activity of poikilothermic mitochondria as a function of temperature. J. Zool. **151**, 277–298 (1967).

Newell, R. C., Northcroft, H. R.: A re-interpretation of the effect of temperature on the metabolism of certain marine invertebrates. J. Zool. **151**, 277–298 (1967).

Nimura, Y., Inoue, M.: Oxygen uptake rate of the Japanese spiny lobster as related to the environmental oxygen concentration. Bull. Japan. Soc. Sci. Fisheries **35**, 852–861 (1969).

Sassaman, C., Mangum, C. P.: Patterns of temperature adaptation in North American Atlantic coastal actinians. Marine Biol. **7**, 123–130 (1970).

Somero, G. N., Hochachka, P. W.: Biochemical adaptation to the environment. Am. Zool. **11**, 157–165 (1971).

Steel, R. G., Torrie, J. H.: Principle and Procedures of Statistics. New York: McGraw-Hill 1960.

Tribe, M. A., Bowler, K.: Temperature dependence of "standard metabolic rate" in a poikilotherm. Comp. Biochem. Physiol. **25**, 427–436 (1968).

Wallace, J. D.: Activity and metabolic rate in the shore crab *Carcinus maenae* (L.) Comp. Biochem. Physiol. **41**, 523–533 (1972).

Some Effects of Temperature and Food on the Rate of Oxygen Consumption by *Mytilus edulis* L.

B. L. Bayne, R. J. Thompson, and J. Widdows

With 7 Figures

Introduction

The effects of temperature on the rates of oxygen consumption by marine intertidal invertebrates have been extensively studied, and the relative temperature independence of the resting, or standard, rates of consumption by some species has aroused much interest (for review, see Newell, 1970). In addition, the role of nutrition in affecting both oxygen consumption and the responses of oxygen consumption to temperature change has been recognised (Davies, 1967; Newell and Bayne, 1973). Recent papers by Thompson and Bayne (1972) and Bayne (1973a) have explored the possibilities for inducing different levels of oxygen consumption rate (Qo_2; ml of oxygen consumed per gram dry tissue weight per hour) by *Mytilus* by manipulating the ration. Widdows (1973) used differences in ration to establish three different levels of Qo_2 in a study of the physiological effects of temperature on *Mytilus*. The purpose of this present paper is to summarise and review some of our recent studies in order to illustrate the importance of knowing the nutritional condition of the animals used in studies on the effects of temperature on metabolism.

Material and Methods

All experiments were carried out with animals from a population at Heacham on the Norfolk coast. The mussels were maintained in an aquarium at Leicester. The flagellate, *Tetraselmis suecica*, was dosed continuously into the experimental tanks as food. Starvation conditions were established by filtering the water through filters with a 90% retention of all particles $>2.0\,\mu m$ in diameter. Temperatures were held constant to within ± 0.5 C.

Rates of oxygen consumption were measured in "open flow" systems using oxygen electrodes as described by Bayne (1971) and Thompson and Bayne (1972).

Filtration rates were determined by measurement of particle concentrations (with a Coulter Counter) in the water before and after it had flowed over the animals. The values for a simple "energy budget" under different environmental conditions were derived from the physiological rate measurements and integrated as described by Widdows and Bayne (1971).

The data are presented largely as weight-specific values, i.e. expressed as ml O_2 consumed (respiration rate) or volume of water cleared (filtration rate) per gram dry weight of flesh per hour. Filtration rate is defined as the volume of water that is cleared of particles $>3.0\mu m$ by the animal in unit time. When the particle retention efficiency closely approached 100%, filtration rate was considered equivalent to ventilation rate, i.e. the volume of water passed through the mantle cavity in unit time.

Results and Discussion

Levels of Oxygen Consumption Rate

We have identified three levels of oxygen consumption rate in *Mytilus* viz. standard, routine and active. This terminology is adopted from Fry (1947) as originally based on experiments with fish. In using these terms it is necessary, on the one hand, to distinguish between "activity" (or "what the organism does", Fry, l. c.) and "metabolism" (or "how the animal works", Fry l. c.) and, on the other hand, to recognise that changes in activity modify the rate of metabolism, and changes in the rate of metabolism determine the scope for activity.

In a sessile suspension feeder, such as *Mytilus*, the animal's "activities" may be listed as follows: a limited locomotion; the secretion and attachment of byssus threads; movement of the shell valves (adduction and abduction); the activity of the gill cilia in passing water through the mantle cavity and in filtering food particles; mechanical activity within the alimentary canal; and activity associated with digestion, excretion and growth. In our experiments activity effects due to locomotion and byssus secretion were largely excluded. We have monitored adduction and abduction over long periods in the laboratory and recorded only slight movements of the shell valves once the animals had embyssed and were left undisturbed. The significant changes in level of activity, therefore, have been due to (a) changes in activity of the gill cilia, which were monitored indirectly as filtration/ventilation rate, and (b) aspects of activity due, directly or indirectly, to the intake of food.

We are unable, at present, to measure independently these two components of "activity" in *Mytilus*, but it is helpful to distinguish conceptually between the "mechanical cost of metabolism" due to filtration and ventilation, and the "physiological cost of metabolism" due to the processes of digestion, excretion and growth.

The metabolic correlates of activity are measured, indirectly, as the rate of oxygen consumption by the intact animal. The standard rate of oxygen consumption is associated with negligible filtration activity, and the active rate of oxygen consumption is associated with the maximal filtration rate (Thompson and Bayne, 1972). In the absence of a particulate food, the rate of oxygen consumption declines over a period that varies with season (Bayne, 1973a) to a new, low steady-state condition, typical of an animal with shell-valves open, with the mantle edge partially extruded, but showing no filtration activity. This is the standard rate (Fig. 1). Confirmation that this rate represents a true resting condition was gained (a) by extrapolation to zero activity on graphs of filtration rate plotted against oxygen consumption rate (Thompson and Bayne, 1972; Widdows, 1973), and (b) from occasional records obtained during continuous monitoring of oxygen consumption and filtration rate, when an individual mussel had ceased to filter and oxygen consumption rate declined as a result.

When a starved mussel that is respiring at the standard level is fed, the animal immediately starts to filter particles from the water and the filtration rate rapidly rises to a maximum. At the same time the rate of oxygen consumption increases to a maximum, and this is identified as the active rate of oxygen uptake. This active rate is maintained for at least six hours if feeding is continued (Fig. 1), but it declines eventually to a new steady-state intermediate between the lower and upper limits set by the standard and active rates of uptake. This intermediate condition is the routine rate of oxygen consumption; it is characteristic of animals that have adjusted their physiological rate functions in the laboratory under conditions of adequate ration. Undoubtedly the animals can regulate their oxygen uptake at any level within the limits of the standard and active rates. Nevertheless, under constant environmental conditions a routine rate is established that is predictable according to season and animal size.

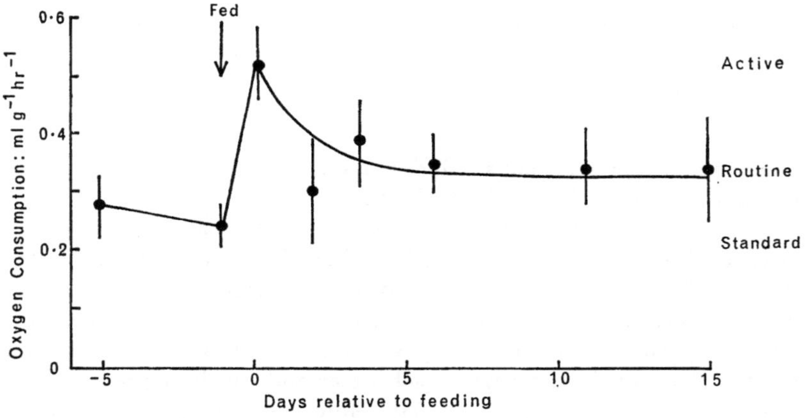

Fig. 1. Rates of oxygen consumption by *Mytilus edulis* when fed following a period of starvation. The three levels of oxygen consumption rate, active, routine and standard, are identified. Vertical bars represent $\pm$ 1 standard deviation throughout. (Adapted from Thompson and Bayne, 1972)

Whereas we induce standard and active rates of consumption by varying the ration level, the routine rate of oxygen consumption is established spontaneously by the animal, once the ration level is above the maintenance requirement.

At this time filtration rate is also intermediate between the fully active condition and the negligible rates recorded during standard oxygen consumption. In the literature on fish studies (Brett, 1964) the routine rate of oxygen consumption is associated with "spontaneous activity". For studies on sessile invertebrates, however, it may be more meaningful to relate routine oxygen consumption to an acclimated activity pattern which is observed when ration exceeds the maintenance level of zero "scope for growth" (see later).

There is some evidence (Bayne, 1973b) to suggest that during prolonged starvation *Mytilus* is capable of a "partial anaerobiosis" in which measured rates of oxygen consumption are lower than values predicted by the absence of both the mechanical and physiological costs of metabolism. These very low rates of uptake probably approach the basal level for the species. Further characterisation of these basal rates is necessary, but in the interim we can define three levels of oxygen consumption rate as follows:

a) Standard: The rate of oxygen consumption by an animal incurring minimal physiological and mechanical costs of metabolism.

b) Routine: The rate of oxygen consumption by an animal fully acclimated at a ration which exceeds the maintenance requirement.

c) Active: The rate of oxygen consumption by an animal incurring the maximum physiological and mechanical metabolic costs.

Relationships between Rates of Oxygen Consumption and Body Weight

The rate of oxygen consumption can be expressed as a power function of weight in the following equation:

$$\frac{d0_2}{dt} = aW^{\mathrm{b}}$$

where a and b are constants, $d0_2/dt$ is the amount of oxygen consumed per animal per unit time, and W is the dry flesh weight.

We have weight-related oxygen consumption data for standard and routine rates for *Mytilus* at 15 C and $33^0/_{00}$ S in winter (December to February inclusive) and summer (June to mid-September), covering a weight range from 0.07 to 3.0 grams dry flesh weight (Table 1). The regression coefficients (the values of "b") relating rate of oxygen consumption to dry weight for these two seasons and four conditions do not differ significantly from each other and can be represented by a mean regression coefficient ot 0.699 ± 0.055. However, the regressions differ significantly in level (i.e. the values of "a"); mean values for oxygen consumption of an animal of 1 gram dry tissue weight are given in Table 2.

Table 1. Regression equations of rate of oxygen consumption to dry weight for *Mytilus edulis*: $y = a \cdot x^{\mathrm{b}}$, where $y = \mu l\ O_2 hr^{-1}$, x = mg dry weight

Season and level of oxygen consumption	Regression equation	Number of determinations	Correlation coefficient	Standard error a) Coefficient	b) Estimate
Winter:					
Standard	$y = 1.76 \cdot x^{0.724}$	44	0.86	0.065	0.089
Routine	$y = 2.64 \cdot x^{0.774}$	41	0.86	0.073	0.087
Summer:					
Standard	$y = 1.87 \cdot x^{0.670}$	42	0.87	0.065	0.137
Routine	$y = 2.64 \cdot x^{0.702}$	68	0.88	0.046	0.123

Table 2. Mean oxygen consumption rates by *Mytilus* weighing 1 gram dry tissue at different seasons, as calculated from the regression equations in Table 1

Season and level of oxygen consumption	Rate of oxygen consumption (ml O_2 hr^{-1} $\pm$ standard error) by animal of 1 gram dry tissue weight	
Winter:		
Standard	0.263	(0.089)
Routine	0.549	(0.087)
Summer:		
Standard	0.164	(0.137)
Routine	0.339	(0.123)

In an earlier paper (Bayne, 1973a) we reported a regression coefficient ("b") of 1.027 when relating routine oxygen consumption in the summer to tissue dry weight. In this earlier series of determinations the weight-range of animals used was relatively small (0.3 to 1.4 gram). Certainly there is considerable variability in the relationship between oxygen consumption and tissue weight, some of which may be due to the seasonal shift in the balance of available energy reserves (Gabbott and Bayne, 1973). More critical measurements of oxygen consumption and nitrogen excretion, taken at the time of this shift in the summer, might help to explain the varied observations in the regression analysis.

The routine rate of oxygen consumption, both at constant temperature of 15 C and also at field ambient temperatures, varies seasonally from high values in the winter to low values in the summer (Bayne, 1973a). There is a similar seasonal variation in standard rate, so that both in winter and summer the standard rate of oxygen consumption is approximately one half of the routine rate.

The Scope for Activity

The scope for activity is usually defined as the difference between the standard and active rates of metabolism (Fry, 1947). This difference represents the maximum energy available to the animal for activity. In an ecological context, however, the difference between standard and *routine* metabolic rate is also of considerable interest as an index of scope for activity, especially in cases where the active rate does not appear to be maintained for long periods of time, but rather declines towards the routine rate after a short period, as in *Mytilus*. The distinction should be made, therefore, between a maximum (active minus standard) and a routine (routine minus standard) scope for activity.

The regression relationships discussed in the last section indicate that the routine scope for activity in *Mytilus* is approximately $\times 2$. Boyden (1972) observed a scope for activity in *Cerastoderma (Cardium)* of $\times 1.9$, which is probably comparable to a routine scope. Our data suggest that the routine scope for activity remains constant during the year.

Our values for maximum scope for activity are more variable than those for routine scope, and there is no apparent consistency in the variability. The relationship between maximum and minimum rates of oxygen consumption in fish is known to be rather variable (Brett, 1964). The overall mean maximum scope for activity in *Mytilus* is calculated as ×2.8 the standard rate of oxygen consumption. This is considerably lower than the ×6.3 calculated for *Cerastoderma* by Newell (1970), but we have some evidence that maximum scope is highest in the winter, when it may approach ×6 the standard rate.

In Fig. 2 (from Widdows, 1973) the maximum scope for activity is plotted against temperature for animals acclimated to 15 C and then subjected to a range of temperatures from 5 to 30 C. The filtration rate, which was measured simultane-

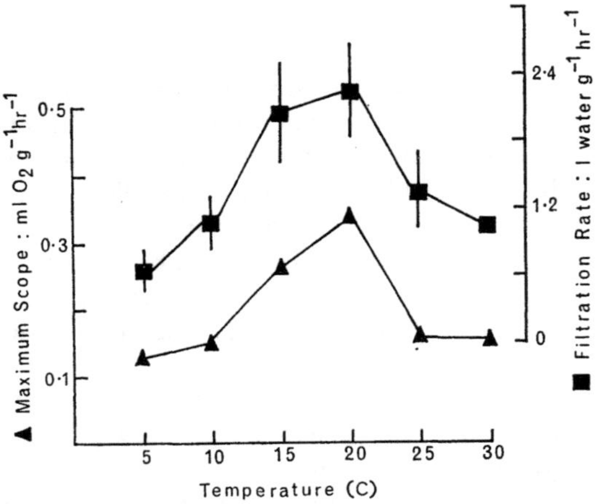

Fig. 2. The immediate (= acute) effects of temperature change on the filtration rate and maximum scope for activity of *Mytilus edulis* acclimated to 15 C. (From Widdows, 1973)

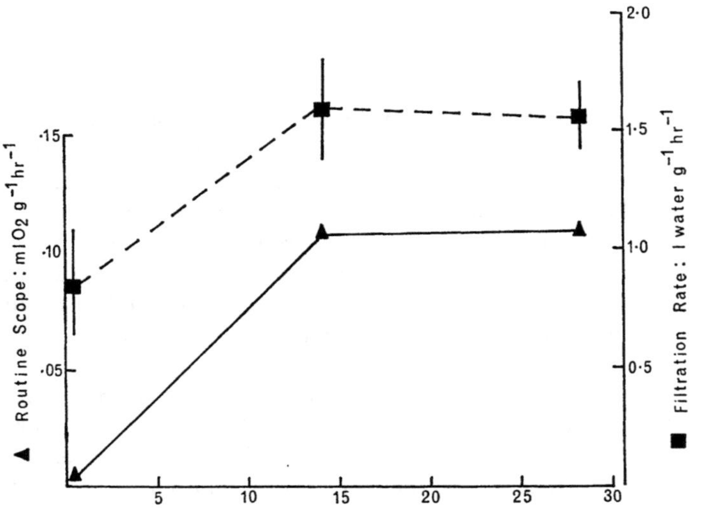

Fig. 3. The acclimation of filtration rate and the routine scope for activity of *Mytilus edulis* subjected on day 0 to a temperature change of 5 C below ambient

ously with the rate of oxygen consumption, is plotted also, and the agreement between the form of the two curves is noticeable.

In Fig. 3 the routine scope for activity is plotted against time for animals acclimating to a temperature change of 5 C below ambient, in June. Cold acclimation here takes the form of an increase in routine scope from low initial values, which resulted from depression of oxygen uptake on sudden exposure to cold, to higher values at a new steady state representing the acclimated condition. Also plotted in Fig. 3 are values of filtration rate, which gradually increased during acclimation to reach a new steady state at 14 days. Once again, the agreement between the activity curve and the estimated scope for activity is good.

The Effects of Temperature Change on the Rate of Oxygen Consumption

Widdows (1973) has recently discussed some of the short and long-term effects of temperature change on the metabolism of *Mytilus*. In this section we will briefly review these and some more recent results.

Mussels were maintained at 15 C and their standard, routine and active rates of oxygen uptake determined at this temperature. The animals were then subjected to a temperature change of either 5 C above or 5 C below the maintenance temperature, and the three levels of oxygen consumption measured immediately (day 0) and on days 14 and 28. The results of one experiment are illustrated in Fig. 4. Two important points emerge. Firstly, in the short-term (acute) response, the standard rate of oxygen consumption is less temperature dependent (mean Q_{10} (10–20 C) for animal 1 gram dry weight = 1.58) than is the routine rate (mean Q_{10} (10–20 C) for animal 1 gram dry weight = 2.70). The active rate of oxygen consumption is intermediate in its temperature response (mean Q_{10} (10–20 C) for animal 1 gram dry

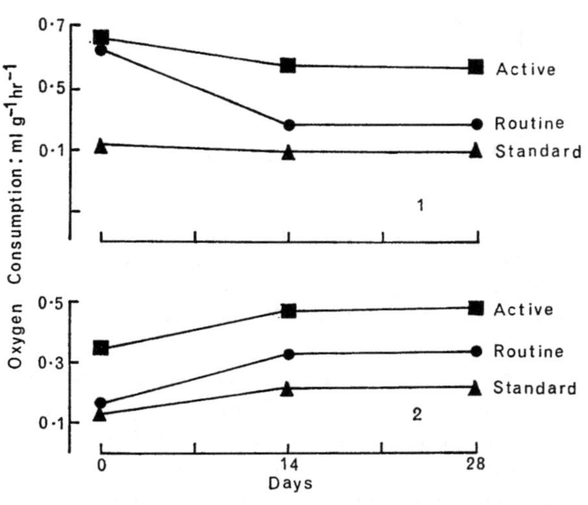

Fig. 4. The change, over a period of 28 days, of active, routine and standard oxygen consumption by *Mytilus edulis* when subjected to temperature changes of 5 C above ambient (1) and 5 C below ambient (2) temperature. (From Widdows, 1973)

weight = 1.76). Secondly, there is little change with time in either standard or active rates of consumption; that is, these rates do not acclimate, or acclimate in part only, to the temperature change. Routine rate, however, acclimates fully within 14 days, as we have described previously (Widdows and Bayne, 1971). As a result, the maximum scope for activity does not change significantly during acclimation, whereas the routine scope is reduced during warm acclimation and increased during cold acclimation.

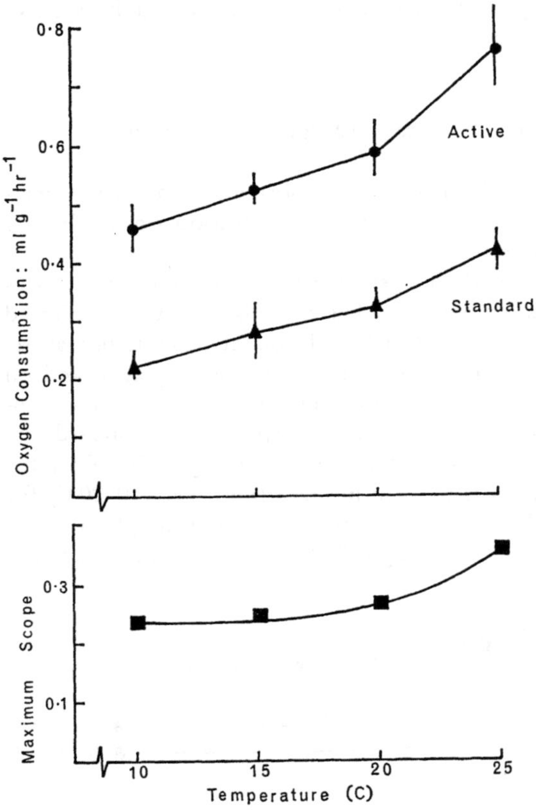

Fig. 5. The active and standard rates of oxygen consumption by *Mytilus edulis* held for 21 to 28 days at 10, 15, 20 and 25 C, and the calculated maximum scope for activity. Measurements were taken at the temperatures at which the animals were held. Notice the constancy of the scope between 10 and 20 C

The maximum scope for activity does not alter with acclimation temperature over a temperature range from <7 to 20 C so that a considerable flexibility of feeding response is maintained. At 25 C, however, the maximum scope for activity is increased, due to a sharp rise in the active rate of oxygen consumption (Fig. 5). In contrast, in the acute response to temperature change (Fig. 2) the maximum scope for activity is decreased above 20 C. The flexibility of the feeding response is therefore limited in the short-term at temperatures above 20 C, but during acclimation this flexibility is apparently restored. The inflation of the maximum scope for activity at temperatures approaching the lethal limits for a species has been observed in fish (Brett, 1964).

Seasonal Acclimation of Routine Rates of Oxygen Consumption

Mussels were maintained at temperatures of 5, 10, 15, 20 and 25 C in February, May, August and November and their routine rates of oxygen consumption determined after 28 days (Fig. 6). On each occasion there was full acclimation of oxygen consumption over the temperature range from 5 to 20 C, with one exception. This was in November, when animals at 5 C showed a significant depression of rate of

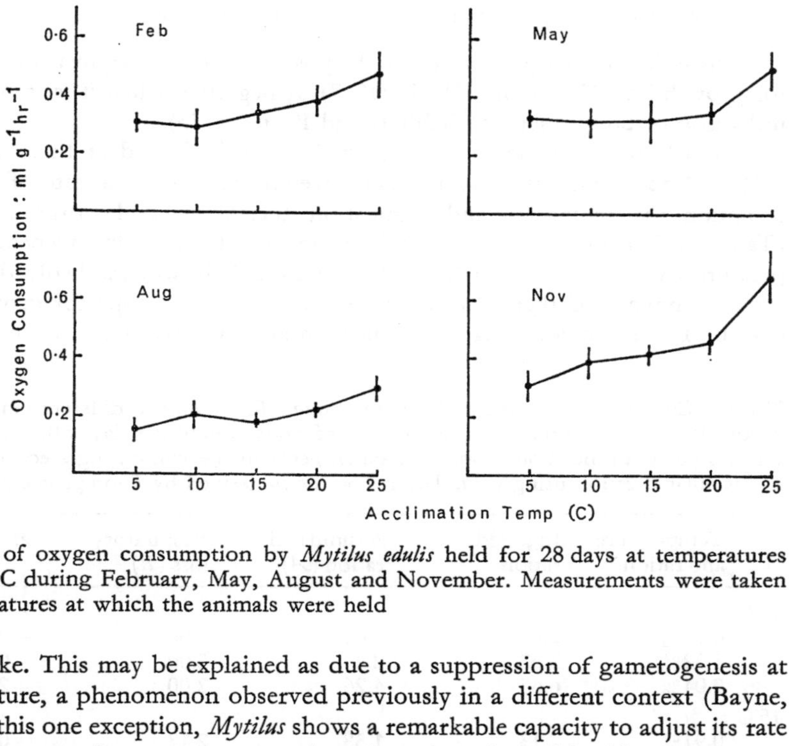

Fig. 6. Rates of oxygen consumption by *Mytilus edulis* held for 28 days at temperatures from 5 to 25 C during February, May, August and November. Measurements were taken at the temperatures at which the animals were held

oxygen uptake. This may be explained as due to a suppression of gametogenesis at this temperature, a phenomenon observed previously in a different context (Bayne, 1965). With this one exception, *Mytilus* shows a remarkable capacity to adjust its rate of oxygen consumption to constant values, in spite of considerable temperature change.

Energy Balance at Different Temperature and with Different Ration

We have found it helpful to bring together measurements of oxygen consumption, feeding rate and assimilation efficiency into a simple "energy balance" equation as an ndex of the animal's response to environmental change (Widdows and Bayne, 1971).

In the discussion that follows we have not considered the calorific value of amino acids lost from the body, although this is an important item in the energy budget (Gabbott and Bayne, 1973). However, the calculations do afford a meaningful "first step" towards understanding the physiological response of the "whole animal" to temperature change.

The index of energy balance is calculated from the calorific equivalents of oxygen consumption and food assimilation, based on the concept of the balanced energy equation of Winberg (1956). Using the IBP terminology as given by Crisp (1971):

$$(P+G)=A-R$$

where P is the energy of growth,
G is the energy released as gametes,
A is the energy assimilated from the ration, and
R is the energy equivalent of oxygen consumption.

This index of energy balance may be positive, which is equivalent to the "scope for growth" of Warren and Davis (1967) or negative, when it is indicative of the utilisation of body reserves (Gabbott and Bayne, 1973).

Animals were maintained at 10, 15, 20 and 25 C, and at three ration levels, 0.91%, 1.52% and 3.04% of the body weight per day. The rate of oxygen consumption, the feeding rate and assimilation efficiency were determined after 21 days (Table 3). The index of energy balance remained relatively independent of temperature over the range 10–20 C (Fig. 7) though it was markedly dependent on ration. Above 20 C there was a marked decline in the scope for growth, and the data in Table 3 indicate that this inflection at 20 C was due to an increase in the

Table 3. Components of the simple energy budget for *Mytilus* at different temperatures and ration. Values for ration given as percent of body weight per day; other values given as calories per gram per hour. Based on experiments in the winter. Ingested ration corrected to assimilated ration using assimilation efficiencies derived by Thompson and Bayne (1972)

Temperature and ration	Ingested ration	Assimilated ration (A)	Respiratory loss (R)	$A-R$
10 C				
0.91%	2.00	1.64	2.00	—0.36
1.52%	3.33	2.26	2.00	+0.26
3.04%	6.65	4.26	2.00	+2.26
15 C				
0.91%	2.13	1.58	2.13	—0.55
1.52%	3.55	2.31	2.13	+0.18
3.04%	7.10	4.26	2.13	+2.13
20 C				
0.91%	2.13	1.43	2.25	—0.82
1.52%	3.55	2.20	2.25	—0.05
3.04%	7.10	4.12	2.25	+1.87
25 C				
0.91%	1.66	1.08	3.45	—2.37
1.52%	2.77	1.66	3.45	—1.79
3.04%	5.54	3.10	3.45	—0.35

rate of oxygen consumption at 25 C, linked with a reduction in filtration, or feeding rate.

The maintenance ration is indicated by the ration level that results in a zero value for the index of energy balance. At 10, 15 and 20 C a ration of 1.52 % of the animal body weight per day is just above the maintenance requirement. At 25 C, however, the animals were in negative energy balance even at a ration of 3.05 % of body weight per day. As with the rate of oxygen consumption, so with the energy balance index, at 25 C the mechanisms of metabolic compensation for temperature appear to be disrupted, resulting in continued stress in the animal.

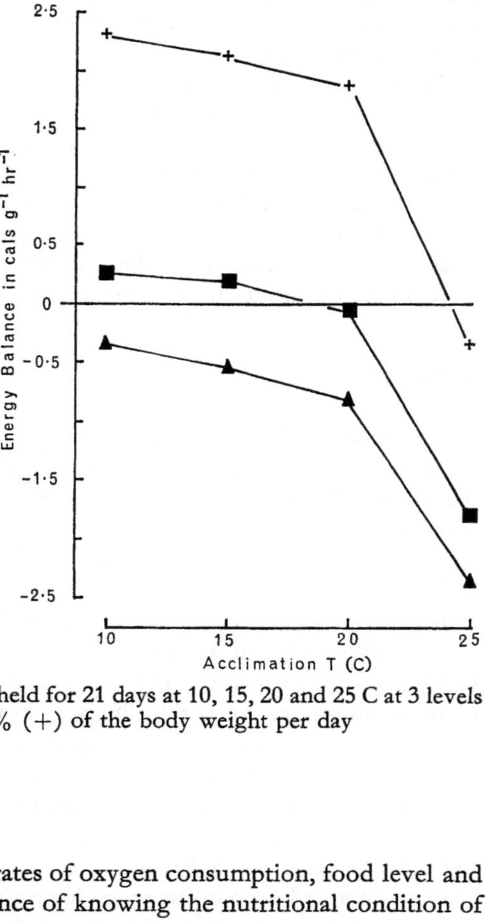

Fig. 7. The energy balance in *Mytilus edulis* held for 21 days at 10, 15, 20 and 25 C at 3 levels of ration: 0.91% (▲), 1.52% (■) and 3.04% (+) of the body weight per day

Conclusions

Our data on the relationships between rates of oxygen consumption, food level and temperature indicate clearly the importance of knowing the nutritional condition of an animal for a proper understanding of the physiological effects of temperature. Starved, inactive animals show a standard rate of oxygen consumption which is relatively insensitive to temperature in the short term, and which does not acclimate to change of temperature in the long term. On the other hand, animals which are fed and are spontaneously active show a routine rate of oxygen consumption which is markedly sensitive to temperature and which is completely acclimated over 14 days of exposure to a temperature change.

In our experiments the main component of the animal's "activity" has been the movement of gill cilia. This activity bears an exponential relationship with oxygen consumption, and as a result, relatively small changes in the former cause large changes in the latter. The acclimation of routine oxygen consumption may therefore be interpreted, at least in part, as resulting from adjustments of filtration rate. The apparent absence of temperature compensation in standard rate may, in turn, be due to the absence of any "activity" change in starved and virtually inactive animals. However, this explanation does not take account of the "physiological costs" of metabolism, or the specific dynamic effect of feeding. Further understanding of many aspects of the eco-physiological effects of temperature change will depend very largely on more precise distinction beween the mechanical and physiological costs of metabolism.

We suggest that integration of the various physiological functions into an index of energy balance is a useful way of visualising the effects of temperature at the level of the intact animal. Acclimation to temperature change can be seen to take the form of apparent control of the energy available for growth. At temperatures above 20 C this control is disrupted as the metabolic requirement for calories is inflated beyond the normal capacity of the feeding apparatus. In order to understand the processes underlying this, and other "inflection points" in the physiological respones to temperature, more information on the specific costs of metabolism is required.

Summary

1. Three levels of oxygen consumption rate are identified and defined, namely active, routine and standard. Active and standard rates can be induced by altering the ration; the routine rate is the acclimated rate when ration exceeds the maintenance requirement.

2. The relationships between standard and routine rates of oxygen consumption and dry tissue weight are quantified for winter and summer. A maximum and a routine scope for activity are identified and discussed.

3. Active and standard rates of oxygen consumption are relatively independent of temperature when compared with the routine rate, and they do not fully acclimate to temperature change within 28 days. Routine rate, however, has a high Q_{10} and fully acclimates within 14 days to temperature change. It is suggested that acclimation of routine rate follows largely from control of filtration rate.

4. Integration of certain physiological rate functions into a simple index of energy balance illustrates a marked temperature independence of this index over the temperature range 10 to 20 C, at levels which are dependent on ration. Above 20 C there is a marked drop in the scope for growth at all rations.

References

Bayne, B. L.: Growth and the delay of metamorphosis in the larvae of *Mytilus edulis* L. Ophelia **2**, 1–47 (1965).

Bayne, B. L.: Oxygen consumption by three species of lamellibranch mollusc in declining ambient oxygen tension. Comp. Biochem. Physiol. **40A**, 955–970 (1971).

Bayne, B. L.: Physiological effects of temperature and nutritive stress in *Mytilus edulis*. J. Marine Biol. Assoc. U. K. **53**, 39–58 (1973a).
Bayne, B. L.: Aspects of the metabolism of *Mytilus edulis* L. during starvation. Proceedings of the 7th European Marine Biology Symposium, 1972 (in press) (1973b).
Boyden, C. R.: Aerial respiration of the cockle *Cerastoderma edule* in relation to temperature. Comp. Biochem. Physiol. **43A**, 697–712 (1972).
Brett, J. R.: The respiratory metabolism and swimming performance of young sockeye salmon. J. Fisheries Res. Board Can. **21**, 1183–1226 (1964).
Crisp, D. J.: Energy flow measurements. In: (Eds.: Holme, N. A., McIntyre, A. D.: Methods for the Study of Marine Benthos, p. 197–279. Blackwell Scientific Publications (1971).
Davies, P. S.: Physiological ecology of *Patella* II. Effect of environmental temperature on metabolic rate. J. Marine Biol. Assoc. U. K. **47**, 61–74 (1967).
Fry, F. E. J.: Effects of the environment on animal activity. Univ. Toronto. Stud. Biol. **68**, 1–62 (1947).
Gabbott, P., Bayne, B. L.: Biochemical effects of temperature and nutritive stress in *Mytilus edulis*. J. Marine Biol. Assoc. U. K. **53**, 287–298 (1973).
Newell, R. C.: Biology of Intertidial Animals, p. 555. London: Logos Press, 1970.
Newell, R. C., Bayne, B. L.: Temperature and metabolic acclimation in marine invertebrates. Proceedings of the 7th European Marine Biology Symposium, 1972 (in press) (1973).
Thompson, R. J., Bayne, B. L.: Active metabolism associated with feeding in the mussel *Mytilus edulis* L. J. Exp. Marine Biol. Ecol. **9**, 111–124 (1972).
Warren, C. E., Davies, G. E.: Laboratory studies on the feeding, bioenergetics and growth of fish. In: Gerking, S. D. (Ed.): The Biological Basis of Freshwater Fish Production, p. 175–214. Blackwell Scientific Publication (1967).
Widdows, J.: The effects of temperature on the metabolism and activity of *Mytilus edulis*. Proceedings of the 7th European Marine Biology Symposium, 1972 (in press) (1973).
Widdows, J., Bayne, B. L.: Temperature acclimation of *Mytilus edulis* with reference to its energy budget. J. Marine Biol. Assoc. U. K. **51**, 827–843 (1971).
Winberg, G. G.: Rate of Metabolism and Food Requirements of Fishes. Fisheries Res. Board Can., Transl. Ser., No. 194, 1960 (1956).

Temperature-dependent Storage Processes in Crayfish and Spiders

KLAUS-G. COLLATZ

With 2 Figures

When, a few years ago, at the Zoological Institute in Berlin systematic attempts to study the body constituents and metabolism of a representative of the decapods were begun, it was soon realised how essential it is to take into account the life cycles of the animals. Analyses of the various organic and inorganic substances in *Orconectes limosus* revealed that in some cases values differed greatly according to season.

Speck (1967) found in crayfish that had been kept from October under relatively constant conditions (water temperature 12–14 C) carbohydrate values up to four times as high as those in outdoor specimens. Further, irregular preparations for moulting were observed: normally the animals moult 2–3 times in quick succession between June and August.

Subsequently, the annual cycle of changes in proteins, chitin, lipids, polysaccharides and "intermediary metabolites" such as sugars, amino acids, lactate and acetate have been studied in the crayfish. Two points deserve mention here:

1. The concentration of most of the substances investigated begins to rise in autumn (from about October), reaches a maximum in early winter and then drops to a spring minimum that is attained in April–May.

2. The degree of change in concentration seems to depend upon the severity of the winter from year to year.

Marked alterations in the annual cycle of events were observed in the cold winter of 1965/66 (carbohydrates: Speck, 1969; organic and inorganic blood components: Andrews, 1967), 1966/67 (lipids: Collatz, 1969) and 1969/70 (proteins, carbohydrates, chitin, intermediary metabolites: Speck *et al.*, 1972). In the very mild winter of 1970/71 the changes were less marked and the individual scatter was particularly high (Speck *et al.*, 1972).

If the temperature of the water from which the crayfish are taken is followed throughout the course of the year it appears that it is always at a minimum when the storage organs of the animals are filled to a maximum. In our latitudes a temperature of 12–14 C seems to be characteristic for the beginning of the storage processes.

In studying the metabolism of spiders, apparently temperature induced storage processes were observed, similar to those already seen in crayfish, and here the influence of temperature could be demonstrated experimentally.

The house spider *Tegenaria atrica* (Agelenidae) was used in the experiments.

It is important in such analyses to study simultaneously as broad a spectrum of material as possible and not, as is so often encountered in the literature, to give priority to one type of material taken more or less at random. The separation process given in Table 1 illustrates this (see Speck and Urich, 1970; Collatz and Speck, 1970).

Table 1. Fractionation scheme of whole spiders; all centrifugation steps are at 3600 rpm/10 min. Sn = supernatant; Sed = sediment

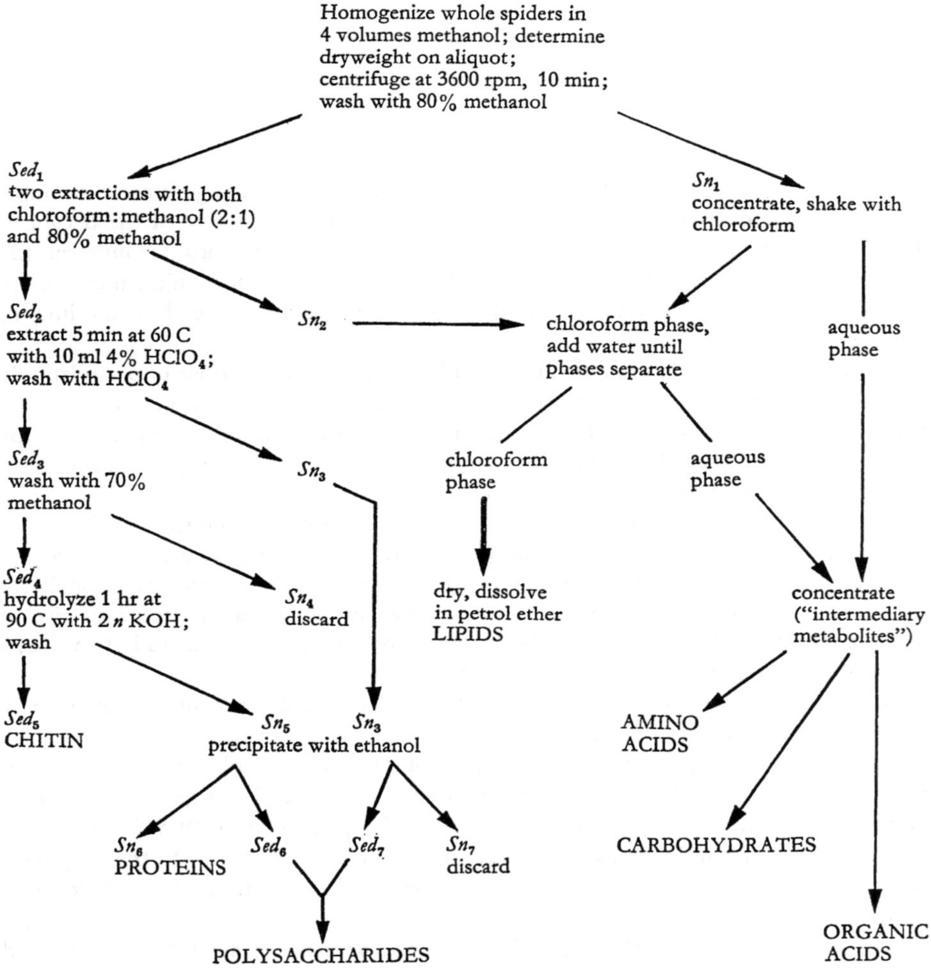

Fig. 1 shows the changes in concentration of the most important organic constituents of female *Tegenaria* over the course of the annual cycle. Most striking, as in crayfish, are the autumn maximum and spring minimum of lipids, and, at a lower level, the corresponding changes of the polysaccharides. The concentrations of chitin and protein, too, vary in a similar manner as in crayfish, although in the latter the differences are more clearly defined. In the case of protein, in particular, this is understandable since it represents the main source of energy for crayfish.

In order to test the influence of temperature on storage of reserve materials spiders were kept 4–5 weeks at 4, 12 and 30 C and fed every 3 days. The results

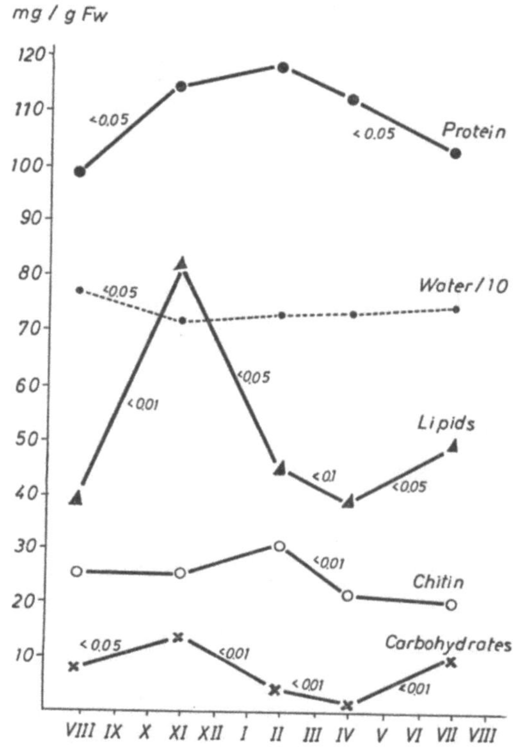

Fig. 1. Seasonal changes of main body constituents of *Tegenaria atrica*. The symbols represent mean values of several determinations. The figures denote levels of significance based on the two-sample t-test of Moore (1957)

are shown in Fig. 2. At 12 C there is a large storage of lipid corresponding to the autumn storage observable in the annual cycle of events. At 30 C the polysaccharides are significantly diminished, which is accompanied by a definite increase in the low molecular carbohydrates contained in the intermediary-metabolite fraction (not shown in the figure).

A speed-up in metabolism such as would explain the above findings is to be anticipated at a temperature of 30 C, which is a high temperature for an ectotherm from these latitudes. If, however, the carbohydrate values at this temperature are compared with the values attained over the course of the year, the high value of 10 mg/g fresh weight points to a relatively small utilization of the polysaccharide stores. It is more likely that the main energy sources are, in fact, the lipids, and this is confirmed by other experiments; the glycerol residues are probably transformed into carbohydrate.

In both crayfish and spiders extensive storage processes occur prior to moulting. Moulting storage and autumn storage are two strictly separate processes in *Orconectes*, whereas *Tegenaria* moults all the year round with the exception of the winter

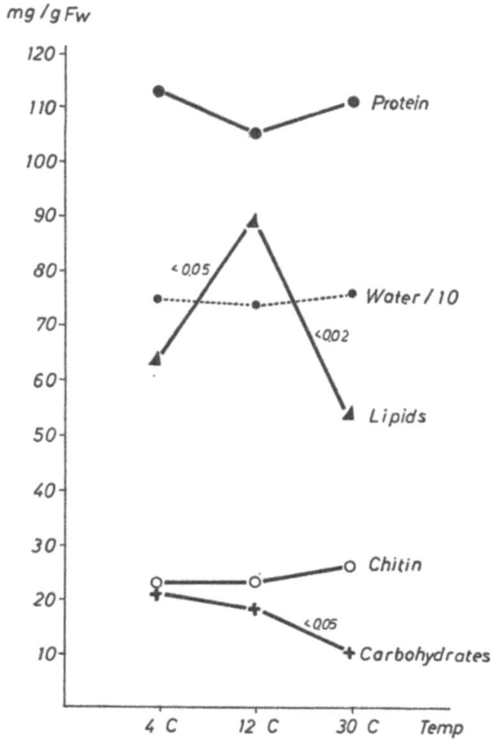

Fig. 2. Changes of main body constituents of *Tegenaria atrica* after 4–5 weeks' acclimation to different temperatures. Symbols and statistics as in Fig. 1

months. It is thus possible that two effects are superimposed so that the height of the moulting lipid peak depends upon the time of year.

It is not yet possible to say at which point temperature acts on metabolism. An answer may perhaps present itself if the pro- and opisthosoma in *Tegenaria* are warmed separately. Since nearly all of the nervous system is contained in the prosoma it might be possible to decide whether the effect of temperature on metabolism is mediated by the central nervous system or is exerted at the level of enzymes.

References

Andrews, P.: Über den Blutchemismus des Flußkrebses *Orconectes limosus* und seine Veränderung im Laufe des Jahres. Z. Vergleich. Physiol. **57**, 7–43 (1967).

Collatz, K.-G.: Das Lipidspektrum des Flußkrebses *Orconectes limosus* und seine jahreszeitlichen Veränderungen. Z. Vergleich. Physiol. **65**, 274–290 (1969).

Collatz, K.-G., Speck, U.: Gesamtbestand an organischen Substanzen der Spinne *Tegenaria atrica* im Vergleich zu *Protophormia terrae novae* (Diptera) und *Orconectes limosus* (Crustacea, Decapoda). Z. Vergleich. Physiol. **70**, 35–44 (1970).

Jantzen, H.: Die Aktivitäten von Enzymen des Glykogen- und Trehalosestoffwechsels in den einzelnen Organen des Flußkrebses *Orconectes limosus* und Kontrollmechanismen des Glykogenstoffwechsels im Schwanzmuskel. Diss. Berlin 1968.

Moore, P. G.: The Two-Sample *t*-Test Based on Range. Biometrics **44**, 482–489 (1957).
Speck, U.: Das Kohlenhydratspektrum in den Organen des Flußkrebses *Orconectes limosus* und seine Veränderungen im Jahresablauf. Z. Vergleich. Physiol. **65**, 51–69 (1969). Diss. Berlin 1967.
Speck, U., Urich, K.: Das Schicksal der Nährstoffe beim Flußkrebs *Orconectes limosus:* Einbau von Glucose-U-^{14}C, Glutamat-U-^{14}C und Palmitat-U-^{14}C in die verschiedenen Stoffklassen. Z. Vergleich. Physiol. **68**, 318–333 (1970).
Speck, U., Urich, K., Hahmann, R.: Der Stoffbestand des Flußkrebses *Orconectes limosus*, Jahreszyklus und Organverteilung. Z. Vergleich. Physiol. **77**, 287–305 (1972).

Effects of Temperature on the Distribution of Turbellarian Triclads

E. Pattee, C. Lascombe, and R. Delolme

With 6 Figures

The aim of this paper is to review our studies on the physiological and ecological effects of temperature on the more common freshwater triclad of southeastern France.

Distribution of Triclads

Three rheophilic triclads succeed each other in clear-cut zones along mountain streams: *Crenobia alpina* above 800–900 m altitude, *Polycelis felina* between 800 and 300 m, and *Dugesia gonocephala* below and in swiftly-flowing cool streams in the plains.

We also studied three limnophilic triclads. The morphologically similar *Polycelis nigra* and *P. tenuis* inhabit stagnant water and sheltered parts of slow streams. Lascombe (1971) mapped their distribution in the Lyons region, showing that both species are found in the plains and on mountain sides. *P. nigra*, however, was never found above 300 m, whereas *P. tenuis* lives as high as 1.000 m, and therefore may be found in ponds, lakes, and in sheltered parts of streams in the *Polycelis felina* zone, whilst the latter lives in the current. *Dugesia tigrina* was introduced into Europe some 50 years ago and still seems to be spreading. It is now quite common in the South of France and was recently found by Russier and Lascombe (1970) in the Lyons region on the warmer side of a sheltered pond and in the stagnant backwater of a river.

Temperature as a Lethal and Limiting Factor

Upper Lethal Temperatures

The animals were exposed to different temperatures and their mean survival time was measured. The curves in Fig. 1, or, at least, the upper parts of the curves which represent long-term exposures, are seen to succeed each other in roughly the same order as the triclad zones in the field.

The usual definition of lethal temperature, i.e. that causing 50% death within 24 hours is not a good one for triclads, since by plotting survival time against temperature a straight line results and there is no justification in selecting a period of 24 hours as a critical interval. The determination of the "incipient lethal limit", according to Fry (1957) (i.e. the limit between resistance and indefinite tolerance), requires up to 6 months in triclads. These limits are listed in Table 1.

In most cases lethal limits do not directly explain triclad zonation since most habitats never become hot enough to kill the animals. There are, however, two instances where such limits might influence distribution.

The first case is that of populations which have been displaced far downstream by floods. Such populations are wiped out by summer temperatures. When reared at outdoor temperatures in the plains, starting in winter, *C. alpina* died in the spring, *P. felina* later, *D. gonocephala* survived one summer but was annihilated by the next, *P. nigra* and *D. tigrina* lived on normally (Pattee 1966, 1968, Russier-Delolme, 1972).

The second case is that of *Crenobia alpina*. Steinböck (1942) and Thienemann (1950) pointed out the apparent contradiction in the ecology of this species which in the mountains is found everywhere, in streams, in large or small lakes, whether

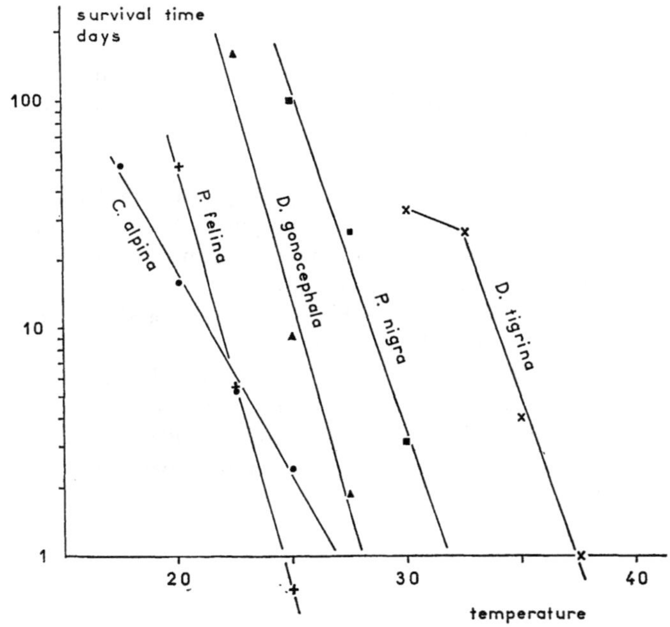

Fig. 1. Semi-logarithmic plot of mean survival time of several species of triclads against temperature. (After Russier-Delolme, 1972)

Table 1. The upper incipient lethal temperature for triclads (50% individuals dead in indefinite time, according to Fry, 1957) accuracy ±1 C

C. alpina	12
P. felina	16
D. gonocephala	21
P. nigra	26
P. tenuis	26
D. tigrina	29

the temperature be stable or unstable, up to 25 C in the sunshine. It thus appears to be eurythermic. At lower altitudes, however, *C. alpina* inhabits only cold springs with a stable temperature and thus appears to be stenothermic. The explanation is that it can tolerate temperatures up to 25 C at midday if cooled down up to 10 C at night (Pattee, 1965). In Fig. 2 each point represents the survival time of the first animal to die in a batch of 10 animals. At 25 C the first deaths occur after 2 days. Thus *C. alpina* runs no danger when subjected to 25 C for one hour daily in the mountains. It may be noticed that the slope of its curve in Fig. 1 is different from that of the others. Resistance to short-term exposure is greater and this difference may be interpreted as a special adaptation to diurnal temperature fluctuations. As has been shown for other animals, the temperature may rise above the upper incipient lethal limit, into the resistance zone, for a short time. But in habitats with a stable temperature such as springs, the temperature must remain below 15 C because at this temperature, as Fig. 2 shows, mortality occurs after a fortnight.

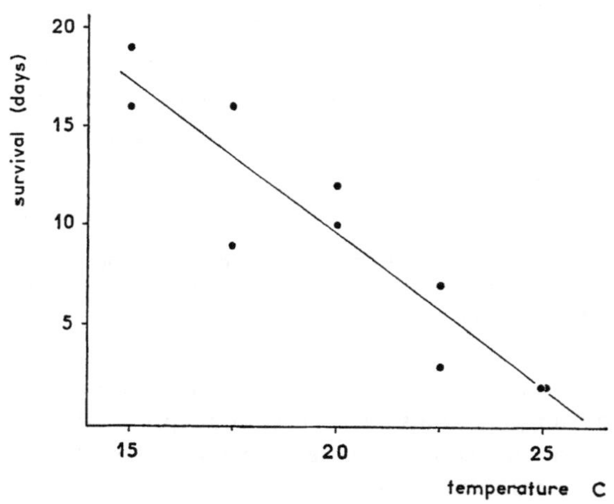

Fig. 2. Minimum survival time of *Crenobia alpina* at different temperatures. Each point represents the survival time of the first animal to die within a batch of 10 animals. (After Pattee, 1969)

C. alpina needs a certain amount of cold, either a stable cool temperature, or an unstable temperature with a cold spell at night. It is a cold-loving, or psychrophilic, animal. These mechanisms account for what Thienemann (1950) called the modification of the ecological valence of a species. *C. alpina* gives the impression of a hardy, eurytopic animal in the centre of its area of distribution (the Alps) but seems to demand more restricted conditions at the periphery of its area, at lower altitudes.

Lower Limiting Temperatures

Individual triclads, whatever their species, show no distinct lower lethal limit. Even *D. tigrina* can be frozen into blocks of ice and thawed out undamaged. The main ecological effect of cold is one of inhibition for which two different limits were observed.

a) *The Lower Limit for General Activity.* At 5 C all species are more or less active and feed and grow, except *D. tigrina*, which again appears as the most thermophilic species. At 5 C it does not feed and the animals shrink accordingly although their survival time is very long: Russier-Delolme (1972) found only 30 % of them to have died after one year. This temperature is thus unfavourable for *D. tigrina* and would tend to eliminate the species if it occurred for long periods each year.

b) *The Lower Limit for Reproduction.* At 5 C only *C. alpina* is capable of reproducing to a significant degree; *P. felina* has a very low reproduction rate and the others are completely sterile. This explains why only *C. alpina* lives in high mountains. At 10 C all species except the thermophilic *D. tigrina* reproduce (Fig. 3).

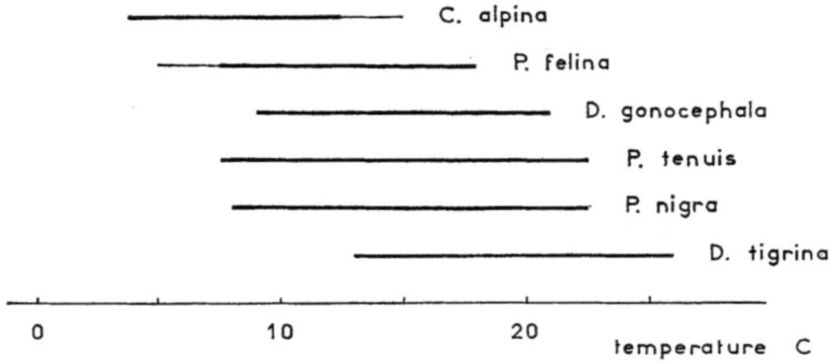

Fig. 3. The temperature range over which sexual or asexual reproduction is possible in triclad species. Thick lines: effective reproduction; thin lines: very low reproduction rate that may be neglected

Below the lower limit for reproduction, the animals feed and grow and thus lay up provisions to be used for reproduction as soon as the temperature rises but the occurence of a species depends on the time during which the temperature exceeds this reproduction threshold.

Temperature as a Controlling Factor

Between the upper and lower limits there is a range of temperature in which life and reproduction are possible. Within this range two sympatric species which occupy the same ecological niche tend to compete for the available resources (food, shelter). According to the competitive exclusion principle, the reproductive capacity of the competing species is one of the parameters which determines the outcome of competition (Hardin, 1960).

We measured the reproductive capacity of our triclads. It depends on the birth rate and the length of a generation, and is computed as the intrinsic rate of natural increase of the population in an unlimited environment. Only by using this parameter can comparisons be made between the sexual reproduction in *C. alpina*, *D. gonocephala*, *P. nigra* and *P. tenuis*, and the asexual reproduction in *C. alpina*, *P. felina* and *D. tigrina*.

Fig. 4 represents the intrinsic rate of increase in rheophilic triclads. The temperature range in which reproductive superiority appears conforms in each species with that recorded in the literature for the animals' habitat. We are at present measuring this range in our region. According to Beauchamp and Ullyott (1932) on the one hand, and Alause (1968) on the other, the mean summer boundary between *C. alpina* and *P. felina* in the field may be set below 9 C when current velocity does not interfere, whereas that between *P. felina* and *D. gonocephala* occurs at about 15 C. Since the possible distribution of each species is curtailed by the presence of other species, competition is likely to be one of the most important mechanisms by which temperature determines triclad zonation.

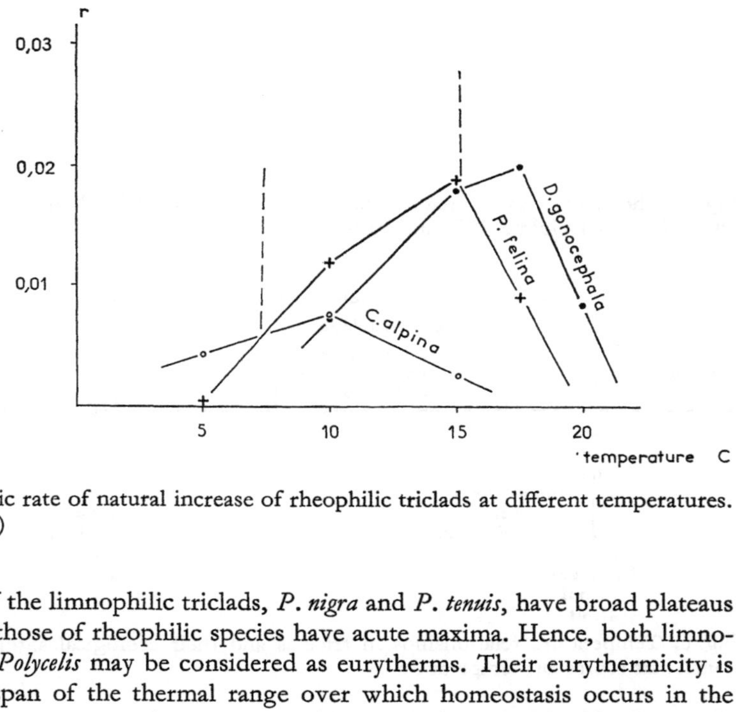

Fig. 4. The intrinsic rate of natural increase of rheophilic triclads at different temperatures. (After Pattee, 1970)

The curves of the limnophilic triclads, *P. nigra* and *P. tenuis*, have broad plateaus (Fig. 5) whereas those of rheophilic species have acute maxima. Hence, both limnophilic species of *Polycelis* may be considered as eurytherms. Their eurythermicity is manifest in the span of the thermal range over which homeostasis occurs in the potential recruitment rate to the next generation. On the other hand, all species which display curves with acute maxima may be termed stenotherms.

One may ask what mechanisms are responsible for this difference and how is homeostasis brought about. A study of the components of the intrinsic rate of natural increase in *P. nigra* and *P. tenuis* shows that several of these components interfere, but that the most effective seems to be cocoon sterility. The proportion of sterile cocoons deposited increases with temperature and plays a part in the levelling-off of the reproduction rate in eurytherms.

The following ecological points may be noted: Rheophilic triclads which have an acute maximum inhabit distinct thermal zones. *P. nigra* which has a maximum over the range 15–20 C inhabits the plains and the lower hills up to the *D. gonocephala* zone. *P. tenuis* has the broadest maximum covering the range 10–20 C and it

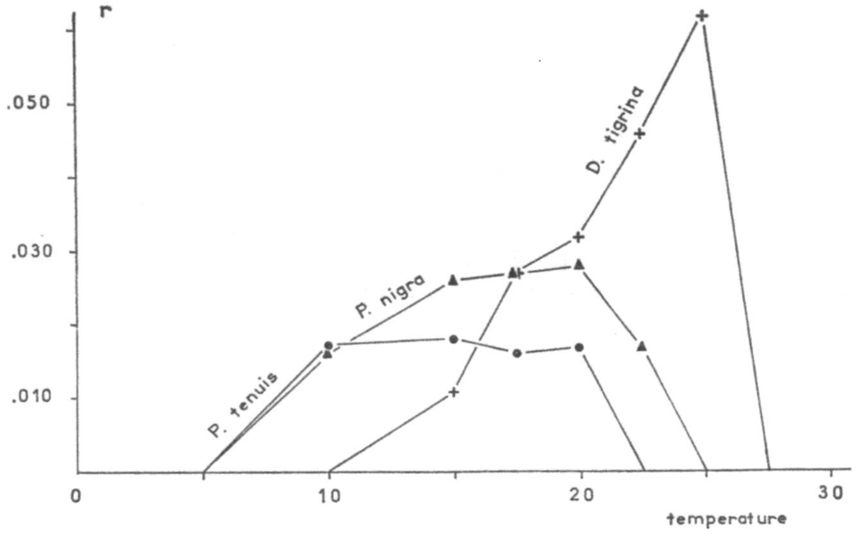

Fig. 5. The intrinsic rate of natural increase of limnophilic triclads at different temperatures. (After Russier-Delolme, 1972 and Pattee, 1972)

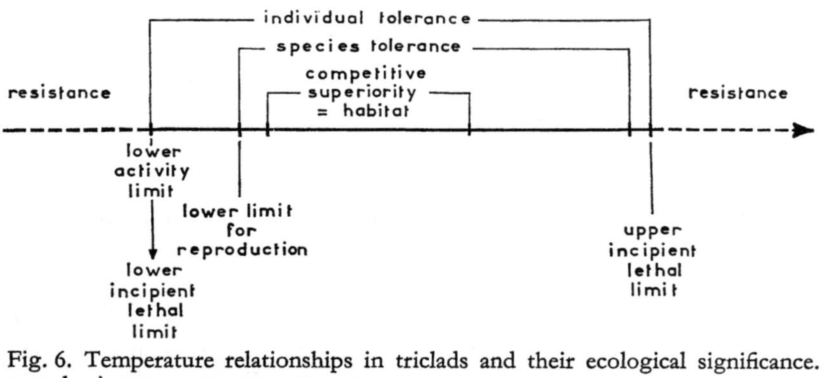

Fig. 6. Temperature relationships in triclads and their ecological significance. The horizontal axis represents temperature

inhabits the plains and the higher land up to and including the *P. felina* zone. The extent of the plateaus in the curves seems to be related to the temperature in the animal's habitat. As Pattee (1972) pointed out, it should be interesting to compare the plateaus of the $R:T$ curves in eurythermic triclads with those of the $R:T$ curves described in intertidal invertebrates by Newell (1967), Newell and Northcroft (1967), Newell and Pye (1971), and in freshwater Trichoptera nymphs by Roux (1969).

According to the preceding conclusions *D. tigrina* is a thermophilic stenotherm. It is found together with exotic plants and fishes in hot houses (Dahm, 1958) and grows more abundant towards the south. According to its $R:T$ curve it becomes competitive at 20 C and above. Its possibility of occupying a habitat apparently depends on:

a) The time available for feeding and storing living material for reproduction (i.e. at temperatures above 5 C).

b) The time during which the species can effectively reproduce at a higher rate than its competitors (i.e. at temperatures above 18 C).

Fig. 6 sums up the authors' present view of the temperature relationships of triclads.

References

Alause, P.: Contribution à l'écologie des Planaires d'eaux courantes: Observations en Languedoc méditerranéen. Le facteur acidité carbonique. Thèse, p. 730, Montpellier (1968).

Beauchamp, R. S. A., Ullyott, P.: Competitive relationships between certain species of fresh-water Triclads. J. Ecol. **20**, 200–208 (1932).

Dahm, A. G.: Taxonomy and ecology of five species groups in the family Planariidae (Turbellaria Tricladida Paludicola), p. 241, Malmö (1958).

Fry, F. E. J.: The lethal temperature as a tool in taxonomy. Année Biol. **33**, 205–219 (1957).

Hardin, G.: The competitive exclusion principle. Science **131**, 1292–1297 (1960).

Lascombe, C.: Recherches écologiques et biogéographiques sur deux espèces jumelles de Planaires d'eau douce dans la région lyonnaise. Thèse, p. 116, Lyon (1971).

Newell, R. C.: Oxidative activity of poikilotherm mitochondria as a function of temperature. J. Zool. (Lond.) **151**, 299–311 (1967).

Newell, R. C., Northroft, H. R.: A re-interpretation of the effect of temperature on the metabolism of certain marine invertebrates. J. Zool. (Lond.) **151**, 277–298 (1967).

Newell, R. C., Pye, V. I.: Quantitative aspects of the relationship between metabolism and temperature in the winkle *Littorina littorea*. Comp. Biochem. Physiol. **38 B**, 635–650 (1971).

Pattee, E.: Sténothermie et eurythermie. Les Invertébrés d'eau douce et la variation journalière de température. Ann. Limnol. **I**, 281–434 (1965).

Pattee, E.: Coefficients thermiques et écologie de quelques Planaires d'eau douce. 1– Tolérance des adultes. Ann. Limnol. **2**, 469–475 (1966).

Pattee, E.: Coefficients thermiques et écologie de quelques Planaires d'eau douce. 2– Tolérance de *Dugesia gonocephala*. Ann. Limnol. **4**, 99–104 (1968).

Pattee, E.: Contribution expérimentale à l'écologie de la Planaire alpine, *Crenobia alpina* (Dana). Bull. Soc. Zool. France **94**, 269–276 (1969).

Pattee, E.: Coefficients thermiques et écologie de quelques Planaires d'eau douce. 4– La reproduction de *Dugesia gonocephala*. Ann. Limnol. **6**, 293–304 (1970).

Pattee, E.: Coefficients thermiques et écologie de quelques Planaires d'eau douce. 5– La reproduction des espèces jumelles *Polycelis nigra* et *Polycelis tenuis*. Ann. Limnol. **8**, 11–30 (1972).

Roux, C.: Les variations de la courbe métabolisme/température de *Micropterna testacea* (Trichoptera Limnophilidae) sous l'influence de divers facteurs écologiques. Ann. Limnol. **5**, 49–60 (1969).

Russier-Delolme, R.: Coefficients thermiques et écologie de quelques Planaires d'eau douce 6– *Dugesia tigrina*. Ann. Limnol. **8**, 119–140 (1972).

Russier, R., Lascombe, C.: La Planaire américaine *Dugesia tigrina* dans la région lyonnaise: écologie et tolérance thermique. Bull. Mens. Soc. Linnéenne Lyon **39**, 197–206 (1970).

Steinböck, O.: Das Verhalten von *Planaria alpina* Dana in der Natur und im Laboratoriumsversuch. Mem. Ist. Ital. Idrobiol. de Marchi **1**, 63–75 (1942).

Thienemann, A.: Verbreitungsgeschichte der Süßwassertierwelt Europas. Binnengewässer **18**, 809, Stuttgart (1950).

The Measurement of the Efficiency of Growth at Different Temperatures and Its Significance

J. N. R. GRAINGER

Introduction

Considerable research has been carried out on the physiological and biochemical adaptations which organisms make at different temperatures (Precht *et al.*, 1955) but little is known of the significance of these adjustments. The present work on bakers yeast was carried out in an attempt to define what is meant by efficiency of growth and reproduction and to try and compare the performance of growing organisms at different temperatures, with a view to applying the results to organisms in the wild. Bakers yeast was chosen because its biochemistry is well known and some work has been carried on the adaptations of yeast to temperature (Christophersen and Precht, 1950). The heat production of yeasts has been studied previously by several authors (for instance Rubner, 1904; Ohlmeyer and Fritz, 1966; Battley, 1960).

Materials and Methods

The experiments were carried out on a locally isolated clone of bakers yeast, maintained on nutrient agar plates in pure culture. The medium used for growth experiments consisted of a broth with buffer and glucose added. The broth consisted of Lab Lemco 5 g, peptone 10 g, NaCl 5 g, distilled water 1,000 ml. The pH of this was adjusted to 5.5 and it was then autoclaved. 100 ml of broth was taken, 100 ml 0.066 M phosphate buffer pH 5.5 was added to it together with 2 g glucose and the mixture sterilised by passing through a Seitz filter. After inoculation with yeast the culture was grown at the appropriate temperature. The medium was continually stirred with a magnetic stirrer and pure O_2 was bubbled through it. Prior to an experiment the cells were harvested by centrifugation, washed in new medium and resuspended in more medium. Samples were taken at the beginning and end of each experiment for determination of N content of cells, cell concentration counts, rates of O_2 consumption and CO_2 outputs. Heat production was measured using the method described in Grainger (1968). The contents of the calorimeter were stirred with a magnetic stirrer and pure O_2 was bubbled through it. The calorimeter was calibrated using the same flow rate of O_2 as was used in the experiments. Heat production was measured after 1 hour and again after a further lapse of about 3 hours. N was determined by taking samples of the medium containing cells, centrifuging, washing twice in tap water and digesting in conc. H_2SO_4. Following dilution of the digest aliquots were put in a Markham still, neutralised and the ammonia was steam distilled into boric acid containing mixed indicator. This was back titrated with 0.01 N HCl. Cell counts were made with a haemocytometer. Rates of O_2 consumption and CO_2 output were measured in a Warburg apparatus. The respiration results are not presented in this paper but will be published elsewhere as they are extensive. Bomb calorimetry was carried out on yeast which had been harvested, washed and dried at 105 C for 2 hours, in a Parr semimicro bomb calorimeter.

I am grateful to Miss Lai Fong Tho for technical assistance and to the Medical Research Council of Ireland for support for some of this work.

Results

Table 1 shows the generation time of yeast grown at three temperatures. This was calculated from the haemocytometer counts and also from the cell N estimations using the usual exponential growth equation:

$$N = N_0 e^{kt}$$

where N_0 is the cell concentration (or N content of the cells) at time zero, N is the cell concentration at time t, k is a constant and e is the base of natural logarithms.

Table 1. Generation time and heat production of growing bakers yeast

Growth temp.	Generation time (mins)			Heat production calories/10^6 cells/hr.			Total heat produced during one generation (cals/10^6 cells)
	Mean	S.D.	n	Mean	S.D.	n	
20.0 C	258.8	34.1	4	0.010[a]	—	—	0.0622
30.0 C	82.3	16.2	11	0.0280	0.013	11	0.0553
40.0 C	452.1	161.0	7	0.0044	0.0035	9	0.0485

[a] Estimated.
S.D. = standard deviation
n = number of experiments.

In all experiments both the cell doubling time and the N doubling time (generation time) was estimated. There was no significant difference between them and so they have been combined together in the results in Table 1. At 30 C the data was: mean doubling time from cell counts = 82.1 mins (standard deviation = 19.7, n = 11), mean N doubling time = 82.5 mins (S. D. = 12.7 n = 11). Two sets of experiments were carried out at 30 C. In one set the yeast had been grown prior to the experiment at 20 C; in the other the yeast was grown at 40 C. There was no significant difference in generation time between them whether measured by N doubling or by cell counts. The generation times based on cell counts were: Mean_{20} = 80.8 mins., S. D. = 17.2, n = 5, Mean_{40} = 83.1 mins., S.D. = 23.1, n = 6. The generation times calculated from N doubling were Mean_{20} = 88.5 mins., S.D. = 9.4, n = 5, Mean_{40}

= 77.5 mins., S.D. = 13.7, $n = 6$. Thus the prior culture temperature does not significantly affect generation time at the experimental temperature.

From the data given in Table 1 it is clear that the shortest generation time is at 30 C. It is longer at 20 C, as might be expected. At 40 C it is also long presumably because the organism is growing close to its lethal limit. At 40 C the respiration measurements (which are not reported here) showed a low rate of oxygen consumption but a high CO_2 output. Since oxygen is said to be essential for multiplication of bakers yeast (Ingram, 1955), the low uptake (perhaps due to some inhibition of the Kreb's cycle) at 40 C may be the reason why growth is so slow at this temperature.

Calorimeter measurements were carried out at two temperatures, 30 C and 40 C. In each experiment the heat production was measured 1 hour after starting and again about 3 hours later. There was no significant difference between the initial and final heat productions (when expressed as calories per 10^6 cells per hour) at 40 C. There was a significant difference between them however at 30 C – the final heat productions being lower. Because of it the initial heat production measurements are given in Table 1. It should be pointed out that the respiration measurements showed that there was still a high O_2 uptake at the end of these experiments.

The value for heat production at 20 C shown in Table 1 is an estimate based on the rate of O_2 consumption and CO_2 output and is put in purely for broad comparison with the 30 C and 40 C results. It is probably an underestimate of the rate of heat production.

The final column on Table 1 is an estimate of the total heat production during one generation. This is calculated as follows. During exponential growth the heat produced by a culture is given by

$$H = H_0 e^{kt}$$

where H = heat production (cal/min) at time t, H_0 = initial heat production at time zero, k = constant, e = base of natural logarithms, t = time (mins.). On integrating between the limits of 0 and t we get:

$$\text{Total heat produced} = \int_0^t H_0 e^{kt} \cdot dt = H_t = \frac{H_0}{k} (e^{kt} - 1).$$

When t = generation time (mins.), and starting at time zero with 10^6 yeast cells this equation can be reduced to

$$H_t = 1.442 \cdot \mathrm{t} \cdot H_0$$

provided H_0 is expressed in calories/min./10^6 cells.

Table 2 gives the heats of combustion of yeast grown at 20 C, 30 C and 40 C. In order to obtain estimates of the heat of combustion per 10^6 cells it was necessary to measure the N content both in terms of dry weight and also in terms of cell concentration. This data is given in the second and third columns. The values given in the final column of Table 2 are no doubt a reflection of differences in cell size. It is generally true in organisms that cell size is greater at lower growth temperatures and decreases as the growth temperature increases until a point is reached when it increases again (Belehradek, 1935).

Table 2. Heats of combustion and N content of bakers yeast grown at different temperatures

Growth temp. C	Heat of combustion cal./g. dry wt.			N content mg N/g. dry wt.			N content mg $N/10^6$ cells			Heat of combustion cal./10^6 cells
	Mean	S.D.	*n*	Mean	S.D.	*n*	Mean	S.D.	*n*	Mean
20.0	4315.3	80.6	7	33.22	1.03	4	.00099	.0002	5	0.128
30.0	4227.8	10.1	6	42.74	0.39	6	.00071	.0001	9	0.0702
40.0	4572.6	81.9	4	29.56	1.23	4	.00109	.0005	6	0.168

The heats of combustion given in this paper are enthalpy measurements. It is probable that the heat production measurements are also predominantly enthalpy estimations (Belaich, 1963). It would seem from the calculations of Morowitz (1955, 1968) that the heat produced as a result of entropy lowering during growth is small.

Discussion

In thermodynamics efficiency is defined as the fraction (or percentage) of the input energy which is used to do useful work. If, during unit time, an organism takes in substances of energy I, gives off H calories of heat and excretes waste products of energy W and adds to itself as growth tissue or cytoplasm of energy G then

$$I - H - W = G.$$

The efficiency of this operation would be defined as

$$E = \frac{G}{I} = \frac{I - W - H}{I}.$$

It is extremely difficult to measure I in biological systems since only a proportion of the total energy of substances entering cells may be available to the organism via its metabolic pathways because of enzyme specificity restrictions etc. Furthermore this amount will vary at different temperatures because of changes in pathway importance and in the activity of many cellular enzymes which are known to occur (Precht *et al.*, 1957). G is also extremely difficult to determine. Bomb calorimetry only measures the heat of combustion which may be an inadequate measure of the energy contained in new cytoplasm or tissue. The nearest approach it is possible to get in the present work to using the above equation would be to replace I by $I - W$, and consider it as the net input of energy and by using the heat of combustion as a measure of G. So we have:

$$E = \frac{G}{I - W} = \frac{G}{G + H}.$$

The values for this at the three temperatures are given in Table 3. In view of what has been said above, however, I consider these values as almost meaningless, in thermodynamic terms, and that it is probably not useful in the present state of ourk nowledge of biological systems to pursue the thermodynamic approach to the question of assessing the efficiency of growth.

It might be thought that a suitable index of efficiency of growth would be the amount of heat expended in construction of an amount of cells or tissue which gives 1 calorie of heat of combustion (Table 2 second column). The difficulty with this however is that the amount of tissue or cells which gives 1 calorie of heat of combustion differs at different temperatures due to differences in cell size (Table 2).

I suggest that the best measure of efficiency of growth that is available at present is the amount of heat which is produced during a generation in the construction of a new cell. It is hoped to investigate the total input energy as an index of growth in a later paper. This data is given in Table 1 and in the last column of Table 3. This is the best measure available of the metabolic work done during the construction and the most efficient situation is that in which this value is smallest. It will be seen from the Table that growth is by this criterion, most efficient at 40 C and least efficient at 20 C.

Table 3. Assessment of efficiency of growth of bakers yeast at different temperatures

Growth temp. C	$\frac{G}{G+H}$	$\frac{H}{G}$	H
20	0.675	0.486	0.0622
30	0.628	0.597	0.0553
40	0.776	0.284	0.0485

G = Heat of combustion (cal./10^6 cells).
H = Total heat produced during one generation (cal./10^6 cells).

This may seem a surprising result especially since the generation time (Table 1) is long at 40 C. It is important however to realise the conditions under which the experiments were carried out. The yeast was growing in pure culture in an "open" situation in which there was no limit on the food supply. Under these circumstances the length of a generation does not matter. In nature however the situation is different. According to evolutionary theory survival is of supreme importance, and it is in terms of survival that the significance of the efficiency of growth must be assessed. This may be clarified by considering some hypothetical situations. We assume for simplicity that the chances of survival are equal with yeast cells grown at 20 C, 30 C and 40 C (which may not be true). Suppose that 10^6 cells are placed in a fixed volume of nutrient which can be completely utilised and that there are no inhibitions to growth by end products, and that the heat of combustion of the dissolved nutrients are 1 cal. At which temperature will most cells be produced before the nutrient is exhausted? The amount of nutrient used in producing one generation may be estimated roughly as the heat of combustion (per 10^6 cells) plus the total heat produced during one generation. This totals 0.19, 0.12 and 0.21 cal. for 20 C, 30 C and 40 C respectively. It will be seen that most cells will be produced at 30 C before the nutrient is exhausted. In this situation then growth is most efficient, when expressed in terms of survival, at 30 C. A second example concerns a similar situation but there is initially 10^6 cells of a competing organism present, in addition to 10^6 cells of yeast, which has the same caloric values as yeast (Tables 1 and 2) but the generation times are 250, 40 and 100 minutes at 20 C, 30 C and 40 C respectively. Calculation shows that in this situation most yeast cells are produced at 20 C, the temperature at which efficiency of growth is lowest in an "open" environment. The calculation in this case includes the generation times.

The above examples illustrate the key point that by assessing the significance of the efficiency of growth in terms of survival in real situations different values are placed on total heat produced during a generation, on heat of combustion of cells, and on generation time in each case. In an open situation such as in the present experiments generation time does not enter the picture since the question of maximising survival does not arise, but immediately constraints of any kind are imposed then the question of maximising survival does arise and the importance of the individual parameters will differ in each real case.

While this paper has been about bakers yeast the principles put forward could be applied to higher organisms. The key parameters to be measured would be the total heat produced from egg until death in an open situation at different temperatures, the heat of combustion of food and faeces, the heat of combustion of the organism, the number of eggs produced during the lifetime, the length of life and the times at which eggs are laid. Efficiency of growth and reproduction could be defined as the heat expended during life divided by the number of eggs produced.

Efficiency would be greatest when this figure was least. Provided the pattern of metabolism is known the rate of oxygen consumption can be a valid measure of heat production. Suitable data in which these parameters have been measured is now becoming available but it is mostly at only one temperature.

The conditions under which efficiency of growth and reproduction are highest could be considered as being optimal for the organism. Most organisms live in changing environments and it is possible that in many species optimal conditions are changing conditions. It must be stressed however that these considerations only apply to "open" environmental situations. In a real situation in nature highest survival (measured initially in terms of fertile eggs laid) may be under conditions in which the organism is growing and reproducing at its least efficient from the energy point of view.

Summary

1. The heat production of cultures of bakers yeast growing in the presence of oxygen was measured at 30 C and 40 C. Generation time was measured at 20 C, 30 C and 40 C and was found to be shortest at 30 C. No significant differences were found between the two methods used to measure this (cell counts and nitrogen content of cells). Heats of combustion of yeast grown at 20 C, 30 C and 40 C were measured and also the nitrogen content.

2. The total heat produced during one generation is suggested as the best measure of efficiency of growth. Growth is considered to be most efficient when this value is smallest. Using this criterion growth was most efficient at 40 C and least efficient at 20 C. It is pointed out that in the experiments carried out the yeast was growing in an open environment in which there were no restrictions of food supply. Under natural conditions survival must be the criterion by which efficiency of

growth is assessed and some hypothetical situations are considered in which different restrictions are applied. Under these conditions efficiency of growth measured in terms of survival may be greatest at 30 C or, under other conditions, at 20 C.

3. The significance of the results are discussed and their application to higher organisms are considered.

References

Battley, E. H.: Enthalpy changes accompanying the growth of *Saccharomyces cerevisiae* (Hanson). Physiologia Pl. **13**, 628–640 (1960).

Belaich, J. P.: Thermogenèse et croissance de *Pseudomonas lindneri* en glucose limitant. Compt. Rend. Soc. Biol. **157**, 316–322 (1963).

Belehradek, J.: Temperature and Living Matter. Berlin: Bornträger 1935.

Christophersen, J., Precht, H.: Untersuchungen über die Temperaturabhängigkeit von Lebensprozessen bei Hefen. Biol. Zentr. **69**, 300–323 (1950).

Grainger, J. N. R.: The relation between heat production, oxygen consumption and temperature in some poikilotherms. In: (Ed.) Locker, A.: Quantitative Biology of Metabolism, 3rd International Symposium, p. 86–90. Berlin-Heidelberg-New York: Springer 1968.

Ingram, M.: An introduction to the biology of yeasts. London: Pitman 1955.

Morowitz, H. J.: Some order-disorder considerations in living systems. Bull. Math. Biophys. **17**, 81–86 (1955).

Morowitz, H. J.: Energy flow in biology. New York: Academic Press 1968.

Ohlmeyer, P., Fritz, U.: Gärungs- und Atmungswärme bei anaerober und aerober Vermehrung von Hefen. Z. Naturforsch. **21b**, 175–180 (1966).

Precht, H., Christophersen, J., Hensel, H.: Temperatur und Leben. Berlin-Göttingen-Heidelberg: Springer 1955.

Rubner, M.: Energieverbrauch im Leben der Mikroorganismen. Arch. Hyg. Bacteriol. **48**, 260–311 (1904).

Life Table Data and Population Dynamics of the Rotifer *Brachionus calyciflorus* Pallas as Influenced by Periodically Oscillating Temperature

Udo Halbach

With 8 Figures

Introduction

The life table data and the population dynamics of *Brachionus calyciflorus* are influenced by a number of abiotic and biotic factors (Halbach, 1969a, b, 1970, 1972). Under constant environmental conditions in a limited system (constant volume and constant daily dose of food) rotifers have very characteristic population dynamics as demonstrated schematically in Fig. 1. After a phase of exponential growth, oscillations are observed around an average density, which is determined by food quantity. The curve is characterized by three parameters: the intrinsic rate of natural increase *r*, the carrying capacity *K* and the duration of one period λ, which is the reciprocal of the frequency. The oscillations are caused by the time lag between food uptake and the resulting production of offspring. The lag leads to primary overshoot and subsequent oscillations. Amplitudes and frequencies of these population curves are drastically affected by temperature (Fig. 2): the fluctuations are more pronounced at high than at low temperatures. The carrying capacity decreases with increasing temperature, whereas the intrinsic rate of increase as well as the frequency of oscillations grow with increasing temperature (Table 1). These differences in the population dynamics are mainly due to changes in the duration of life and in fertility (Fig. 3): both life expectancy and immaturation time are reduced at increased temperatures. Fertility increases with increasing temperature, but the duration of the fertile phase becomes shorter.

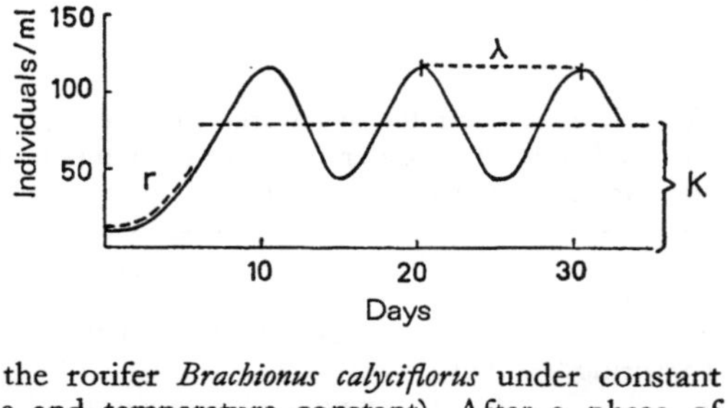

Fig. 1. Typical population dynamics of the rotifer *Brachionus calyciflorus* under constant conditions (volume, ration, light regime and temperature constant). After a phase of exponential growth, oscillations about a mean population density are observed. The curve is characterized by three parameters: the intrinsic rate of natural increase (*r*), the carrying capacity (*K*) and the wave-length of the oscillations (λ). λ is a function of the time lag *T*

In the field, too, there are fluctuations in population density, but they are less regular than in the experiments. Therefore amplitude and frequency are not useful quantities for describing the changes. The absolute exponential growth rate has proven to be a better and more general measure for characterizing the severity of the fluctuations. This parameter increases with increasing temperature, but at all temperatures the fluctuations are much larger in the field than in the laboratory (Fig. 4).

One of the major differences between laboratory and natural conditions is the daily temperature fluctuation in the ponds. There is much evidence for an ecological significance of fluctuating temperatures (e.g. Precht *et al.*, 1955; Hagstrum and Hagstrum, 1971). It might be that the more severe population fluctuations in the field are due to the diurnal temperature changes. Fig. 5 gives an example of the periodicity of the temperature in one of the ponds under study. To test this hypothesis life table data and population dynamics of animals kept under changing temperature conditions were compared with data of "control" animals at an intermediate constant temperature.

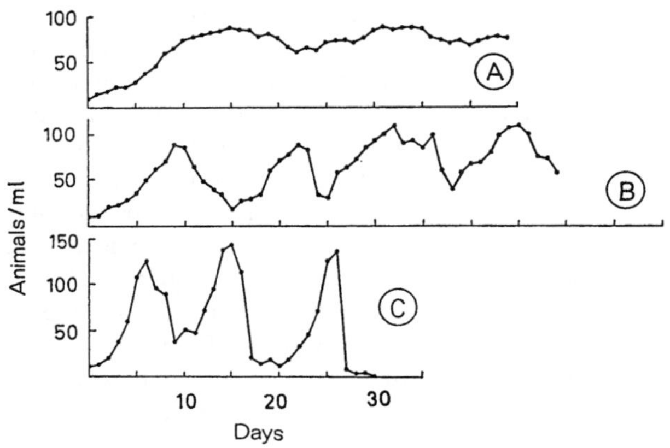

Fig. 2 A–C. Experimental population dynamics of the rotifer *Brachionus calyciflorus* at three different constant temperatures: 15 C (A), 20 C (B), and 25 C (C). With increasing temperature the fluctuations become stronger

Table 1. Intrinsic rates of natural increase (r), capacities (K), and frequencies of the oscillations (f) at three different constant temperatures (15, 20, and 25 C). The error is the standard error of the mean ($S_{\bar{x}}$)

	15 C	20 C	25 C
r (offspring/♀ × hr)	0.34 ± 0.06	0.48 ± 0.07	0.82 ± 0.07
K (animals/ml)	74.7	66.8	57.3
f (oscillations/day)	0.050	0.088	0.118

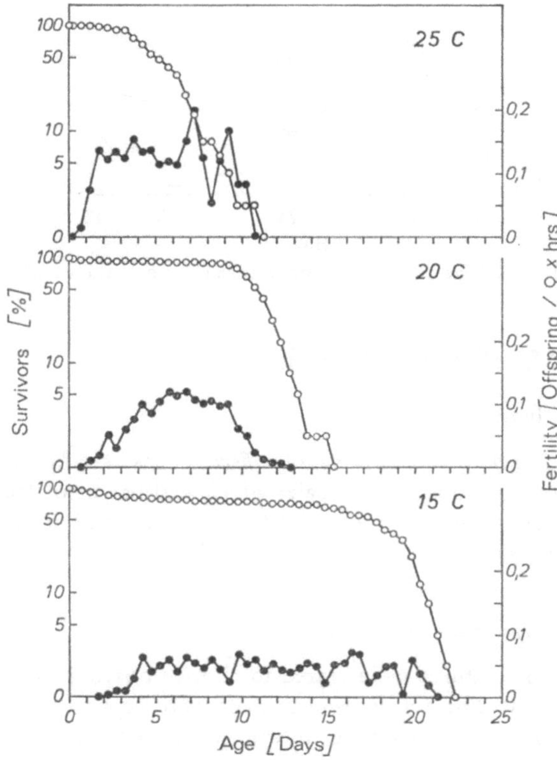

Fig. 3. Life table data of *Brachionus calyciflorus* at three different constant temperatures (15, 20 and 25 C). White symbols: survivorship curves; black symbols: fertility curves

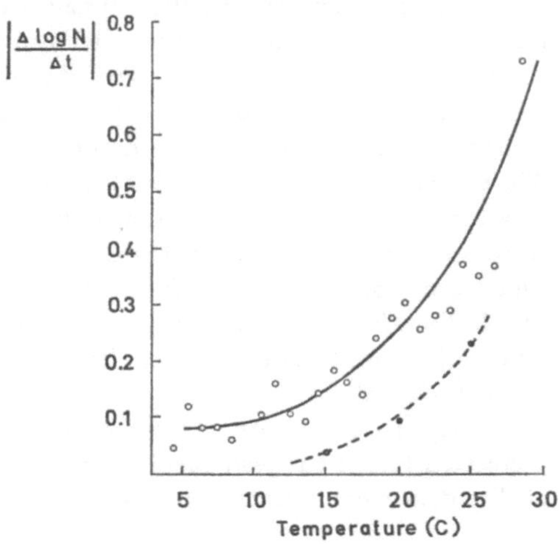

Fig. 4. The absolute exponential growth rate $|\Delta \log N/\Delta t|$ (N = population density, t = time) as a measure of the severity of density fluctuations at different temperatures. Each symbol represents the average of one temperature class (range 1 C). The field data (white symbols, solid line) are always higher than the corresponding experimental values (black symbols, broken line). The curves are fitted by eye

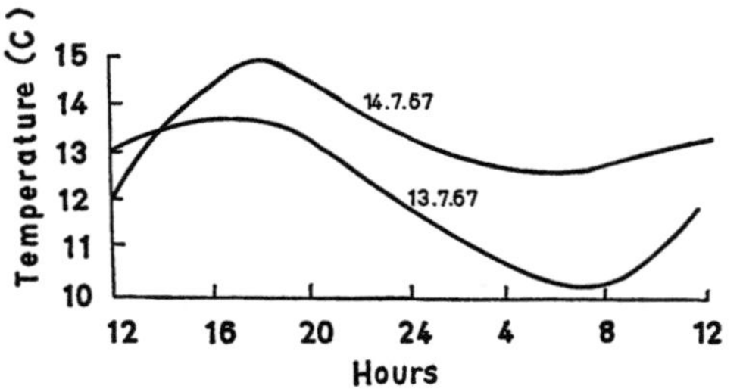

Fig. 5. The course of temperature in one of the ponds under study on two consecutive days

Material and Methods

Animals

The clone of *Brachionus calyciflorus* Pallas used in this study was derived from an amictic (parthenogenetic) female, which had been isolated from a fishpond near Würzburg (Germany) in January 1970. Under the experimental conditions the bisexual resting eggs do not emerge. Therefore the cultures were probably genetically homogeneous.

Experimental Conditions

The cultures were reared in climate boxes in the dark:

a) The *experimental* animals were kept in an incubator with a programmed sine-shaped diurnal temperature curve. The temperature ranged from 15 C to 25 C, with a mean of 20 C.

b) The *control* animals were reared in a constant temperature room at 20 $\pm$ 1 C.

Before each experiment the animals were allowed to acclimatize to the special temperature conditions for about 10 days.

Boiled and filtered tap water served as medium. Culture vessels of 100 ml volume were used for the population experiments. In order to determine life table data, isolated animals were reared in depression slides of 1 ml volume in wet chambers. The unicellular green alga *Kirchneriella lunaris* served as food; it was grown aseptically in a synthetic anorganic medium with a light-dark-cycle of 16:8 hours. The algae were used only during the rather dilute exponential growth phase. They were centrifuged and redispersed in rotifer medium. The algal density was determined nephelometrically with a spectrophotometer; the quantitative dosage was made with an automatic constriction pipette. Every 12 hours the rotifer cultures received 1×10^6 cells/ml.

A population census was made every 24 hours: using hair pipettes all animals were transferred to fresh medium containing 10^6 algae/ml. During this procedure all life data were counted or measured.

Simulations

In order to simulate the population dynamics, the logistic growth function with simple time lag (Halbach and Burkhardt, 1972) was used. The simulation language was DSL/90 with a special extension for using variable time lags (Burkhardt, in preparation). The calculations were made on the IBM 7094 of the Deutsches Rechenzentrum Darmstadt. Numerical integrations were made by using Runge-Kutta methods with variable strides.

Results

Experiments

Life Table Data

Comparisons of the survivorship and the fertility curves are given in Fig. 6. The average life durations are 10.6 ± 0.3 days for constant and 12.0 ± 0.4 days for changing temperatures (difference significant with $p < 0.01$).

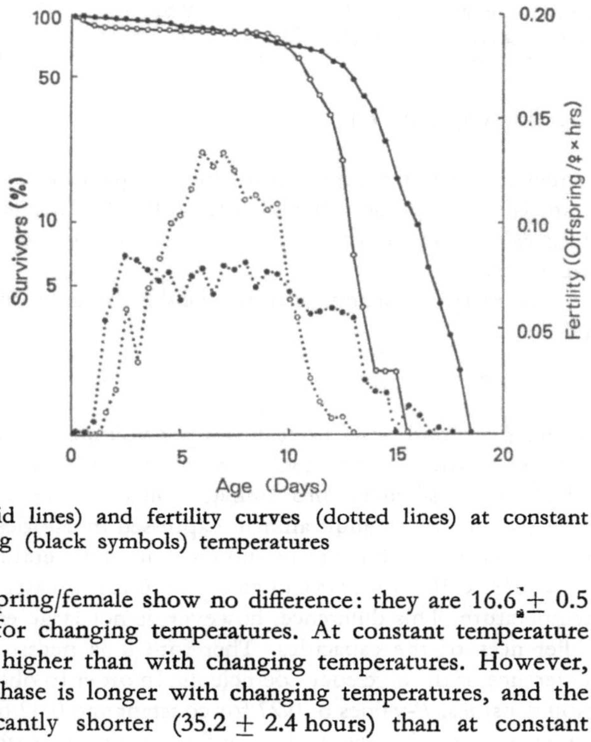

Fig. 6. Survivorship curves (solid lines) and fertility curves (dotted lines) at constant (white symbols) and at changing (black symbols) temperatures

The average values of offspring/female show no difference: they are 16.6 ± 0.5 for constant and 16.7 ± 0.8 for changing temperatures. At constant temperature the fertility rate is obviously higher than with changing temperatures. However, the duration of the fertility phase is longer with changing temperatures, and the immaturation time is significantly shorter (35.2 ± 2.4 hours) than at constant temperature (51.9 ± 0.8 hours).

Intrinsic Rate of Natural Increase (r)

From the survivorship and fertility curves it is possible to construct the intrinsic rate of increase from a number of models. In our case the Edmondson model was used (Edmondson, 1968; Halbach, 1970). In Table 2a a comparison is made between the r-values obtained from the model and the r-values empirically determined by the culture of unlimited populations. The agreement between model and empirical values is fairly good. In both cases the intrinsic rates of increase are higher with changing temperatures. It can be shown that this is almost completely due to the shorter immaturation time.

Table 2. Intrinsic rates of natural increase r under constant temperature (20 C) and changing temperatures (15–25 C). Comparison of the empirical values and the values calculated from the life table data by the Edmondson model. The differences between the empirical and the calculated values are not significant. The difference between the empirical values for constant and for changing temperatures, however, is significant ($p < 0.02$). The error is the standard error of the mean

	intrinsic rates of increase r	
	calculated values (Edmondson model)	empirical values
constant temperature (20 C)	0.55	0.48 ± 0.07
changing temperatures (15–25 C)	0.63	0.60 ± 0.06

Carrying Capacity (K)

Under changing temperatures the mean population density at equilibrium (88.5 ± 4.4 animals/ml) is significantly higher ($p \ll 0.01$) than at constant temperature (66.8 ± 5.2 animals/ml). These values may be assumed to represent the respective carrying capacities.

The carrying capacity may also be determined formally by the equation (Halbach 1970):

$$K = \frac{F \cdot P \cdot L}{B},$$

where F = food dose, i.e. calories per volume and day, put into the system in the form of nutrients; P = coefficient of food utilization, i.e. the percentage of food which is transformed into animal biomass; L = mean duration of life in days; B = biomass of an adult animal, expressed in calories. In our experiments F and B were equal for both constant and changing temperatures ($F = 4.125 \times 10^{-2}$ cal/ml; $B = 1.78 \times 10^{-3}$ cal). As shown in Fig. 6 L is greater for changing than for constant temperature. This difference, however, is not large enough to explain the observed differences of the capacities. Therefore it is necessary to postulate an additional difference of the efficiency coefficients. In order to obtain the observed mean population densities, P-values of 0.27 for constant and 0.32 for changing temperatures have to be assumed. It appears that food utilization is better under changing temperature conditions.

Wave-length of the Oscillations (λ)

Fig. 7 demonstrates the population dynamics for constant (the parallels A, B, C) and changing temperatures (the parallels D, E, F). The average duration of one period is 11.4 days for constant and 10.0 days for changing temperatures (see Halbach and Burkhardt, 1972). This difference does not seem to be highly significant.

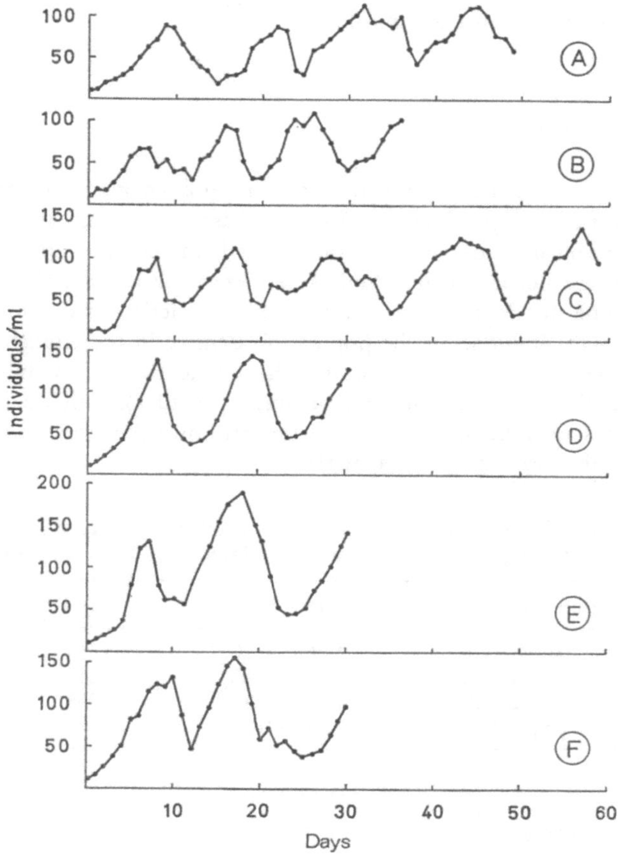

Fig. 7 A–F. Experimental population dynamics at constant (A, B, C) and changing (D, E, F) temperatures

Mean Absolute Exponential Growth Rate ($|\Delta \log N/\Delta t|$)

In the field the fluctuations of the population density are much greater but not as regular as in the laboratory experiments (see Halbach, 1970, 1972). Therefore amplitudes and frequencies are no adequate measurements for characterizing the fluctuations. A more general measurement of the degree of the fluctuations is $|\Delta \log N/\Delta t|$, the mean absolute exponential growth rate, where $N =$ population density and $t =$ time. Applying this expression to the experiments, it is found that the temperature oscillations lead to more severe fluctuations of the population density: with changing temperatures, the values were significantly ($p < 0.01$) higher (0.126 $\pm$ 0.012) than at constant temperature (0.103 $\pm$ 0.012). However, these values still lie far below the field data (0.287 ± 0.069 is the average value of the class between 19 and 21 C; see Fig. 4). Therefore the enormous fluctuations in the field cannot be explained by temperature changes alone. Other factors such as competition and predation must play a role!

Mictic Ratio

Monogonont rotifers normally reproduce parthenogenetically with diploid eggs (amictic reproduction). In most cases, however, there is a certain percentage of bisexual animals in the population which produce haploid eggs (mictic reproduction); these become males (unless fertilized) or thick-shelled resting eggs (if fertilized). The mictic ratio (the percentage of mictic females in the population) could be a factor influencing the population density in the equilibrium (Seitz and Halbach, in preparation). Table 3 shows the mictic ratio with different constant temperatures and with a changing temperature (15–20 C). Obviously the changing temperature has no influence on the mictic ratio and the latter can be excluded as a cause of differential carrying capacities.

Table 3. Mictic rate at different constant temperatures and under changing temperatures (15–20 C) (unpublished data by H. Rauscher). The error is the standard error of the mean

Temperature	Mictic rate (%)	*n* (number of samples)
constant (15 C)	4.56 ± 0.20	1314
constant (17.5 C)	2.79 ± 0.28	249
constant (20 C)	1.67 ± 0.18	355
changing (15–20 C)	3.21 ± 0.39	189

Simulations

Using life table data it is possible to simulate the population dynamics with computers. For our purpose, the logistic growth function with simple time lag was used as simulation model (see Halbach and Burkhardt, 1972). The three curves at the top of Fig. 8 show simulations of populations at constant temperatures: 15 C (*A*), 25 C (*B*), and 20 C (*C*). The agreement with the experimental populations is fairly good. The curves *D*, *E*, and *F* of Fig. 8 are all simulations of populations exposed to changing temperatures (the temperature course is shown in curve *G*) employing different simulation techniques: In curve *D* the three parameters (intrinsic rate of increase *r*, carrying capacity *K* and time lag *T*) oscillated exactly according to the temperature oscillations (curve *G*). The parameters were intrapolated at each integration step. In curve *E* only *r* and *K* oscillated, whereas the time lag *T* was kept constant (the integrated mean value was taken). In *F* all three parameters were kept constant using the empirical values of the experiments with changing temperatures. Simulation *F* shows by far the best resemblance to the experimental curves with regard to amplitudes and frequencies of oscillation and – most important –

with the capacity. This means that the higher capacity with changing temperatures cannot be explained by the oscillations of the parameters in accordance with the temperature course. Rather, it must be assumed that with changing temperatures there are some stimulating effects which lead to a better utilization of food.

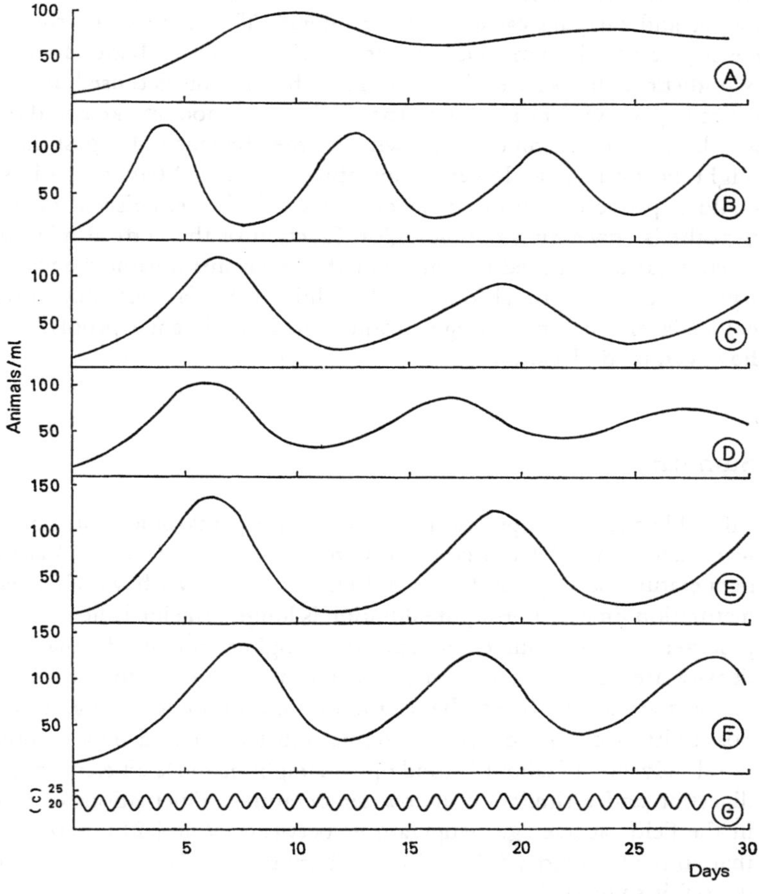

Fig. 8 A–G. Computer simulations of the experimental population dynamics using the logistic growth curve with simple time lag:

$$\frac{dN_{(t)}}{dt} = r \cdot N_{(t)} \frac{K - N_{(t-T)}}{K}$$

(K = carrying capacity, N = population density, r = intrinsic rate of natural increase, t = time, T = time lag). A, 15 C constant temperature ($r = 0.35$, $K = 75.0$, $T = 3.0$); B, 25 C constant temperature ($r = 0.82$, $K = 57.3$, $T = 1.9$); C, 20 C constant temperature ($r = 0.5$, $K = 55.0$, $T = 3.0$); D, changing temperatures (r, K, and T oscillate according to curve G; intrapolations were made at each integration step); E, changing temperatures (only r and K oscillate according to curve G, T is constant: 3.0); F, changing temperatures (empirically determined constant parameters, $r = 0.63$, $K = 89.0$, $T = 3.0$); G, sine-shaped course of temperature during experiments with changing temperatures (D, E, and F)

Discussion

In spite of the fact that the mean temperature is the same with constant and with changing temperatures, significant differences in the life data and in the population dynamics have been observed. With changing temperatures immaturation time was shorter (and as a consequence the intrinsic rate of increase higher), life duration was longer and carrying capacity was greater)*. The increase of the mean population density at equilibrium must at least partly be due to better food utilization. One hypothetical but likely explanation might be that this is caused by periodical changes of feeding activity. For instance, the efficiency of food uptake could have its optimum at relatively high temperatures, whereas the efficiency of digestion and assimilation might be optimal at lower temperatures. Such a "temporal division of labour" could improve the mean efficiency and would be reminiscent of McLaren's (1963) hypothesis concerning the biological function of the vertical migration of plankton organisms; it is argued that at night the organisms inhabit the warmer upper water layers to ensure optimal food uptake while in the daytime they move to deeper and colder layers for optimal egg production. Experimental proof for this hypothesis, however, is still lacking.

Summary

Life table data and population dynamics of the planktonic rotifer *Brachionus calyciflorus* were determined (a) at a constant temperature of 20 C and (b) at daily sine-shaped temperature oscillations between 15 C and 25 C. With variable temperatures, immaturation time is shorter, life duration is longer, the intrinsic rate of natural increase is larger, the environmental capacity is higher and the fluctuations of population density are more severe than at an intermediate constant temperature. These differences are mainly due to better food utilization under changing temperatures. The possibility is discussed that this higher efficiency could be accomplished by a temporal "division of labour": food uptake might be more efficient at high temperatures, digestion and assimilation at low temperatures. Fluctuations of population density in the field (at average temperatures between 19 and 21 C) are much more severe than in the laboratory (15–25 C). The importance of biotic factors (competition, predation) is suggested.

Acknowledgements

The values of the mixis rate were kindly supplied by H. Rauscher, Univ. of Munich (unpublished data). The calculations for the population simulations were carried

* A reduction of developmental time and a prolongation of life and thus of the intrinsic rate of increase r by changing temperatures has been demonstrated often with insects (Headlee, 1914; Peairs, 1927; Shelford, 1927; Parker, 1930; Gösswald, 1936; Herfs, 1936; Buchanan, 1940; Johnson, 1940; Stellwaag, 1940; Maksomovit, 1958; Strong and Sheldahl, 1970).

out by H. J. Burkhardt, Deutsches Rechenzentrum Darmstadt. I wish to thank Prof. A. Duncan, Univ. of London, and Prof. A. Ruttner-Kolisko, Biologische Station Lunz, Austria, for helpful discussions. For reading and improving the manuscript, the help of Prof. J. Jacobs, Univ. of Munich and of Prof. W. Wieser, Univ. of Innsbruck, is gratefully acknowledged.

References

Buchanan, J. W.: Developmental rate and alternating temperatures. J. Exptl. Zool. **83**, 235–248 (1940).

Edmondson, W. T.: A graphical model for evaluating the use of the egg ratio for measuring birth and death rates. Oecologia **1**, 1–37 (1968).

Gösswald, K.: Physiologische Untersuchungen über die Einwirkung ökologischer Faktoren, besonders der Temperatur und Luftfeuchtigkeit, auf die Entwicklung von *Diprion (Lophyrus) pini* L. zur Feststellung der Ursachen des Massenwechsels. Z. Angew. Entomol. **22**, 331–384 (1936).

Hagstrum, D. W., Hagstrum, W. R.: A simple device for producing fluctuating temperatures, with an evaluation of the ecological significance of fluctuating temperatures. Ann. Entomol. Soc. Am. **63**, 1385–1389 (1971).

Halbach, U.: Das Zusammenwirken von Konkurrenz und Räuber-Beute-Beziehungen bei Rädertieren. Zool. Anz., Suppl. **33**, 72–79 (1969a).

Halbach, U.: Räuber und ihre Beute: Der Anpassungswert von Dornen bei Rädertieren. Naturwiss. **56**, 142–143 (1969b).

Halbach, U.: Einfluß der Temperatur auf die Populationsdynamik des planktischen Rädertieres *Brachionus calyciflorus* Pallas. Oecologia **4**, 176–207 (1970).

Halbach, U.: Zum Adaptivwert der zyklomorphen Dornenbildung von *Brachionus calyciflorus* Pallas (Rotatoria). I. Räuber-Beute-Beziehung in Kurzzeit-Versuchen. Oecologia **6**, 267–288 (1971).

Halbach, U.: Einfluß der Nahrungsqualität und -quantität auf die Populationsdynamik des planktischen Rädertieres *Brachionus calyciflorus* im Labor und im Freiland. Verhandl. Deut. Zool. Ges. **65**, 83–88 (1972).

Halbach, U., Burkhardt, H. J.: Sind einfache Zeitverzögerungen die Ursachen für periodische Populationsschwankungen? Vergleich experimenteller Untersuchungen an *Brachionus calyciflorus* Pallas (Rotatoria) mit Computer-Simulationen. Oecologia **9**, 215–222 (1972).

Headlee, T. J.: Some data on the effect of temperature and moisture on the rate of insect metabolism. J. Econ. Entomol. **7**, 413–417 (1914).

Herfs, A.: Ökologisch-physiologische Studien an *Anthrenus fasciatus* Herbst. Zoologica **34**, 90 (1936).

Johnson, C. G.: Development, hatching, and mortality of the eggs of *Cimex lectularius* L. (Hemiptera) in relation to climate, with observations on the effect of preconditioning to temperature. Parasitology **32**, 127–173 (1940).

Maksomović, M.: Experimental researches on the influence of temperature upon the development and the dynamics of populations of the Gypsy Moth (*Liparis dispar* L.). Biol. Inst. Pos. izd. **3**, Beograd (1958).

McLaren, I. A.: Effects of temperature on growth of zooplankton, and the adaptive value of vertical migration. J. Fisheries Res. Board Can. **20**, 685–727 (1963).

Parker, J. R.: Some effects of temperature and moisture upon *Melanoplus mexicanus* Saussure and *Camnula pellucida* Scudder (Orthoptera). Bull. Univ. Montana Agr. Expt. Sta. **223**, 132 (1930).

Peairs, L. M.: Some phases of the relation of temperature to the development of insects. Bull. West. Va. Agr. Expt. Sta. **208**, 1–62 (1927).

Precht, H., Christophersen, J., Hensel, H.: Temperatur und Leben. Berlin-Göttingen-Heidelberg: Springer 1955.

Shelford, V. E.: An experimental investigation of the relations of the Codling Moth to weather and climate. Bull. Illinois Nat. Hist. Surv. **16**, 307–440 (1927).

Stellwaag, F.: Die Einwirkung schwankender Freilandtemperaturen auf Insekten. Anz. Schädlingskunde **16**, 109 (1940).

Strong, F. E., Sheldahl, J. A.: The influence of temperature on longevity and fecundity in the bug *Lygus hesperus* (Hemiptera: Miridae). Ann. Entomol. Soc. Am. **63**, 1509–1515 (1970).

The Role of Daily Temperature Rhythms in Brood Development of Ants of the Tribe Leptothoracini (Hymenoptera; Formicidae)

ALFRED BUSCHINGER

With 3 Figures

A number of species of ants belonging to the tribe Leptothoracini, subfamily Myrmicinae, provide good subjects for laboratory experiments on behaviour and ecology. The populations are small, consisting as a rule of approximately 200–500 adult individuals and about the same number of larvae. Biological and ethological peculiarities are much more difficult to investigate in species with larger populations. The Leptothoracini, for example, include species exhibiting various forms of monogyny and polygyny (Buschinger, 1967, 1968), workerless social parasites (Kutter, 1969; Buschinger, 1965, 1971), slave keeping species and guest ants.

Most Leptothoracini nest in exposed places such as superficial cracks in rocks, in tree bark, tree stumps and twigs lying on the ground. Only a few species inhabit the denser deciduous forests. Nests are not left in strong sunshine in summer nor in severe frosts in winter.

In the course of many years of breeding and keeping these insects it has been observed that the independent species *Leptothorax acervorum* (Fabr.), *L. muscorum* (Nyl.), *L. gredleri* (Mayr), the workerless social parasites *L. kutteri* Buschinger and *Doronomyrex pacis* Kutter, as well as the slave keeping species *Harpagoxenus sublaevis* (Nyl.) not only tolerate the daily temperature variations occurring at their nesting sites (Fig. 1) but even require them for certain phases of larval development.

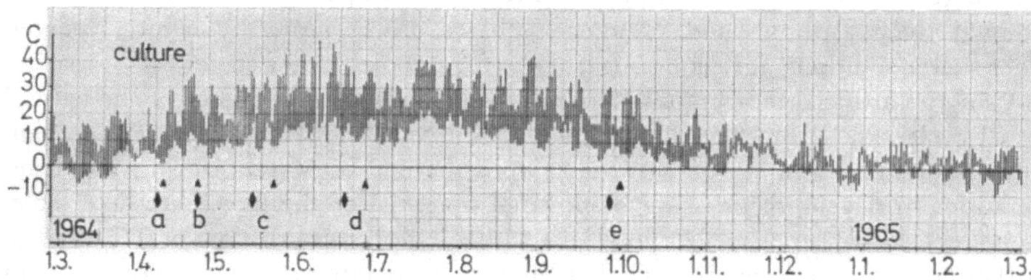

Fig. 1. Daily temperature variations during the course of one year in a simulated nest in the open. The different stages of development of *Leptothorax acervorum* (*a* feeding of larvae after winter; *b* shrinkage of larvae); *c* first pupa; *d* hatching of sexuals; *e* shrinkage of larvae) occur more or less simultaneously in the artificial and natural nests. The annual amplitude tolerated by the ants is 58 C (—10 to +48 C), the maximum daily variations were 33 C (+15 to +48 C)

The larvae of workers and of males overwinter once, those of queens usually twice. In constant or only slightly varying temperatures in the laboratory (± 2 or

3 C) the larvae die upon reaching the final prepupal stage. If, at this point, a daily temperature rhythm of about 8h at 0 C to 16 h at +20 C is offered, all larval stages present shrinkage whithin 6–8 days and take on the appearance of outdoor populations in winter (Fig. 2). Constant low temperatures or photoperiods do not have this effect.

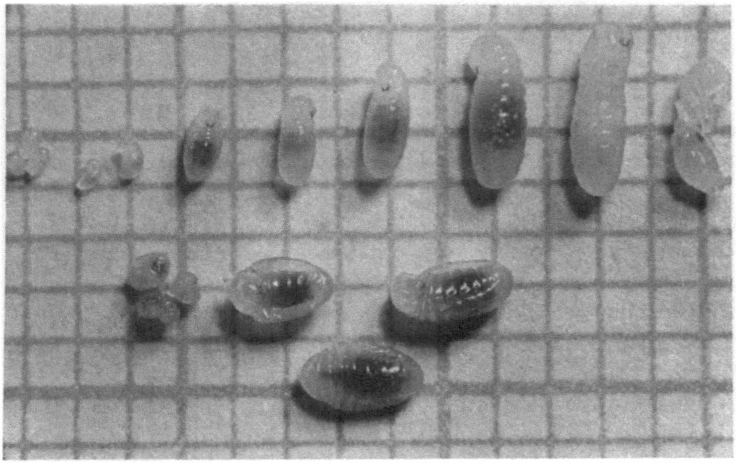

Fig. 2. Larval development of *Leptothorax acervorum*. Upper row: eggs, various larval stages, pre-pupa worker pupa. Below: shrunk winter larvae. The pre-pupal stage is only achieved after a phase of shrinking followed by a phase of feeding

Following shrinkage, the larvae can be induced to feed at laboratory temperatures (20–24 C) or under a rhythm of about +10 to +24 C and later +20 to +35 C, and the grown larvae then form pupae. Although a temperature rhythm is advantageous at these "spring" and "summer" phases, lower mortality being observed under such conditions, it is not obligatory as for the pre-winter stage.

Using an artificial temperature programme (Fig. 3), certain phases of the natural annual cycle can be drastically shortened. The five months of winter in the open, from about the beginning of November until the end of March, during which no alterations in development or developmental potential of the larvae are detectable, can be replaced by a period of two weeks with a temperature rhythm of 0 C to 20 C. Furthermore, the long periods of relatively low temperatures occurring outdoors in spring and autumn can also be reduced. Thus the entire breeding cycle which, in the open, is completed once each year, can be condensed into 3–4 months (Fig. 3). Using the above-mentioned species and the programme described it was possible to rear up to four sexual broods within one year (Buschinger, 1972).

These observations do not hold true for all Leptothoracini: Plateaux (1968, 1970), using *Leptothorax nylanderi* (Foerst.), only succeeded in producing males, queens and workers from larve following winter conditions lasting at least three, and usually five, months at constant +5 to +8 C. *L. nylanderi* is a species choosing less exposed nesting sites.

The optimum daily temperature regimes at various phases of the breeding and annual cycles have not yet been investigated in our experiments. If it is borne in mind that in Europe *L. acervorum*, for example, is found from sea level up to more than 3000 m, a certain spectrum of variation is to be expected from species to species and among the individual populations.

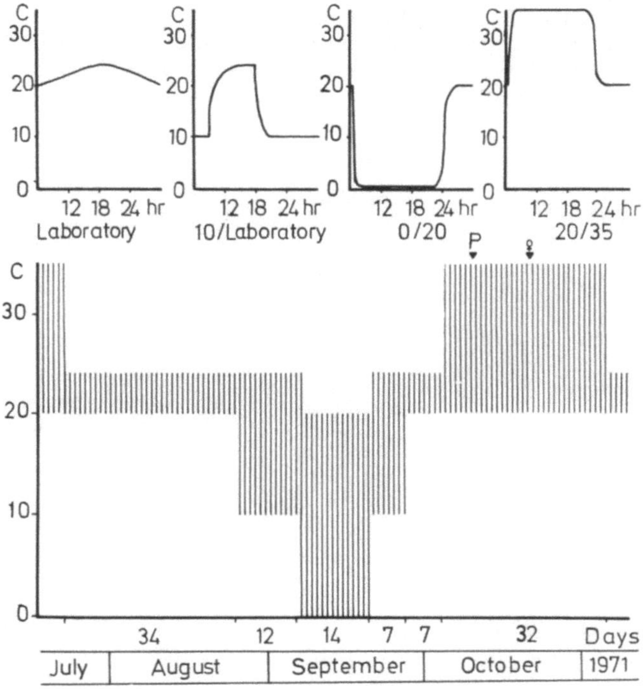

Fig. 3. Temperature programme for breeding experimental animals. Below: Daily temperature variations in an artificially shortened annual cycle. The arrows "*P*" and "♀" mark the appearance of the first female pupa and imago in the "summer period" of any one brood. Above: Daily temperature variations in the various "artificial seasons". The laboratory temperature varies between 20 and 24 C. To prepare for winter the populations are placed in a refrigerator at +10 C for 16h daily (10/laboratory). The "winter" is spend in a refrigerator at 0 C, which is switched off for 8h daily during which time room temperature is attained (0/20 C). The summer rhythm is achieved using a thermostat at +35 C, also switched off for 8h daily so that room temperature is reached (20/35 C). (From Buschinger 1972)

References

Buschinger, A.: *Leptothorax (Mychothorax) kutteri* n. sp., eine sozialparasitische Ameise (Hymenoptera, Formicidae). Insectes Sociaux **12**, 327–334 (1965).

Buschinger, A.: Verbreitung und Auswirkungen von Mono- und Polygynie bei Arten der Gattung *Leptothorax* Mayr (Hymenoptera, Formicidae). Inaugural-Dissertation, Würzburg (1967).

Buschinger, A.: Mono- und Polygynie bei Arten der Gattung *Leptothorax* Mayr (Hymenoptera, Formicidae). Insectes Sociaux **15**, 217–226 (1968).

Buschinger, A.: Kreuzung zweier sozialparasitischer Ameisenarten *Doronomyrmex pacis* Kutter und *Leptothorax kutteri* Buschinger (Hym., Formicidae). Zool. Anz. **189**, 169–179 (1972).

Kutter, H.: Die sozialparasitischen Ameisen der Schweiz. Neujahrsbl. Naturf. Ges. Zürich, p. 62 (1969).

Plateaux, L.: Sur le polymorphisme social de la fourmi *Leptothorax nylanderi* (Foerster). Thèse, Fac. Sci., Paris (1968).

Plateaux, L.: Sur le polymorphisme social de la fourmi *Leptothorax nylanderi* (Foerster). I. Morphologie et Biologie comparées des castes. Ann. Sci. Nat., Zool., Paris. 12. Sér. **12**, 373–478 (1970).

The Role of Temperature in Controlling Obligatory Diapause

H.-J. BRAUNE

With 2 Figures

Periods of arrested growth and development which are frequent in the ontogenesis of insect species inhabiting areas with fluctuating ecological conditions are usually termed collectively *dormancy*. The dormant state is recognized as an adaptive mechanism enabling species to survive seasons of adverse environmental conditions such as cold, heat, or drought. Dormancy may interrupt embryonic, larval, pupal or adult reproductive stages. Many authors, especially Andrewartha (1952), Lees (1955), Müller (1970) and Mansingh (1971) have pointed out the ecological aspect of dormancy in synchronizing the life cycle with seasonal fluctuations of climate so as to ensure an abundance of food for all active stages.

The spectrum of dormancy depends upon the nature and extent of deviation of environmental factor(s) from the optimum (Mansingh, 1971). Because of the different mechanisms involved a distinction is usually made between *quiescence* and *diapause*. However, certain cases cannot be fitted into the narrow framework of these two terms and many attempts have been made by ecologists and physiologists to introduce a more exact terminology and classify the different mechanisms of the dormant state. Adequate reviews are given by Müller (1970), Mansingh (1971) and Laudien (in Precht *et al.*, 1973).

Quiescence is regarded as the simplest and primary mechanism of dormancy. It is defined as the direct result of deviation of an environmental factor from its optimum. The dormant state begins and ends without a time lag. Most individuals of a population withstand short-term adversity by means of their natural tolerance. The development of a special biochemical adaptation is unnecessary. This kind of dormancy can occur at any time of the year in any or all of the embryonic, larval, pupal, or adult stages of the life cycle. Since arrest of growth is caused directly by an unfavourable environment, Müller (1970) proposes the term *Konsekutive Dormanz*. Quiescence is a very common mechanism in insects inhabiting geographical zones with only slight and short-term fluctuations of ecological factors from the optimum.

On the other hand, long-term deviation of ecological factors from the optimum occur in geographical areas with cyclic seasonal changes such as winter and summer. Insects of such habitats do not experience quiescence but long-term dormancies which ensure that the most cold-hardy developmental stages occur during winter, and the active stages under optimum ecological conditions. The dormant stage is genetically determined and therefore characteristic for a species. Since these long-term dormancies set in long before the occurrence of adverse environmental conditions, Müller (1970) proposes the term *Prospective Dormanz*.

Diapause has been defined by Beck (1968) as a genetically determined state of suppressed development, the manifestation of which may be induced by environmental factors. The distinction between simple cold torpor, like quiescence, and diapause is based on the relative speed of their termination. Quiescence is terminated immediately upon adverse environmental conditions coming to an end, whereas in diapause there is a period of prolonged developmental arrest, which cannot be terminated at all stages of the dormant state. Once diapause has set in, specific physiological processes must occur and be completed before a resumption of development is possible. Diapause starts long before conditions become adverse and is maintained for some time whatever the state of the environment. The most important ecological advantage of diapause is that it renders insects less dependent on the influence of adverse environmental conditions.

Diapause is either facultative or obligatory. The former case holds for many polyvoltine insects which diapause only in the last generation before winter. Sensitive stages in early development perceive the oncoming seasonal changes as changes in photoperiod. Short day often serves as an ecological stimulus and induces diapause in hibernating insects. If day length falls below a critical value and thus becomes a diapause inducing factor, dormancy occurs in a later developmental stage. Species of wide distribution with facultative diapause can produce varying numbers of generations per year dependent upon climatic region. In certain habitats they are univoltine and thus appear to have an obligatory diapause, but their dormancy must be called facultative since they are capable of continuous development if day length is maintained above the critical value either experimentally or under natural conditions.

On the other hand, there are insects with obligatory diapause, the initiation of which is independent of external factors. Such species are always univoltine or have a 2-year life cycle, and diapause in every generation at one and the same genetically determined development stage. In this case dormancy cannot be prevented either in the laboratory or under natural conditions.

In most reviews on dormancy cases of obligatory diapause are insufficiently considered, the existence of such phenomena being either neglected or questioned. Recently, however, some striking results have proved the existence of obligatory diapause. The aim of this paper is to emphasize such instances of diapause as occur irrespectively of external factors. Only the duration and termination of obligatory periods of arrested growth are controlled and influenced by temperature.

Three distinctly different mechanisms of obligatory diapause will be discussed.

The first type is exemplified by *Mitopus morio* which is a singlebrooded opilionid and hibernates in the egg stage (Tischler, 1967). The egg deposition of this species takes place at the end of July. Embryogenesis proceeds at summer temperatures until at a genetically determined developmental stage, development is arrested. Hatching of the nymphs occurs in the following spring. The experiments of Tischler show that dormancy in *Mitopus morio* is an obligatory diapause, based on a genetically determined alteration of temperature requirements in successive phases during embryonic development. Three phases can be distinguished. The developmental processes of the first and the last phase can take place only at high temperatures, whereas the middle one demands a significantly lower range of temperature; making it impossible to rear *Mitopus morio* in the laboratory at constant temperatures. With respect to the conditions in the natural habitat diapause in this species may be interpreted as follows: Dormancy is initiated under the high temperatures of summer which inhibit developmental processes of the second embryonic phase. Development remains arrested until the low temperatures of winter gain influence and set off the second phase of embryogenesis. These low winter temperatures cause a further period of arrested growth due to the fact that the last embryonic phase, too, requires a higher temperature. Thus in the natural habitat embryogenesis can be completed only when the temperature increases in the spring. For this type of obligatory diapause Müller (1970) introduced the term *Parapause*. The mechanism discussed prevents untimely hatching in the autumn and winter, and ensures the eggs' hatching in the spring, when adequate food is available for the nymphs.

Still other mechanisms of obligatory diapause have evolved which are equally efficient in synchronizing the insect's life cycle with seasonal fluctuations of environmental conditions. A further way in which temperature may control obligatory diapause is demonstrated in *Gryllulus mitratus* (Masaki, 1962). This species has a wide range of distribution and lives in a variety of climatic regions. Nevertheless, the species always retains its univoltine life cycle, hibernating in the egg stage. Diapause is obligatory and genetically fixed, but the intensity or duration of diapause and the range of temperature required for terminating dormancy depends on the temperature which the animal experienced during early embryogenesis.

High temperatures during this early embryonic period intensify diapause, whereas lower ones cause only a weak diapause. During the long reproductive season of this cricket, which lasts in Japan from the end of July to the middle of October, temperature usually attains its annual maximum at the beginning and declines towards the end. It follows that eggs exposed to warm weather, i.e. eggs laid in the early season, have a relatively deep diapause and require lower temperatures for its termination. This guarantees that the eggs remain in the resistant state until the following spring, and untimely hatching is thus prevented. The weaker diapause in the later season, however, enables the eggs to hatch as soon as adequate environmental conditions prevail in spring.

There is, finally one other way of preventing inappropriate timing of hatching. This is realized in hibernating eggs of the aphid *Aphis fabae* (Behrendt, 1963) and in mirid bugs (Braune, 1971), of which the species *Leptopterna dolobrata* will be discussed as an example: Larvae and adults of this single-brooded mirid bug can be found for $2^1/_2$ months of the year only, whilst the hibernating egg stage lasts over 9 months.

Dormancy during embryogenesis always occurs before unfolding of the embryo. In the field this stage is reached at the end of July (about 20 days after deposition of the eggs). The initiation of dormancy does not depend on external factors and therefore belongs to the obligatory type of diapause. Studies on the influence of temperature and photoperiod on the duration of the egg period and on termination of diapause were carried out in the laboratory.

The eggs used in the experiments were deposited by females kept at 20 C and long-day photoperiod (18L/6D).

The different samples were at first given a general treatment of 31 days at +20 C and long-day photoperiod, thus guaranteeing that the eggs had developed up to the dormant state before they were transferred to various experimental conditions. No part of development is influenced by photoperiod.

At constant temperatures between —1 and +23 C hatching of the eggs occurred only within the small range from +10 to +16 C, but the percentage of eggs that hatched in these samples was very low.

Development at temperatures above +16 C, and a significant increase in the success of hatching, was obtained by exposing the eggs to a period of chilling at +5 C for 30 to 360 days and then incubating them at higher temperatures (Fig. 1). At least 210 days' chilling at +5 C were required to terminate diapause in more than half of the eggs at +23 C.

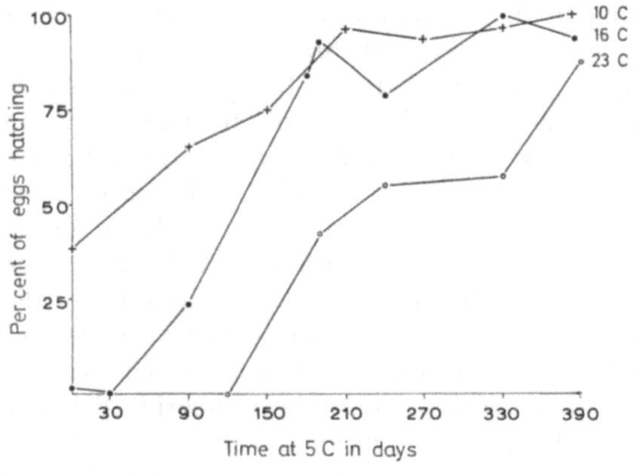

Fig. 1. Per cent of eggs hatching at different temperatures, plotted against length of chilling period at +5 C (in days)

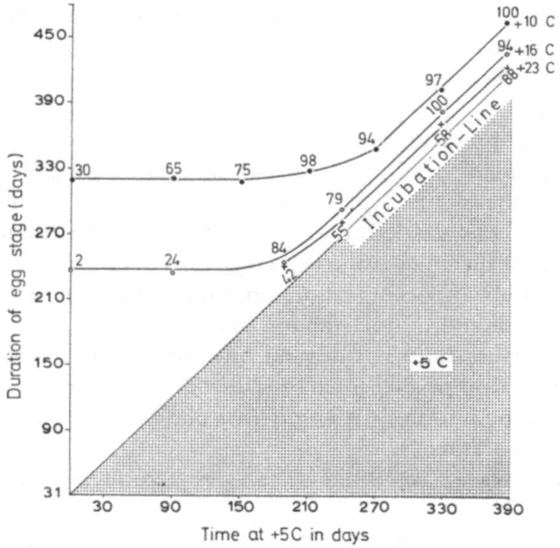

Fig. 2. Duration of egg stage (in days) plotted against length of chilling period at +5 C. After chilling the eggs were incubated at +10, +16 and +23 C. The phases of chilling (stippled part) and of incubation are separated by the incubation line. The figures above the curves indicate hatching success in per cent

Fig. 2 shows the duration of the egg stage at +10, +16 and +23 C related to the time of chilling at +5 C. The egg period is divided into three parts: 1. the general 31 days' treatment at +20 C, 2. the period of chilling at +5 C and 3. the time of incubation at +10, +16 or +23 C. In the graph the last two phases are separated by the incubation line (time at +5 C below this line, time of incubation at +10, +16 or +23 C above it). Considering the hatching lines at +10 and +16 C one can see that chilling periods of up to 180 days have no influence on the duration of the egg period. A prolongation of the chilling period, however, causes a proportional increase of the egg period. As periods of up to 180 days at +16 C can be replaced by a treatment at +5 C without changing the duration of the egg period, morphogenesis cannot have taken place during this time since developmental processes always depend on temperature.

The above results suggest that within the egg period four phases have to be distinguished: *prae-*, *meso-*, *meta-* and *postdiapause* (for details see Braune, 1971): In praediapause, embryogenesis takes place up to the dormant stage. During meso- and metadiapause development is arrested irrespective of environmental conditions. In postdiapause development recommences and continues until the emergence of the larvae. These four phases react differently to temperature treatment (Table 1).

Only the periods of prae- and postdiapause, during which morphogenetic processes take place, show "normal" temperature reactions, i.e. acceleration of development with increasing temperature. Mesodiapause, which lasts for about 6 months, is a peculiarity of the embryonic period. Whatever the temperature, this phase requires exactly the same period of time for its completion, indicating that its duration is genetically determined and thus independent of temperature.

Table 1. Phases of Embryonic Development (in Days) at Different Temperatures

Temperature C	Morphogenesis Praediapause	Diapause Mesodiapause	Metadiapause	Morphogenesis Postdiapause
— 1	—	180	?	—
+ 5	?	180	~150	—
+10	>40	180	72	45
+16	21	180	18	21
+20	16	180	—	?
+23	?	180	—	12

Metadiapause is an interphase before normal development is resumed in postdiapause. The unknown physiological changes which take place in the embryo during this phase can only be completed at low temperatures. From the ecological point of view the mechanism of dormancy in *Leptopterna dolobrata* may be interpreted as follows: The deposition of eggs in this mirid bug occurs at the end of July. As the first larval instars have very specific requirements concerning food and since adequate food is not available during autumn and winter, the eggs must be prevented from hatching under these adverse conditions. This is guaranteed by the phases of meso- and metadiapause. The long-lasting temperature-independent mesodiapause (6 months) prevents any influence of environment on developmental processes before winter. On the other hand, a resumption of development with rising temperatures in the spring is only possible when metadiapause is completed. This is guaranteed by the low temperatures of winter. Thus the protracted egg diapause in *Leptopterna dolobrata* is a good means of synchronizing the appearance of larvae and adults with that of their food plants.

As far as is known, obligatory diapause occurs mainly in the embryonic and sometimes in the pupal stage, whereas facultative diapause is to be found mainly in adult reproductive or larval stages. The examples discussed in this paper demonstrate that obligatory as well as facultative diapause are useful mechanisms for synchronizing the life cycles of the insects with seasonal changes in environmental conditions.

References

Andrewartha, H. C.: Diapause in relation to the ecology of insects. Biol. Rev. **27**, 50–92 (1952).

Beck, S. D.: Insect photoperiodism. New York-London: Academic Press 1968.

Behrendt, K.: Über die Eidiapause von *Aphis fabae* Scop. (Homoptera, Aphididae). Zool. Jahrb., Abt. Allgem. Zool. Physiol. Tiere **70**, 309–398 (1963).

Braune, H. J.: Der Einfluß der Temperatur auf Eidiapause und Entwicklung von Weichwanzen (Heteroptera, Miridae). Oecologia (Berl.) **8**, 223–266 (1971).

Lees, A. D.: The physiology of diapause in arthropods. Cambridge: Univ. Press 1955.

Mansingh, A.: Physiological classification of dormancies in insects. Can. Entomologist **103**, 983–1009 (1971).

Masaki, S.: The influence of temperature on the intensity of diapause in the eggs of the Emma Field Cricket (Orthoptera: Gryllidae). Kontyu **31**, 9–16 (1962).

Müller, H. J.: Formen der Dormanz bei Insekten. Nova Acta Leopoldina, N. F. No **191**, 35, 7–27 (1970).

Precht, H.: Christophersen, J., Hensel, H.: Temperature and Life. 2nd ed. Berlin-Heidelberg-New York: Springer 1973 (in press).

Tischler, W.: Zur Biologie und Ökologie des Opilioniden *Mitopus morio* F. Biol. Zentr. **86**, 473–484 (1967).

Critical Minimum Temperatures of Reptiles

Ian F. Spellerberg

With 2 Figures

Introduction

Reptile biological studies have been most productive at centres near deserts. This may have led to the considerable amount of research into the upper temperature relations of reptiles. Reptile cryobiological research has received little attention although it is commonly known that some reptile species inhabit temperate biomes, high latitudes and high altitudes.

Suggested reptile adaptations to the cold are vivipary (Hock, 1964) and body size and form (Heatwole *et al.*, 1969; Spellerberg, 1971). Some seasonal changes in physiological aspects (Aleksiuk, 1971 a, b) as well as thermoregulatory mechanisms are clearly important for temperate biome species because reptiles have a low oxidative metabolism, compared to endotherms, and their body temperature is lowered at night and during the winter. Viitanen (1967) has shown that seasonal activity and distribution of vipers (*Vipera berus* – Viperidae) is limited by low ground and sub-surface temperatures. Another detailed report by Prieto and Whitford (1971) described the importance of low temperature tolerances in the distribution of high altitude Iguanid species.

There are but few other studies (Bailey, 1949; Saint-Girons and Saint-Girons, 1956) which attempt to explain reptile distribution and speciation in cold regions. This situation may be due in part to a lack of suitable criteria which could be used in comparative ecological studies.

Many casual reports describe low temperature responses of reptiles but it was not until about three decades ago that a serious attempt was made to clarify criteria used for the examination of reptile low temperature responses. Cowles and Bogert (1944) examined the thermal requirements of desert reptiles and noted ecologically important thermal reactions. Amongst the major categories suggested was the Critical Minimum temperature which was defined as: "From the ecological standpoint this is equivalent to the lethal (minimum) under certain conditions, since it is the temperature that causes a cold narcosis and effectually prevents locomotion. At this temperature the animals are helpless to escape enemies or to remedy their thermal impasse if they have not already sought sanctuary." It is now generally accepted that the Critical Minimum temperature of reptiles is the lower body temperature at which the animal has just lost its righting reflex, along with locomotory ability, after the specimen has been cooled (Jacobson and Whitford, 1970; Spellerberg, 1971).

The status of the Critical Minimum temperature and its usuage in reptile ecology is discussed here.

Literature Review

There are few accounts of low temperature responses of reptiles at the whole animal level and it is rare when the ecological significance is discussed or when reference is made to the categories used by Cowles and Bogert (1944). Commencing with Cowles and Bogert's own report, there is one reference to the Critical Minimum temperature. That is, the Yucca Night Lizard (*Xantusia vigilis* – Xantusiidae) was reported to be capable of righting itself from the supine to the normal position at 7 to 10 C. Herbst (1949) determined the "first righting response on stimulation" of several lizards and snakes by first cooling them and then warming them. Fitch (1954) noted that at a body temperature of 1.4 C a Five Lined Skink (*Eumeces fasciatus* – Scincidae) was able to turn over slowly after several seconds. Pearson (1954) made a general reference to the low temperature behaviour of *Liolaemus multiformis multiformis* (Iguanidae) and stated that at a temperature of 1.5 C it could walk and right itself very slowly. Fitch (1956) reported that a Collared Lizard (*Crotaphytus collaris* – Iguanidae) could right itself at 13.9 C. In the same year Bogert and Martin del Campo (1956) noted that the righting response of *Heloderma suspectum* (Helodermatidae) began at 4.6 C. In a report by Cunningham (1966) there is mention of a righting response and the term "immobilized" is used in descriptions of the behaviour of the Alligator Lizard (*Gerrhonotus multicarinatus webbi* – Anguidae) but there is no mention of the Critical Minimum temperature. More recently Jacobson and Whitford (1970) found that the Critical Minimum of the snake *Thamnophis proximus* (Colubridae) was lower than the Critical Minimum of *Natrix rhombifera* (Colubridae) and that the distribution of *T. proximus* extended further north than that of *N. rhombifera*. In the same year Kour and Hutchison (1970) reported the Critical Minima for four lizard species at each of three acclimation temperatures. Hall (1971) determined the Critical Minimum of the Great Plains Skink *(Eumeces obsoletus)* by first cooling then warming the specimens. A good ecological evaluation of the Critical Minimum is given by Prieto and Whitford (1971). In their paper the Critical Minimum of *Phrynosoma cornutum* (Iguanidae) was reported to be 9.4 C and this species is found in open desert scrub habitats. In contrast the Critical Minimum of *Phrynosoma douglassii* was found to be 2.7 C and this species has a distribution range which extends to altitudes as high as 3494 m. Recently Spellerberg (1972) evaluated the ecological importance of the Critical Minimum temperature for 33 Australian reptile species.

Variables Affecting the Critical Minimum

Laboratory Procedures. With a decrease in the reptile's body temperature the movements of the animal are protracted and there is a diminished response to touch as the body temperature approaches a level where loss of locomotion and loss of the righting reflex is imminent. It has been important therefore to ensure that instrumentation of the specimen does not impede body movements. Further, the stimulus strength required to induce a righting response (stimulating the pelvic region with a fine brush – Spellerberg, 1972) is markedly different from specimen to specimen. Consequently the experimenter must be quite satisfied that the righting response has just been lost before the Critical Minimum temperature is noted. Temperature tolerance studies reported in Spellerberg (1972) and here were performed with the reptile instrumented for the duration of the experiment. That is the animal was not handled apart from quickly turning the specimen over by the tail to test the righting response. In addition the thermistor or thermocouple wires were very fine and were attached so that the reptile had maximum freedom of movement.

One way to estimate the accuracy of these methods was to determine the Critical Minimum several times on the same specimen. Data were taken from observations on the daily rhythmicity of the Critical Minimum where each lizard was tested at intervals of five to six days. The time of day for each test was constant for each specimen and the holding cage temperature was kept at 26 C. Altogether the Critical Minimum was determined 85 times using 27 adult male *Lacerta sicula* (Lacertidae). The mean intra-individual difference was 0.5 C with a range of 0.0 C to 1.3 C for the 85 tests.

Reptile Body Form. Intra-specific variation of the Critical Minimum level can be increased when the animal is damaged. The tail of some species is used as a lever when the animal rights itself and similarly the limbs may also be employed but to a lesser extent. For example the mean summer Critical Minimum for *Lacerta agilis* with tails complete was 3.0 C ($N = 6$) and the mean for those without tails was 4.7 C ($N = 3$). Similarly the mean summer Critical Minimum for intact *L. sicula* was 7.2 C ($N = 18$) and the mean for animals without tails was 8.0 C ($N = 5$).

Rate of Cooling. A very fast or a very slow cooling rate may also alter the Critical Minimum value (see discussion of standard cooling rates in Spellerberg, 1971, 1972). When the cooling rate was increased, values for the Critical Minimum were greater as a result of differential cooling of the body and subsequent immobilization of the surface tissues. An example of this effect was evident when eleven adult *Sphenomorphus tympanum* – cool temperate form (Scincidae) were placed in the cooling chamber with the ambient temperature set at 0.0 C. The body temperature of each lizard was 25 C when first placed in the chamber and this temperature then rapidly decreased. A Critical Minimum of 10.8 C (Range = 9.4 C to 11.2 C) was obtained as compared with a mean Critical Minimum of 2.8 C (Range = 2.2 C to 3.8 C, $N = 65$) determined by the standard procedure. A subcutaneous probe in the dorsal pectoral region gave values between 0.0 and 3.0 C when the righting reflex was lost. During cooling at this rapid rate the lizards did not show any basking postures whereas they are present during the normal cooling procedure (Spellerberg, 1971). Struggling and burrowing were persistent and when the body temperature reached about 20.0 C both anterior tarsi were turned palmar surface uppermost.

If the rate of cooling was very slow then the Critical Minimum values are lower than those recorded in the standardized conditions. For example when a summer sample of eight adult *S. tympanum* (cool temperate form) were cooled at about one third the rate of their standard cooling, a mean value of 2.2 C (Range = 1.8 C to 2.4 C) was obtained for the Critical Minimum temperature (compare with standard value above).

Altitude. The Critical Minimum temperature and its variation with altitude has been examined for the *Sphenomorphus quoyi* species complex of southeast Australia (Spellerberg, 1971). A small variation was found between the mean Critical Minima of different samples of one species collected from different altitudes in one day during the summer. These differences (between mean values) ranged from 0.1 C *(S. quoyi)* to 0.4 C (*S. tympanum* – cool temperate form) but none were significant. However two winter populations of *S. tympanum* (cool temperate form) collected from 400 and 1200 m had Critical Minimum temperatures that were significantly different. The Critical Minimum range for each population did not overlap (Fig. 1).

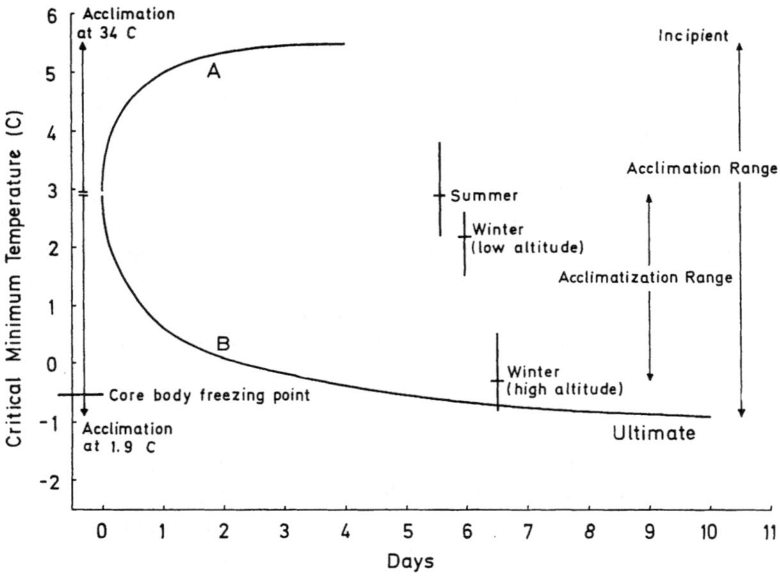

Fig. 1. Critical Minimum acclimation and acclimatization for temperate reptile species. *A* Shows a generalized high temperature acclimation and *B* shows a generalized low temperature acclimation. The acclimation range encompasses the mean Critical Minimum temperatures obtained in the summer and in the winter

Seasonal Differences. Evidence for seasonal changes (acclimatization) has been found for several temperate lizard species (Spellerberg, 1972). Significant differences were found between the mean Critical Minimum temperature obtained in the summer and that obtained in the winter. A probability of 0.05 between means was considered significant.

Daily Variation. Circadian rhythms in the Critical Minimum temperature of reptiles are currently being investigated (Spellerberg and Hoffmann, 1972). There was a circadian rhythm in the lizard Critical Minimum temperature during natural temperature and light conditions as well as in laboratory conditions where temperature was constant (26.0) and light-dark cycles and continuous light was used. In the natural conditions and the LD conditions there was a significant difference between the mean Critical Minimum determined at 0600 hours and that determined at 1800 hours (P less than 0.001). In the LL conditions, there was the same significant difference between the mean Critical Minimum determined at the onset of activity and at the end of activity.

Laboratory Acclimation. This aspect is discussed by Spellerberg (1972). A rapid acclimation over a period of 48 hours was found where lizards were held at temperatures just below the level of the Critical Minimum that was first determined immediately following collection in the summer months. A similar rapid rate of acclimation for the Critical Maximum of lizards has been reported by Ballinger and Schrank (1970).

Critical Minimum Acclimation Range

Alterations to the Critical Minimum level by acclimation and acclimatization have not previously been examined in detail and consequently the value of this aspect in reptile ecological studies has not been discussed.

Data based on records for *S. tympanum* (cool temperate form) are shown in Fig. 1 (Spellerberg, 1971) and this could well be considered as a generalization for temperate reptile species. The two graphs (A, B) show the acclimation rate and range for the Critical Minimum temperature when the animal is held at high temperatures (A) and at low temperatures (B). Low temperature acclimation was at 1.9 C which is just below the mean summer Critical Minimum level. High temperature acclimation was at 34 C which approximates the mean Voluntary Maximum (Spellerberg, 1972) for this species. Equivalent graphs for acclimatization might be difficult to obtain because it would necessitate repeated recapture of the members of a lizard population. However the mean and range for the Critical Minimum is shown here for a sample of lizards collected in mid-summer. Two levels are shown for lizards collected in mid-winter: one taken from high altitudes (1000–1200 m) when there was a light snow cover and low air temperatures had prevailed for two to three weeks; one taken from a lower altitude (400 m) following a nine day period of fine weather (cloud cover less than the daily average for that month). A combination of high altitudes, light snow cover and low air temperatures probably resulted in the very low winter Critical Minimum temperatures shown in Fig. 1. The terms "incipient" and "ultimate" are discussed at length in Spellerberg (1971) and here they indicate the highest and lowest levels respectively for the Critical Minimum temperature following acclimation.

Various aspects of Fig. 1 might be used in comparative studies and of the more meaningful levels the mean summer Critical Minimum and the ultimate Critical Minimum may be two. That is for each species examined there should be a value for the Critical Minimum temperature of the species, immediately following collection in the field, then a further value after low temperature acclimation has been applied. The acclimation temperature would best be just below the mean summer Critical Minimum because this is where the greatest acclimation rate and range occurs (Spellerberg, 1971). The results of this approach would give a range for the Critical Minimum which could be used as an index for the comparison of temperature tolerances in different species. The same approach might also be used for those animals collected in the winter months.

Fig. 2 shows a possible ecological application of the Critical Minimum acclimation range in a comparison of the temperature tolerances of four allopatric *Sphenomorphus* species from southeast Australia. The distribution of these four congeneric species shows an interesting altitudinal zonation with little or no overlap at the species distribution contact zones. Data for Fig. 2 are taken from Spellerberg (1971) and are rearranged here to show the acclimation range (incipient level to ultimate

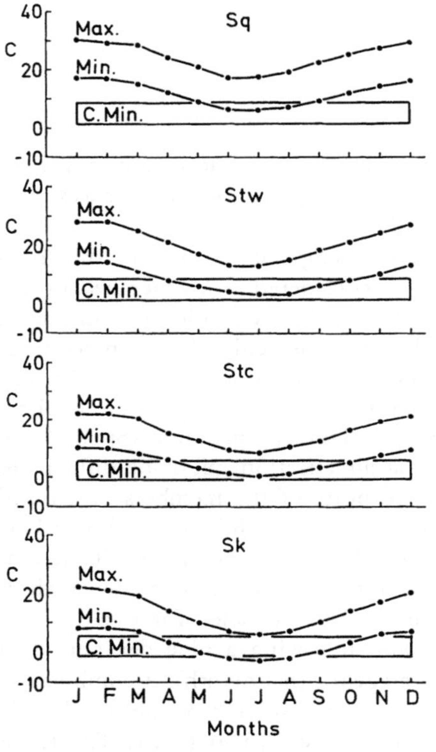

Fig. 2. Critical Minimum acclimation range (incipient to ultimate level) for four *Sphenomorphus* species and the mean daily maximum and minimum air temperatures in each of the four species distribution areas. Sq = *Sphenomorphus quoyi*, Stw = *S. tympanum* (warm temperate), Stc = *S. tympanum* (cool temperate), Sk = *S. kosciuskoi*

level) of the four species. The mean daily maximum and minimum air temperatures (meteorological screen) in each of the four species distribution areas are also shown on the same figure. The fully acclimated level (ultimate) of the Critical Minimum remained above the level of the winter daily minimum air temperature except in the case of *S. kosciuskoi*. Air temperatures may often be well below the temperatures found at sub-surface levels and below the thermal levels experienced by these lizards. Snow cover ameliorates sub-surface temperatures and it is reasonable to assume that the ultimate Critical Minimum temperature of *S. kosciuskoi* is sufficiently low to allow locomotory ability throughout the winter period.

Conclusions

Compared to endotherms the terrestrial ectotherm vertebrates are found in a diverse but a limited number of localities and obviously dispersion is limited largely by their dependence on external heat sources for thermal homeostasis and normal activity. Terrestrial reptiles are poikilothermic (variable body temperature) but have an ectothermic temperature regulation (body temperature determined by heat obtained from the environment). They are well equipped with specialized thermoregulatory mechanisms for thermal homeostasis and, in some cases, a wide range of temperature tolerances which has enabled populations to reach elevations as high as 4000 and 5000 m (Strel'nikov, 1944; Pearson, 1954). Snakes are found as far north as 67 N in Europe (Mertens, 1968) and lizards occur in the southern regions of New Zealand to about 46 S. Appreciation of reptile ecology in temperate regions therefore requires in part thermoregulatory studies which, when augmented by results from behavioural and physiological research, should indicate some reptile adaptations to the cold.

Temperature relations of reptile behaviour responses have long been studied. It is clear that many of these behavioural responses are stereotyped as well as discrete and that they are elicited over a characteristic and sometimes limited range of body temperatures. The range of body temperatures associated with each response is greater at low body temperatures than at high body temperatures as illustrated by the Critical Minimum and Critical Maximum acclimation range (Fig. 6 in Spellerberg, 1972). In addition and following laboratory acclimation, the intraspecific range for the Critical Minimum may be quite large, because it seems that there may be genetic divergence (in contrast to phenotypic adaptations) amongst some populations which together form a geographically widespread species. If weather conditions at the time of collection were markedly different on subsequent field excursions, this would further spread the values obtained for the Critical Minimum of any species. The laboratory acclimation experiments indicate that acclimatization may be very rapid and over a few summer days, where air temperatures were unusually low, the Critical Minimum level could change significantly. Clearly the climate (together with topography and climax vegetation) should be carefully examined and weather conditions should be noted so that the Critical Minimum temperatures may be suitably qualified.

I suggest that reporting Critical Minimum temperatures as unqualified and isolated measurements (mean value for a species) has little value although I have (Spellerberg, 1972) reported these for 33 Australian reptile species and have attempted ecological evaluations. The danger of using an unqualified Critical Minimum temperature can be illustrated in the following way. The lower temperature tolerances of several species may have been compared using the mean Critical Minimum values. If acclimation, acclimatization and endogenous rhythms are not considered then several similar Critical Minimum values might well be obtained although the acclimation range for each species could be different but still overlap. That is while the Critical Minimum acclimation range is markedly different in each of the hypothetical species (and this is not known by the investigator) the recorded Critical Minimum temperatures could be very similar. This would lead to the erroneous conclusion that the lower temperature tolerances of the species were similar whereas the ultimate Critical Minimum and acclimation range would show otherwise.

For use in comparative studies the Critical Minimum temperature must be qualified and the pre-experimental conditions must be similar before conclusions of a comparative and ecological nature can be made. Examination of the Critical Minimum acclimation range of a number of species from diverse habitats may assist in an understanding of the thermal ecology of each species. Further value is added to this approach when the results are linked with physiological studies of the intact animal and behavioural studies of the individual and the population in the natural environment.

Summary

The Critical Minimum temperature of reptiles is defined as the lower body temperature at which locomotory ability is disorganized and which produces a cold narcosis so that the animal loses its ability to escape from conditions that may lead to its death. The loss of the righting reflex may be used to indicate this temperature.

The Critical Minimum temperature can be altered by acclimatization and acclimation. That is altitude, season, time of day and laboratory temperature levels are important variables which can raise or lower the Critical Minimum temperature of reptiles.

Without accurate information on pre-experimental conditions the mean Critical Minimum temperature has little value in comparative ecological studies. Rather than attempting to monitor pre-experimental conditions it is suggested that the Critical Minimum acclimation range be used. Application of the Critical Minimum acclimation range in ecological research is demonstrated in four allopatric lizard species.

Acknowledgements

This paper was prepared while I held an Alexander von Humboldt-Stiftung Research Fellowship. Further assistance was provided by the Alexander von Humboldt-Stiftung to attend the Temperature Symposium at Obergurgl, Austria. It is a pleasure to thank Professor J. Aschoff for the use of research facilities at the Max-Planck-Institut für Verhaltensphysiologie, Erling-Andechs. I thank Dr. K. Hoffmann for the many valuable discussions which led to some of the conclusions reached here.

References

Aleksiuk, M.: Temperature-dependent shifts in the metabolism of a cool temperate reptile, *Thamnophis sirtalis parietalis.* Comp. Biochem. Physiol. **39 A**, 495–503 (1971 a).

Aleksiuk, M.: An isoenzymic basis for instantaneous cold compensation in reptiles: Lactate dehydrogenase kinetics in *Thamnophis sirtalis.* Comp. Biochem. Physiol. **40 B**, 671–681 (1971 b).

Bailey, R. M.: Temperature toleration of Gartersnakes in hibernation. Ecology **30** (2), 238–242 (1949).

Ballinger, R. E., Schrank, G. D.: Acclimation rate and variability of the Critical Thermal Maximum in the lizard *Phrynosoma cornutum*. Physiol. Zool. **43** (1), 19–22 (1970).

Bogert, C. M., Martin del Campo, R.: The Gila Monster and its allies. Bull. Mus. Nat. Hist. **109**, 1–238 (1956).

Cowles, R. B., Bogert, C. M.: A preliminary study of the thermal requirements of desert reptiles. Bull. Am. Mus. Nat. Hist. **83**, 265–296 (1944).

Cunningham, J. D., Thermal relations of the Alligator Lizard *Gerrhonotus multicarinatus webbi*. Herpetologica **22** (1), 1–7 (1966).

Fitch, H. S.: Life history and ecology of the Five-Lined Skink *Eumeces fasciatus*. Univ. Kans. Publ. Mus. Nat. Hist. **8**, 1–156 (1954).

Fitch, H. S.: Temperature responses in free-living amphibians and reptiles of north-eastern Kansas. Univ. Kans. Publ. Mus. Nat. Hist. **8**, 417–476 (1956).

Hall, R. J.: Ecology of a population of the Great Plains Skink *(Eumeces obsoletus)*. Univ. Kansas Sci. Bull. **49** (7), 357–388 (1971).

Heatwole, H., Lin, T. H., Villalon, E., Muniz, A., Matte, A.: Some aspects of the thermal ecology of Puerto Rican anoline lizards. J. Herpetol. **31**, 65–77 (1969).

Herbst, L. C.: A preliminary study of the reactions of snakes and lizards at low temperatures. M. Sc. thesis, Univ. Cal., Los Angeles (1949).

Hock, R. J.: Terrestrial animals in cold: reptiles, p. 357–359. In: Dill, D. B. (Ed.): Handbook of physiology. Sect. 4: Adaptation to the environment. Washington, D. C.: Am. Physiol. Soc. (1964).

Jacobson, E. R., Whitford, W. G.: The effect of acclimation on physiological responses to temperature in the snakes *Thamnophis proximus* and *Natrix rhombifera*. Comp. Biochem. Physiol. **35** (2), 439–449 (1970).

Kour, E. L., Hutchinson, V. H.: Critical thermal tolerances and heating and cooling rates of lizards from diverse habitats. Copeia **1970** (2), 219–229 (1970).

Mertens, R.: Kriechtiere und Lurche, p. 98. Stuttgart: Franckh'sche Verlagshandlung (1968).

Pearson, O. P.: Habitats of the Lizard *Liolaemus multiformis multiformis* at high altitudes in southern Peru. Copeia **1954** (2), 111–116 (1954).

Prieto, A. A., Whitford, W. G.: Physiological responses to temperature in the Horned Lizards, *Phrynosoma cornutum* and *Phrynosoma douglassii*. Copeia **1971** (3), 498–504 (1971).

Saint Girons, H., Saint Girons, M. C.: Cycle d'activité et thermorégulation chez les reptiles (lézards et serpents). Vie Milieu **7**, 133–226 (1956).

Spellerberg, I. F.: Thermoregulation and temperature tolerances in the *Sphenomorphus quoyi* species complex (Lacertilia: Scincidae) of southeast Australia. Ph. D. thesis, La Trobe Univ. Australia (1971).

Spellerberg, I. F.: Temperature tolerances of southeast Australian reptiles examined in relation to reptile thermoregulatory behaviour and distribution. Oecologia (Berl.) **9**, 23–46 (1972).

Spellerberg, I. F., Hoffmann, K.: Circadian rhythm in lizard Critical Minimum temperature. Naturwissenschaften **59**, (11), 517–518 (1972).

Strel'nikov, I.: Importance of solar radiation in the ecology of high mountain reptiles. Zool. Zh. **23**, 250–257 (1944).

Viitanen, P.: Hibernation and seasonal movements of the viper, *Vipera berus berus* (L.), in southern Finland. Ann. Zool. Fennici **4**, 472–546 (1967).

Resistance Adaptations of Marine Invertebrates and Fish to Cold

Hans Theede

With 11 Figures

Introduction

Upper and lower temperature limits are contributing factors in determining the vertical and geographical distribution of many marine animals. The lower temperature limits, however, have been studied much less than the upper ones. (For literature see e.g. Precht *et al.*, 1955; Gunter, 1957; Kanwisher, 1966; Kinne, 1963, 1970; Newell, 1970; Theede, 1972). Damaging effects of low temperatures may be distinguished by their connection with formation of ice crystals or not. Under the more general term cold resistance a distinction can be drawn between chilling resistance and freezing resistance. Chilling resistance is the ability to survive at low temperatures without the formation of ice, whereas freezing resistance is the ability to withstand the direct and indirect effects of extra- or intracellular ice formation. Many animals avoid severe cooling or freezing by appropriate behaviour and other protective mechanisms. Other organisms can be completely frozen under natural conditions and resume their former activity after thawing. In such forms a special tolerance of single cell and plasma structures to direct and indirect influences of freezing may play an important role. Many invertebrates of the boreal coasts (e.g. mussels, clams, periwinkles and barnacles), which are regularly left dry at ebb tide, are partially able to withstand the formation of ice in cold winters. In the following, I intend to demonstrate what adaptations enable marine animals to endure different degrees of cold and freezing.

Invertebrates

Chilling Resistance

The primary sources of damage by chilling in invertebrates are to be sought in the defective integration of the nervous system and of the metabolic processes, in insufficient energy supply to the tissues and in disturbances in the electrolyte economy (for further literature consult Luyet and Gehenio, 1940; Gunter, 1957; Kinne, 1963, 1972). It is well known from ecological studies that tropical marine species may die at temperatures far above zero centigrade when chilled. Vernberg and Vernberg (1970) have demonstrated that numerous tropical-subtropical crustacea and molluscs die within 48 hours when subjected to a bath temperature of 10 C. In contrast, most of the boreal marine species endure water temperatures around or below 0 C. Only after continuous exposure to cold for a period of weeks or months will some of these animals die, especially those living in the boreal region near their northern distribution limit. In nature it was observed that the population densities of several marine invertebrates were reduced during long winters, e.g. 1962/63 (Crisp, 1964a, b; Courtney and Webb, 1964; Ziegelmeier, 1964). Along the English coast the mortality rate was greater among the lusitanian forms with a more southern distribution than among the celtic species. Arctic-boreal forms showed the smallest losses (Crisp, 1964b). In the waters near Helgoland, lamellibranchs especially suffered high losses whereas polychaetes and crustaceans survived better during this winter. Juvenile forms were more tolerant than the adults (Ziegelmeier, 1964).

Experimentally it was possible to obtain reproducible values for the chilling resistance of isolated tissue pieces when the temperature of the sea water was gradually lowered below the freezing point without ice formation (Schlieper *et al.*, 1967; Theede, 1972). In short-term chilling experiments, specific differences could be demonstrated in the onset of rigidity of cilia of isolated gill pieces of different species (Fig. 1). Accordingly, the cellular chilling resistance of lower marine invertebrates can be applied in the ecophysiological characterization of species in the same way as the cellular heat resistance (e.g. Vernberg *et al.*, 1963; Schlieper,

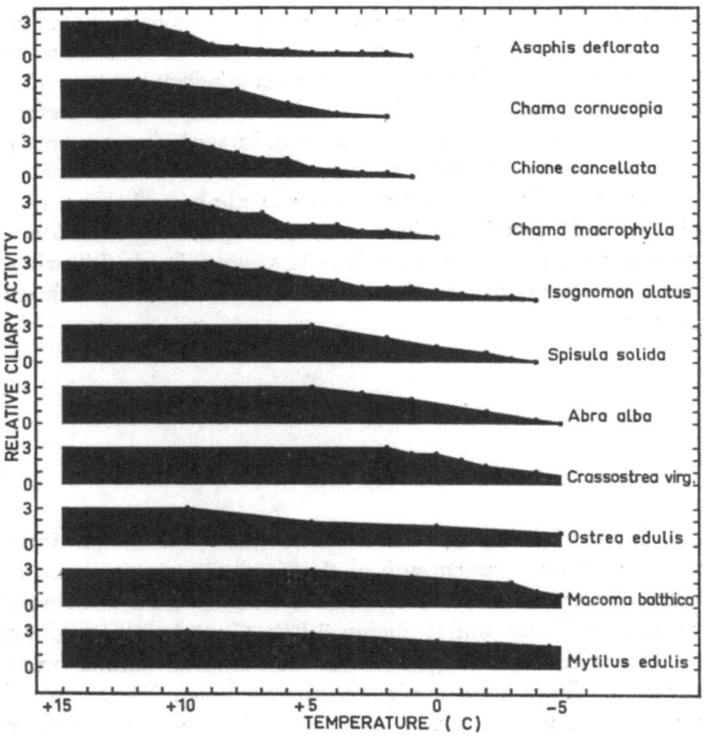

Fig. 1. Different types of reaction of gill cilia of marine bivalves during cooling (1 C/min). (After Theede, 1972)

Ushakov, Zhirmunsky, in Troshin, 1967). If the temperature is lowered gradually rigidity commences in gill pieces of tropical and subtropical lamellibranchs at temperatures above or around 0 C (Fig. 1). One such species having a high sensitivity to cold is *Chama cornucopia*. It lives attached to the rocky ground of the upper tidal region of the Farasan Islands in the southern Red Sea. Two other bivalves of the Red Sea, *Avicula savignyi* and *Barbatia helblingii*, exhibit a relatively high chilling resistance. But they cannot have acquired this in the Red Sea, where even at the bottom the temperature never falls below 20–18 C. Therefore it may be assumed that these species originated in the colder deep-bottom waters of the Indian Ocean before migrating to the Red Sea. Here, they obviously have retained the genetic adaptation to cold of their ancestors (Schlieper *et al.*, 1967).

In species from coastal areas of the North Sea that suffer considerable losses during occasional severe winters – *Spisula solida*, *Abra alba* (Ziegelmeier, 1964; Kühlmorgen-Hille, 1963) – the activity of gill cilia stops before the onset of ice formation. On the other hand, gills of bivalves such as *Mytilus edulis* or *Macoma balthica* show distinct cellular activity until ice forms. These animals also occur in the tidal zone and are among the few lamellibranchs in the sublittoral region that survived the long European winter of 1962/63 with little or no damage. Similar experimental results involving cellular chilling resistance were obtained by exposure of excised gill-tissue pieces to supercooled sea water held at constant temperature, —3 C for example (Theede, 1972).

In shallow-water forms, the effects of individual temperature adaptation on chilling resistance were observed. The cells of the subtropical-tropical species *Chione cancellata* exhibit adaptation of heat and chilling resistance when the acclimation temperature is changed in the range of 10 to 23.5 C. However, no concurrent change in cellular freezing resistance occurs. This suggests that different mechanisms are involved in the adaptation of chilling and of freezing resistance.

Another example which demonstrates that chilling and freezing resistance need not be influenced by one factor in the same way is the different action of an addition of calcium ions to the external medium (Theede, 1972). After 3 days pretreatment of the animals with increased Ca concentrations in the external medium, the cellular chilling resistance of *Abra alba* was enhanced. In contrast, the cellular freezing resistance of Baltic *Mytilus edulis* was not significantly influenced and the cellular freezing resistance of cold acclimated North Sea *Mytilus* was reduced after the same pretreatment.

Freezing Resistance

Observations in Nature

Natural freezing resistance in marine invertebrates seems to be restricted to species that are able to survive in the supra- or eulittoral of temperate and colder regions. In winter, this may be a determining factor for the upper limit of distribution in the tidal zone, especially in northern latitudes. In the Arctic, sessile species of macrofauna which are limited to the tidal zone do not occur. A few mobile types of macrofauna can be found in isolated places: in winter they withdraw into deeper waters and in summer they temporarily recolonize the upper littoral areas (Madsen, 1940; Gerlach, 1965).

In severe winters mass mortality of littoral macrofauna species has been observed in shore areas of the temperate region (Blegvad, 1929; Brongersma-Sanders, 1957; Crisp, 1964a, b; Tiedke, 1967). In this way frost has a selective effect on marine littoral animals. A progressive impoverishment of the tidal fauna is observed on approaching the Arctic. Only a few species survive over the whole year in the exposed upper littoral in the subarctic (or boreo-arctic) region. The barnacle *Balanus balanoides*, the periwinkle *Littorina saxatilis* and the mussel *Mytilus edulis* are the last typical representatives of the temperate zone tidal fauna in the southwest of Greenland and on Spitzbergen. On the West-Greenland coast these species are reported to survive cold winters in crevices of rocks or enclosed in the ice (Madsen, 1940; Petersen, 1962a, b). Kanwisher (1955) reported that exposed intertidal animals of the subarctic region may be frozen for as long as 6 months. The survival of such forms depends on their ability to tolerate the formation of ice in a great part of their body water. Kanwisher (1955) measured the temperature within frozen intertidal specimens and obtained values as low as —20 C. Such animals were alive after thawing. Gerlach (1965) mentions several species of meiofauna which survive the winter in the eu- and supralittoral of the subarctic region of Spitzbergen. Lorenzen (1968) found species of nematodes, halacarides, small oligochaetes, harpacticoides and collemboles surviving in the frozen boreal salt marsh. The heat capacity of the substratum, the insulating property of ice and snow, as well as the so-called greenhouse effect of an ice cover may be of great significance for protection from extreme cold. Under arctic and antarctic conditions representatives of the meio- and microfauna, such as protozoa, rotifers, tardigrades and nematodes, as well as the larval and adult stages of some insect species, may be the only invertebrates which successfully survive repeated freezing and thawing (Scholander *et al.*, 1953; Tilbrook, 1970).

Experimental Studies

Experimental data about freezing resistance of marine invertebrates are rare (for references see: Smith, 1958; Kinne, 1963; Kanwisher, 1966; Theede, 1967, 1972). Kanwisher (1955) found that several bivalves and gastropods from the tidal region are able to endure lenghty freezing, whilst species occurring below this region fail to survive. Similar results for the resistance of intertidal molluscs have been obtained by Asahina (1958), Sömme (1966), Tanno and Wada (1966) and Williams (1970) (Fig. 2). According to Zhyubikas (1968), eulittoral specimens of *Mytilus edulis* proved to be more resistant to —10 C than those from deeper tidal ponds. In addition, a high freezing resistance was found for *Balanus balanoides* (Sömme, 1966; Crisp and Ritz, 1967) and also in the periwinkles *Littorina littorea* and *L. rudis*. On the South English coast Southward (1958) found species of exposed sessile barnacles to possess greater resistance than the mobile "top-shells". The gastropods also exhibit correlations between their resistance and geographical and vertical distribution. The resistance to high as well as to low temperatures is greater in those forms which are better able to endure long exposure to air. In this way the more exposed *Monodonta lineata* and *Gibbula umbilicalis* have a greater resistance to low temperatures (—5 C) than the less exposed *Gibbula cineraria* and *Calliostoma zizyphinum*, although the range of distribution of the two first-mentioned species reaches farther south (Southward, 1958). Marine oligochaetes (*Enchytraeus albidus*) from the supralittoral attain a high freezing resistance when kept in the cold and at high salinities (Kähler, 1970). In freezing experiments at —4 C, nematodes, oligochaetes, small polychaetes (*Manayunkia*), and halacarides survived for several hours, whereas harpacticoids, ostracods and turbellarians were sensitive to freezing (Lorenzen,

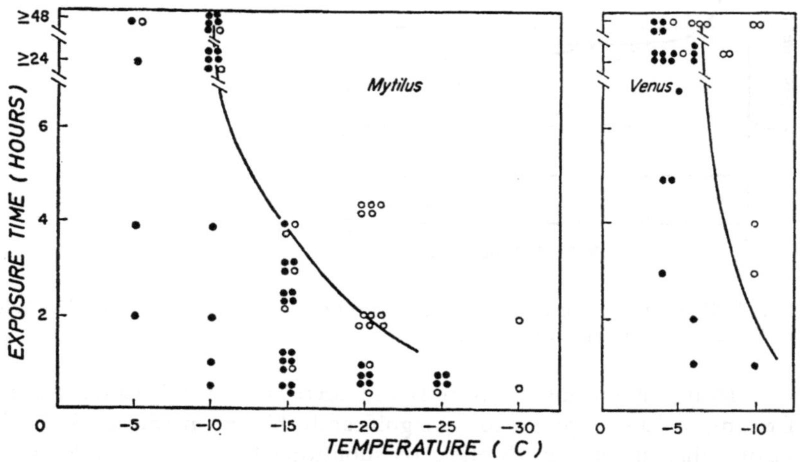

Fig. 2. Survival of two species of intertidal mollusks in cold air. ● Animal survived, ○ animal died. (After Williams, 1970)

1968). On the whole, species from the salt marsh proved to be more resistant than those from the bottom of the Wadden Sea (Lorenzen, 1968).

Nair and Leivestad (1958) demonstrated a considerable freezing resistance in wood-boring crustaceans on the Norwegian coast. *Limnoria lignorum* requires four hours at —18 C to reach 50% mortality. At —8 C 8 hours are needed by this species and 4 hours by the more sensitive *Chelura terebrans*. To protect themselves from extreme cold, these animals migrate in autumn and winter to a lower level of the same pile or to a neighbouring one. Consequently, after the winter months, empty burrows are observed in the wood piles above the water line. Animals that in winter remain too high up the pile will die. Hence, an upper limit of their winter distribution is established. This limit may be extended in the spring when population pressure drives the animals to unoccupied regions of the pile.

When studying the freezing resistance of lamellibranchs, it is noticed that under unfavourable external conditions these animals close their shells. By this action a lowering of temperature in the interior occurs more slowly in larger specimens with a greater store of water than in smaller ones (Fig. 3).

In order to measure cellular freezing resistance, isolated pieces of gill tissues of marine bivalves were frozen in small amounts of liquid. Most of the experiments were performed at the lethal temperature of —10 C. The decisive criterion for survival under freezing conditions was the continuation of the ciliary beat after subsequent thawing and a short recovery period at room temperature. Fluctuations of cellular activity rates observed at different intervals of time subsequent to freezing and thawing may reflect a balance between processes of regeneration and repair on the one hand and secondary freezing injuries (e. g. autolysis) on the other.

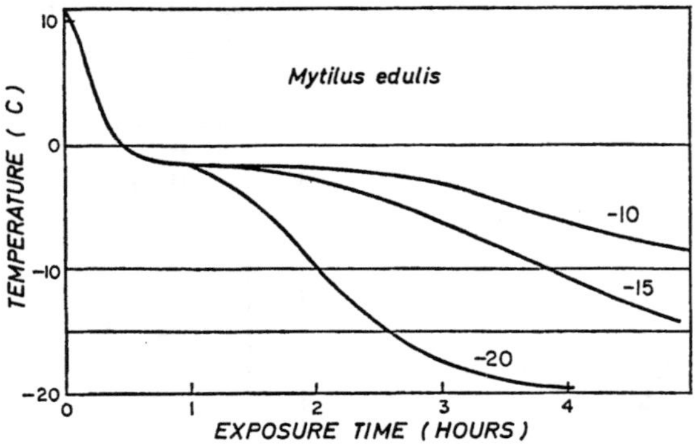

Fig. 3. Rate of cooling and freezing in specimes of *Mytilus edulis* (10–11 cm long) exposed to different low temperatures in air. (After Williams, 1970)

Among marine animals from the subarctic and boreal region, values for cellular freezing resistance prove to be significantly higher in species living in the upper littoral than in bottom dwellers or sublittoral forms. On the North German Sea coast *Mytilus edulis* possesses the greatest cellular freezing resistance, which is less pronounced in the sediment-inhabiting and therefore more protected forms *(Cardium edule, Macoma balthica, Mya arenaria)* and least in species limited to deeper water layers *(Spisula solida)* (Table 1). Similar results were obtained with lamellibranchs from the American Atlantic coast. In order of decreasing resistance to

Table 1. Cellular freezing resistance of isolated gill pieces of different lamellibranchs from the North Sea. The ciliary activity as a criterion for the degree of freezing injuries was observed after freezing of the isolated tissue pieces at —10 C, subsequent thawing and a period of recovery of 10 min at room temperature of about 20 C. (After Theede, 1965)

Species	50% of normal ciliary activity was found after tissue exposure to —10 C, for the periods of time (in minutes) given below	Complete cessation of ciliary activity was found after tissue exposure to —10 C for the periods of time (in minutes) given below
Tissues tested at 30 ‰ S		
Mytilus edulis	120	330
Cardium edule	65	210
Mya arenaria	42	90
Macoma balthica	35	100
Modiolus modiolus	3,5	6
Ostrea edulis	3	5,5
Spisula solida	2	4

freezing after acclimation to low temperature (5 C), some littoral bivalves from America's Atlantic Coast show the following sequence: *Mytilus edulis*, *Modiolus demissus*, *Crassostrea virginica*, *Macoma balthica*. Relationships exist between the freezing resistance of eulittoral species, the northward range of distribution, and the degree of exposure in the region of their northern distribution boundaries.

Whereas high cellular freezing- and chilling-resistance are characteristic for boreal lamellibranchs surviving in the tidal region in winter, only few species in subtropic-tropic eulittoral areas with distribution limits further northward possess similar resistances (e. g. *Crassostrea virginica* and *Brachidontes exustus*). Populations of oysters from the coast of Florida and from Cape Cod do not significantly differ in their cellular heat- and cold-resistance after acclimation to the same ambient temperature (Theede, 1972).

Effects of Environmental Factors on Temperature Resistance

Season

The freezing resistance is not developed to the same extent throughout the year. In nature animals exhibit pronounced seasonal fluctuations of freezing resistance on the organismic level, as has been shown by different authors, in, for instance, some littoral bivalves and gastropods (Kanwisher, 1955) and other forms. Sömme (1966) studied the seasonal changes of freezing resistance in three intertidal species from the Oslofjord. The common periwinkle *(Littorina littorea)*, the rough periwinkle *(L. rudis)* (Fig. 4) and the common barnacle *(Balanus balanoides)* survive

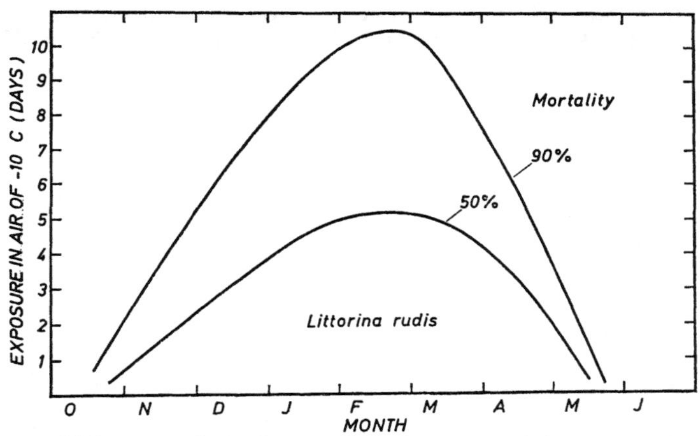

Fig. 4. Freezing resistance of *Littorina rudis* Jeffreys at different times of the year. (After Sömme, 1966)

several days in air at —10 C in mid-winter, whereas they are less resistant during other seasons. At —20 C, all specimens are killed by short exposure. On the cellular level, seasonal fluctuations of freezing resistance are apparent, too, as has been shown in bivalve tissue (Theede, 1965, 1972) (Fig. 5). These are more pronounced in animals from normal sea water than in brackish-water specimens. According to

Crisp and Ritz (1967), the middle lethal temperature (50% mortality after 18 hrs exposure to cold) for *Balanus balanoides* in summer is —6 to —8 C and in winter —14 to —18 C. In autumn, during the phase of rapid increase of freezing resistance, the animals at the higher levels of the shore acquire tolerance sooner than individuals at lower levels. The remarkable freezing resistance in winter specimens of *Balanus balanoides* is not only found in the adult, but also in the incubating eggs and nauplius stages. The planctonic stages, however, particularly the cyprids, are considerably less tolerant than the winter-resistant sessile adults (Fig. 6) (Crisp and Ritz, 1967).

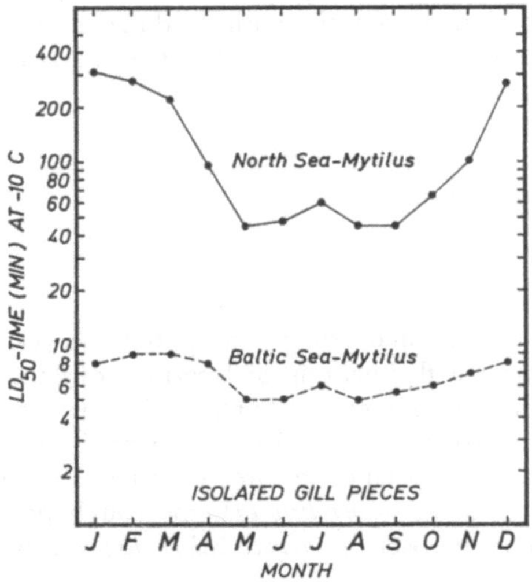

Fig. 5. Seasonal variations of cellular freezing resistance of *Mytilus edulis* from the North Sea (● — ●) and from the Baltic Sea (● --- ●). (After Theede, 1972)

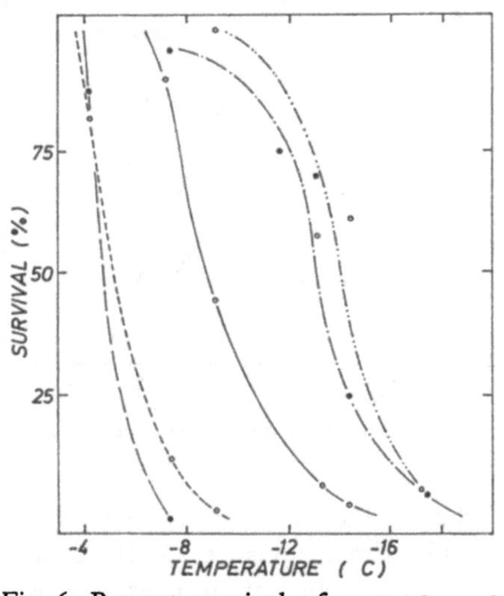

Fig. 6. Percent survival of eggs ○-···-○, late stage nauplii ●-·-●, cyprids ○--- ○, settled unmetamorphosed cyprids ●--●, and young spat ○— ○ of *Balanus balanoides* after exposure to low temperatures. (After Crisp and Ritz, 1967)

These seasonal resistance changes are clearly distinguished from the seasonal course of heat resistance which, according to Dregolskaya (1963), Crisp and Ritz (1967) and Friedrich (1967), is related mainly to the reproductive cycle.

The mechanisms leading to increases of freezing resistance are not yet fully understood. According to Crisp and Ritz (1967), the appearance of freezing tolerance in adult barnacles in winter coincides with a period of physiological hibernation involving relative metabolic inactivity. Cook and Gabbott (1970) found in adult specimens of *B. balanoides* that the glycerol content increases in winter when the barnacles are most cold tolerant and decreases in summer when the animals are more sensitive to freezing. The glycerol level in January amounts to 0,036% of the dry weight corresponding to body tissue concentrations of 1.06 mM. This value is about 1000-fold lower than concentrations of glycerol in the pupal haemolymph of some insects. It is not known whether these low glycerol levels are high enough to account for the freezing resistance of *Balanus* by a supercooling effect. Determinations of the glycerol content and of the supercooling points of the haemolymph of barnacles would answer this question. According to Cook and Gabbott (1972) the degree of unsaturation (iodine number) of the total body lipids of *B. balanoides* increases during autumn and remains high throughout the winter when the barnacles are most cold tolerant. During spring the iodine number falls and remains low in the summer. The high iodine values of the body lipids during winter are attributed to the phospholipid fraction, and it is suggested by the authors that a high degree of unsaturation of the fatty acid components of membrane-bound phospholipids may be an important factor in the acquisition of cold tolerance of barnacles. A similar mechanism has also been suggested by Dawson and Barnes (1966) for the cold tolerance of the eggs of *B. balanoides* in winter. Such seasonal changes of unsaturated lipoprotein contents have not yet been studied in bivalves in connection with varying cold tolerance.

Salinity and Salinity Adaptation

Marine invertebrates living in brackish water show a very small cellular freezing resistance when compared to specimens of resistant populations of the same species from normal sea water. "Survival times" of the order of not more than a few minutes have been obtained with tissue of specimens from the Baltic Sea at —10 C. An interesting aspect for the physiologist is that other resistances (to dehydration by exposure to air, to high salinities, heat, hydrostatic pressure, O_2-deficiency and H_2S) are also reduced (Theede *et al.*, 1969; Schlieper, 1971). The loss of freezing resistance, therefore, could be considered to be, at least partly, the expression of a "non-specific" reduction of resistance.

Marine lammelibranchs from the Gulf of Finland in a salinity of 6‰ (near their extreme limit of distribution) have almost completely lost the ability to survive freezing: exposure to —10 C kills the tissue within 2 to 3 minutes. The loss of freezing resistance has ecological consequences. In the inner Baltic and in certain parts of estuaries, marine macrofauna survive in winter only if the specimens remain covered by water. Animals which in summer have chosen localities too close to the shore or the water line on rocks and piles, die at low water levels (Theede and Lassing, 1967). Altogether the reduced resistance to abiotic influences such as heat, freezing, and dehydration may be a contributory factor to the "upper submergence" of several marine species living in brackish waters.

Furthermore, it was demonstrated that there are only slight seasonal fluctuations in the cellular freezing resistance of brackish-water forms (Fig. 5). In the laboratory, acclimation to low temperatures in conjunction with high salinities results in high freezing resistance. This holds for whole animals (*Enchytraeus:* Kähler, 1970) and for isolated tissues (*Mytilus:* Theede, 1969, 1972). In both cases acclimation to high salinity increases the degree of resistance to temperature extremes. The effect of salinity on cold resistance decreases with rising adaptation temperature (Fig. 7).

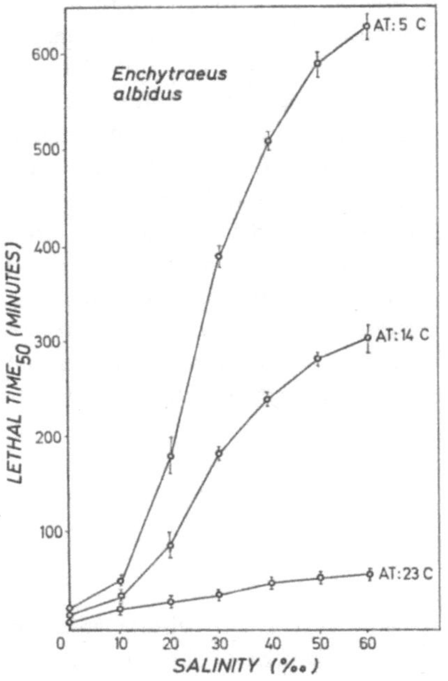

Fig. 7. Freezing resistance of the oligochaete *Enchytraeus albidus* after pretreatment at different temperature-salinity combinations. (After Kähler, 1970)

The magnitude of this effect is genetically fixed. The time course of adaptation was first observed after transfer of adult specimens of *Mytilus edulis* from the Baltic into normal sea water by Theede (1965) (Fig. 8). Both the cellular resistance to high salt concentrations and the cellular freezing resistance of these mussels increase in animals from normal sea-water during a slow process of adjustment of about 4 weeks. On the other hand, the level of salinity resistance and freezing resistance of North Sea animals decreases sharply after transfer into brackish water. These results suggest that the differences of resistance found in marine invertebrates from the North Sea and the Baltic are mainly due to individual non-genetic adaptation.

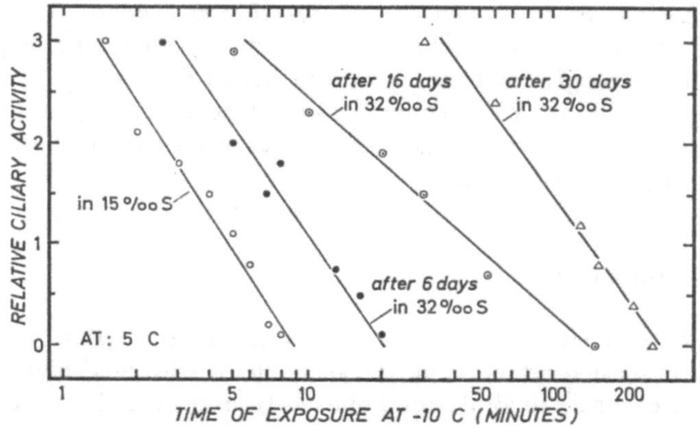

Fig. 8. The time course of change of cellular freezing resistance of *Mytilus edulis* from the Baltic Sea (~15‰ S) after transfer of the animals to sea water of 32‰ S (at 5 C). (After Theede, 1965)

Temperature and Temperature Adaptation

Adaptation of freezing resistance is known in different types of invertebrates e.g. in the ciliate *Zoothamnium hiketes* (Vogel, 1966), the oligochaete *Enchytraeus albidus* (Kähler, 1970), the barnacle *Balanus balanoides* (Cook and Lewis, 1971; Cook and Gabbott, 1972), the mussels *Mytilus edulis* and some other bivalves (Theede, 1972). After transfer from 15 C to 5 C mussels from the North Sea need about 3 weeks to acquire their ultimate high cellular freezing resistance. The corresponding loss of freezing resistance, when cold-acclimated animals are transferred to a higher temperature, occurs within a much shorter time (Fig. 9). Experiments with other bivalve species from the American east coast demonstrate smaller, but pronounced, effects of acclimation temperature on freezing resistance *(Crassostrea virginica, Macoma*

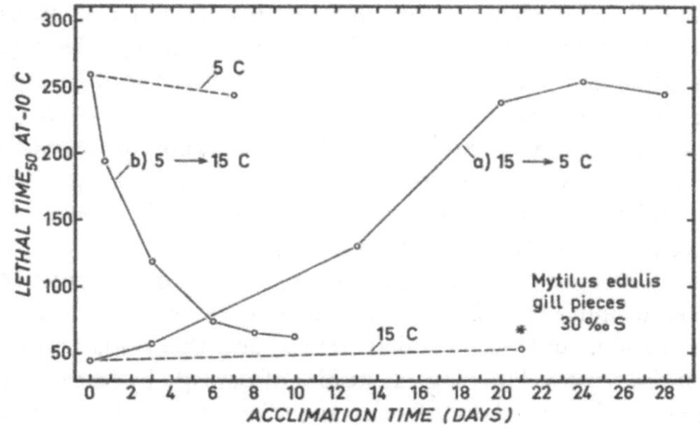

Fig. 9. Influence of acclimation temperature on the freezing resistance of isolated gill tissue of *Mytilus edulis*. (After Theede, 1972)

balthica, Modiolus demissus and *Brachidontes exustus)*. The degree of adaptation of freezing resistance in different species may be considered as genetically fixed, although it is strongly modified by other external factors (e.g. salinity, ionic relations, exposure to air, food supply). The influence of non-genetic temperature adaptation seems to account mainly for the seasonal fluctuations of freezing resistance in bivalves (Theede, 1972).

Other Factors

When considering the resistant littoral invertebrates, it is interesting to note that the most resistant forms are also able to withstand high salinities and dehydration by exposure to air. Littoral mussels from the North Sea are able to survive exposure for longer than a week. The animals close their shells, reduce heart beat and ciliary activity and liberate energy by glycolytic processes. Experimental studies demonstrated a strong increase of freezing resistance over the course of some days when mussels were exposed to air. An increase in osmotic concentration by evaporation, and in metabolites from the more or less glycolytic metabolism, may account for this effect. When animals are kept in oxygen-deficient water, the cellular freezing resistance is somewhat increased, too, although this effect is much smaller than during exposure to air (Theede, 1972).

Laukner (1972) points out that the age of animals and infection by parasites may influence their tolerance. Small specimens of *Cardium lamarcki* Reeve proved to be more resistant to freezing than large ones. The "stress effects" of a short freezing exposure caused a higher mortality rate in those specimens which were under the additional stress of a parasitic infestation.

Mechanisms of Freezing Injury and Freezing Resistance

Literature on damaging effects of freezing and types of cryoprotection has recently been discussed by Ashwood-Smith (1970), Farrant (1970), Farrant and Woolgar (1970) and other authors in Meryman (1966) and Wolstenholme and O'Connor (1970). Intracellular ice may have very damaging effects (Chambers and Hale, 1932) especially due to physical (mechanical) disruption. In experiments with the hydroid polyps *Laomedia flexuosa* and *L. loveni* Jankowsky *et al.* (1969) were able to induce, selectively, extra- and intra-cellular freezing or extracellular freezing alone. The animals withstood extracellular freezing much better than the formation of intracellular ice. About 20% of the specimens of *Laomedea loveni* survived intracellular ice formation at —4 C for up to 10 min.

The first step in slow freezing under natural conditions is the formation of extracellular ice. With growth of the ice crystals the osmotic concentration of the remaining unfrozen fluid increases, and this leads to greater or lesser dehydration of the tissues. Histological studies of damage by freezing and thawing reveal shrinkages and distortions of cells and disruptions of membranes (Luyet and Gehenio, 1940; Kanwisher, 1955, 1966). Experiments by Lovelock (1953) with red blood cells indicate that the increasing electrolyte concentration during freezing is a destructive factor. Meryman (1970, 1971), however, found some evidence that the absolute concentration of electrolytes is not the main reason for membrane damage, but that the osmotic pressure, by increasing the concentration of non-penetrating solutes, may be an even more decisive factor in producing freezing injuries. The idea of this author is that the cell membrane is damaged when the osmotic stress during extracellular freezing causes the cell volume to decrease below a specific minimum value. Steps during this process are increase and later loss of semipermeability of the cell membrane, and finally irreversible disruptions.

The most sensitive organelles within the cells seem to be the lysosomes (Kohn, 1960; Chilson *et al.*, 1965). In bivalve gills the importance of lysosomal injuries due to freezing was demonstrated by Theede (1972). Quick release of acid phosphatase from lysosomes could be measured after freezing and thawing of the gills. The effect was stronger in the less resistant tissues and indicates early damages to the organelle and cell membranes. Persidsky (1971) pretreated bone marrow cells with trypan blue which concentrates selectively in lysosomes and inhibits their enzymes. Such pretreated cells showed better recovery after freezing than the untreated ones. It was concluded from these experiments that activation of lysosomal enzymes by freezing and thawing is an important cause of cryoinjury. The hypothesis of this author is that cellular cryopreservation might be improved by specific protection of lysosomes, inhibition or depletion of their enzymes. The unanswered question is whether and how similar phenomena are already realized in resistant forms in nature.

Besides these cellular effects Scholander *et al.* (1953) showed a sharp drop in oxygen consumption concurrent with ice formation (in chironomid larvae). Q_{10} values as high as about 50 resulted in the temperature range below freezing. Similar results were obtained in frozen periwinkles (*Littorina littorea*) by Kanwisher (1959). This author concluded that the salinity increase during freezing is largely responsible for the high Q_{10}. Further causes for this effect were sought in increased viscosity of cellular and body fluids and in the fact that ice may act as a diffusion barrier to gases. The drop of oxygen consumption during freezing does not seem to have harmful effects within short exposure times. But one can imagine that the formation of gas bubbles during ice crystallization could have deleterious consequences.

The conditions within animals and their cells, to which the ability to tolerate freezing may be attributed, may be characterized according to my own and other investigations as follows: Kanwisher (1955) found by calorimetric studies that in some littoral molluscs a large percentage of body water can freeze extracellularly without lethal consequences. According to Williams (1970) death in some bivalves occurred when more than 64% of their water content was frozen. In *Mytilus edulis* this occurred on cooling below —10 C; in *Venus mercenaria* (a species more sensitive to freezing) this amount of water already froze at —6 C. During freezing, 20% of the water content of *Mytilus* is osmotically inactive (Fig. 10). This water seems to be important for the freezing resistance because a higher degree of hydration is maintained at low temperatures, and consequently the cell volume does not readily fall below a certain critical value, as Meryman (1970) has pointed out. How this tissue water becomes osmotically inactive is not yet known. Williams (1967, 1968) suggests the occurrence of organic intracellular cryoprotective substances. According to Williams (1970), the content of free amino acids within the cells does not seem to be responsible for the high freezing resistance of *Mytilus* tissue. Protecting substances such as glycerol and DMSO could not be detected in bivalves (Kanwisher, 1966). Therefore, it might be concluded that water is bound to fine cellular structures or macromolecular substances. In this connection it may be of interest that Antarctic limpets *(Patinigera polaris)*, sometimes found in near-shore ice, protect themselves by an envelope of mucus which inhibits ice penetration into the tissue. The ice inhibiting components in limpet mucus are in the macromolecular fraction (Hargens, personal communication).

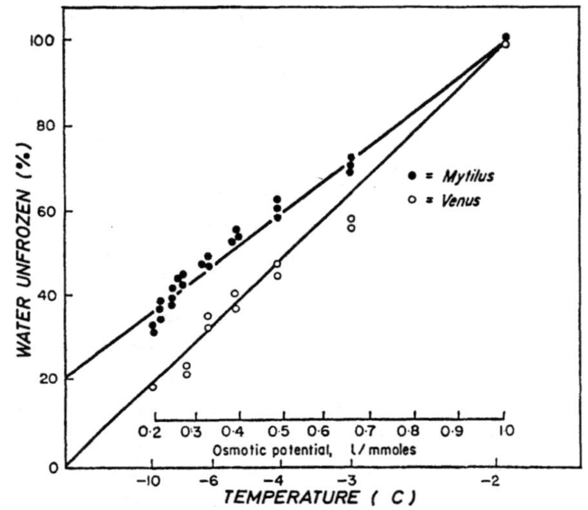

Fig. 10. Boyle-van't Hoff plot comparing the amount of frozen water as a function of temperature in two species of mollusks. *Mytilus* shows about 20%, *Venus* shows no osmotically inactive water. (After Williams, 1970)

Heber and Santarius (1971) found that in plant cells a great number of high- as well as low-molecular compounds are able to protect isolated membranes of plastides from the harmful effects of freezing. The protecting effect can even be increased by combination of different organic and inorganic substances. These authors speculate that high freezing resistance need not be due to a single factor (increase of the content of one protecting substance), but can result from the complex action of a number of different cellular constituents.

It is interesting to note that those littoral marine invertebrates that are distinguished by high cellular freezing resistance also possess a high cellular salinity resistance (osmotic resistance) (Theede, 1965). Tissues of these marine animals have to withstand dehydration and osmotic stress effects associated with ice formation if they are to survive. Consequently, resistance to high osmotic concentrations is a prerequisite for high cellular freezing resistance.

When mussels *(Mytilus edulis)* from brackish water are transferred to normal sea water, not only the cellular freezing resistance but also the cellular salinity resistance (to high salt concentrations) increases (Theede, 1965). The acclimation time required for reaching the final level of freezing resistance is definitely longer than for the osmotic adjustment to the changed salinity. The long duration of adjustment and the effect of the acclimation temperature seem to point to a special mechanism responsible for the development of high cellular freezing resistance. One important aspect seems to be that the conditions within the cells, influenced by the salinity of acclimation, do not only directly affect the cellular freezing resistance but also strongly modify the effects of other environmental factors (air-exposure, O_2-deficiency, protecting agents, acclimation temperature) on this resistance. These results and the above-mentioned relatively "nonspecific" influence of salinity on different resistances suggest that structural stabilities of cellular components may be involved.

Fish

Ecological aspects of temperature adaptation in marine fish were recently discussed by Brett (1970). In this chapter I restrict myself to problems concerning the survival of fish under supercooled conditions. From earlier investigations, especially of Scholander and coworkers (Scholander *et al.*, 1957; Gordon *et al.*, 1962), it is known that polar fish species may occur under layers of ice at temperatures near the freezing point of sea water. The survival of these fish depends upon their ability to prevent their body fluids from freezing. A few arctic bottom fish are able to survive under supercooled conditions, at which time they avoid any contact with ice. Polar shallow-water species appreciably raise the osmotic value of their blood measured in terms of freezing-point depression. They are able to prevent ice propagation in their body fluids to an even greater extent. Hargens (1972) recently determined plasma melting points, freezing points, temperatures of initial ice formation and rates of ice propagation in polar fish by use of a modified micro osmometer. According to his findings, arctic and antarctic shallow-water fish, living in contact with sea water at —1.8 to —1.9 C are able to prevent the onset of ice propagation in their blood down to a temperature of about —2.1 C although their equilibrium freezing point (the temperature where the edges of a seed crystal of ice sharpen) lies at about —1.1 C to —1.2 C (Fig. 11). The great disparity between equilibrium freezing point (as a measure for the number of osmotically active molecules and ions) and the temperature of initial ice formation is correlated with a high colloid osmotic pressure caused by components of the protein fraction of the plasma. Scholander *et al.* (1957) found that the freezing-point depression of these fish species is subject to seasonal variations. Gordon (1957), Gordon *et al.* (1962) and Raschack (1967, 1969) confirmed this and demonstrated additionally that the osmotic concentration of the blood of *Myoxocephalus scorpius* is raised during acclimation to low temperatures by increasing, in the first place, the content of organic compounds. The amount of added organic substances during cold acclimation is smaller when the fish (*M. scorpius*) are kept in brackish water (Raschack, 1969). According to Hargens (1972) the protein content and the colloid osmotic pressure are mainly involved. DeVries and Wohlschlag (1969) found that it is the high content of serum glycoproteins which inhibits the ice crystal growth in some antarctic fish. DeVries (1971) determined the molecular structure of a subunit of a serum glycoprotein. His findings indicate that cis-hydroxyl groups on the galactose residues of these subunits are the active components which prevent ice crystal growth. When a low temperature is attained, where ice propagation can start, the crystallization pattern also will be influenced (Hargens, 1972). Dendritic forms of ice crystals were observed in blood samples of polar deep-bottom fish. In the blood of polar shallow-water fish sharp monoclinic ice spears formed when the samples were cooled to 0.8–1.2 C below the freezing point.

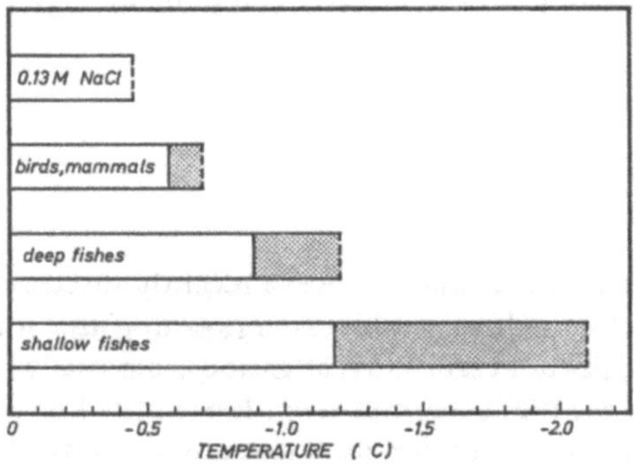

Fig. 11. Equilibrium freezing point (|) and temperature of initial ice propagation (¦) in different solutions. (No difference of both values in 0,13 M NaCl-solution, small differences in plasma of birds, mammals and polar deep-bottom fishes, great difference in polar shallow-water fishes). (After Hargens, 1972)

A temperate zone species, the killifish *Fundulus heteroclitus* increases its glucose level in the blood at subzero temperatures (Umminger, 1967, 1969, 1974). In this form the hydroxyl-rich compound glucose is suggested to act as a supercooling stabilizer by avoiding spontaneous nucleation in the absence of ice crystals. This substance, however, is ineffective in promoting supercooling when ice crystals are present in the water. Therefore supercooled killifish avoid contact with sea ice. Glycoproteins present in the blood of several polar species are much more effective. For this reason, e. g. supercooled specimens of antarctic *Trematomus* can actually tolerate contact with ice (DeVries, 1971).

Umminger (1974) summarizes the chemical changes and regulatory mechanisms involved in the acclimation of the killifish to subzero temperatures: The hyperglycemic response is produced by breakdown of liver glycogen, which is mainly catalyzed by hepatic glycogen phosphorylase. The hormonal control of these cold-induced changes in carbohydrate metabolism results from a hypertrophy and degranulation of glucagon-containig alpha cells and an atrophy of insulin-producing beta cells in the pancreatic islets. Also a possible neural involvement is discussed. Umminger (1974) points out that the manner in which carbohydrate metabolism is altered by low temperature in the killifish should not be considered as a generalization to be applied to all fish. Different responses of single species to extreme low temperatures will have to be interpreted in terms of their phylogenetic histories, ecologies and physiological demands.

Summary and Conclusions

Different degrees of resistance to chilling and freezing in littoral marine invertebrates can be distinguished on the organismic and on the cellular level. They correspond to the degree of exposure of the species to low temperatures in winter and to their geographic distribution. Considerable seasonal variations of freezing resistance in marine littoral invertebrates of the temperate and colder regions are induced mainly by temperature adaptation. Specimens living in brackish waters are very sensitive to freezing in comparison with resistant individuals of the same species from normal sea water. Seasonal variations of their freezing resistance are only very small. Long-term adaptation experiments have shown that these differences are mainly induced by individual acclimatization to the salinity of the external medium. The reduction of cellular freezing resistance, caused by adaptation of whole animals to brackish water, is related to the similarly reduced resistance to other abiotic environmental factors (high salinity, heat, hydrostatic pressure). Altogether high freezing resistance results from adaptation to low temperature combined with high salinity. Additional increases of cellular freezing resistance caused by exposure to air or to oxygen-deficient water or to different organic compounds can be strongly modified by temperature and salinity conditions during pretreatment. Results of adaptation experiments suggest that different mechanisms are involved when the cellular resistances to heat, chilling or freezing are changed by individual adaptation. High cellular freezing resistance of some intertidal marine invertebrates requires high resistances to dehydration and high salinities. In addition to this, structural peculiarities (high degree of unsaturation of the fatty acid components of membrane-bound phospholipids) and intracellular antifreeze components (which inhibit intracellular ice propagation) seem to play an important role.

In contrast to some marine littoral invertebrates, which are able to withstand partial freezing over a certain period of time, polar marine fish species have developed mechanisms that prevent freezing by adding organic compounds to the blood.

Further studies of the properties of the cellular and intracellular fluids of freezing-resistant marine invertebrates, comparable to those recently performed with the blood of fish, seem to be required to elucidate the protecting mechanisms in these forms.

References

Asahina, E.: On a probably freezing process of molluscan cells enabling them to survive at a super-low temperature. Low Temp. Sci., Ser. B **16**, 74–75 (1958). (Japanese with English summary.)

Ashwood-Smith, M. J.: Effects of low temperatures on microorganisms, plants, and cold-blooded animals. In: Smith, A. U.: Current trends in Cryobiology, p. 5–42. London and New York: Plenum Press 1970.

Behrisch, H. W.: Molecular mechanisms of adaptation to low temperature in marine poikilotherms. Some regulatory properties of dehydrogenases from two arctic species. Marine Biol. **13**, 267–275 (1972).

Blegvad, H.: Mortality among animals of the littoral region in ice winters. Reports Danish Biol. Station **35**, 49–62 (1929).

Brett, J. R.: Temperature: animals – fishes. In: O. Kinne (Ed.): Marine Ecology. Vol. 1. Environmental factors, Part 1. London and New York: Wiley Interscience 1970.

Brongersma-Sanders, M.: Mass mortality in the sea. In: Hedgpeth, J. W. (Ed.): Treatise on marine ecology and paleoecology, Vol. 1. Ecology, p. 941–1010. (Mem. Geol. Soc. Am. 67.) Baltimore: Waverly Press 1957.

Chambers, R., Hale, H. P.: The formation of ice in protoplasm. Proc. Roy. Soc. (Lond.) **B 110**, 336–352 (1932).

Chilson, O. P., Costello, L. A., Kaplan, N. O.: Effects of freezing on enzymes. Federation Proc. Federation. Am. Soc. Exp. Biol. **24** (Suppl. 15), 55–65 (1965).

Cook, P. A., Gabbott, P. A.: Seasonal changes in the free glycerol level in the body parts of the adult barnacle Balanus balanoides. Marine Biology **7**, 11–13 (1970).

Cook, P. A., Gabbott, P. A.: Seatonal changes in the biochemical composition of the adult barnacle, Balanus balanoides, and the possible relationships between biochemical composition and cold-tolerance. J. Marine Biol. Assoc. U.K. **52**, 805–825 (1972)

Cook, P. A., Lewis, A. H.: Acquisition and loss of cold-tolerance in adult barnacles *(Balanus balanoides)* kept under laboratory conditions. Marine Biol. **9**, 26–30 (1971).

Courtney, W. A. M., Webb, J. E.: The effects of the cold winter of 1962/63 on the Helgoland population of Branchiostoma lanceolatum (Pallas). Helgoländer Wiss. Meeresuntersuch. **10**, 301–312 (1964).

Crisp, D. J. (Ed.): The effects of the severe winter of 1962–63 on marine life in Britain. J. Animal. Ecol. **33**, 165–210 (1964).

Crisp, D. J.: The effects of the winter of 1962/63 on the British marine fauna. Helgoländer Wiss. Meeresuntersuch. **10**, 313–327 (1964a).

Crisp, D. J., Ritz, D. A.: Changes in temperature tolerance of Balanus balanoides during its life-cycle. Helgoländer Wiss. Meeresuntersuch. **15**, 98–115 (1967).

Dawson, R. M. C., Barnes, H.: Studies in the biochemistry of cirripede eggs. II. Changes in lipid composition during development of *Balanus balanoides* and *B. balanus*. J. Marine Biol. Assoc. U. K. **46**, 249–261 (1966).

Devries, A. L.: Glycoproteins as biological antifreeze agents in antarctic fishes. Science **172**, 1152–1155 (1971).

Devries, A. L., Wohlschlag, D. E.: Freezing resistance in some antarctic fishes. Science **163**, 1073–1075 (1969).

Dregolskaya, J. N.: Heat resistance of ciliated epithelium of gills of *Mytilus galloprovincialis* L. from the Black Sea. (Russ., Eng. Summary.) In: Ushakow, B. P. (Ed.): Problems of cytoecology of animals, p. 43–50. Moscow: Acad. Sci. 1963.

Farrant, J.: Mechanisms of injury and protection in living cells and tissues at low temperatures. In: Smith, A. U. (Ed.): Current trends in Cryobiology, p. 139–152. London and New York: Plenum Press 1970.

Farrant, J., Woolgar, A. E.: Possible relationships between the physical properties of solutions and cell damage during freezing. In: Wolstenholme, G. E. W., O'Connor, M. (Eds.): The frozen cell, p. 97–119. London: J. & A. Churchill 1970.

Friedrich, L.: Experimentelle Untersuchungen zum Problem zellulärer nichtgenetischer Resistenzänderungen bei der Miesmuschel *Mytilis edulis* L. Kiel. Meeresforsch. **23**, 105–126 (1967).

Gerlach, S. A.: Über die Fauna der Gezeitenzone von Spitzbergen. Proc. 5th Marine Biol. Symp. Botanica Gothoburgensia **3**, 81–92 (1965).

Gordon, M. S.: Observations on osmoregulation in the arctic char (*Salvelinus alpinus* L.). Biol. Bull. **112**, 28–33 (1957).

Gordon, M. S., Amdur, B. H., Scholander, P. F.: Freezing resistance in some northern fishes. Biol. Bull. **122**, 52–56 (1962).

Gunter, G.: Temperature. In: Hedgpeth, J. W. (Ed.): Treatise on marine ecology and paleoecology, Vol. 1: Ecology, p. 159–184. (Mem. Geol. Soc. Am. 67.) Baltimore: Waverly Press 1957.

Hargens, A. R.: Freezing resistance in polar fishes. Science **176**, 184–186 (1972).

Heber, U., Santarius, K. A.: Empfindlichkeit und Resistenz der Zelle gegen Frost. Umschau, Heft 25, 930–936 (1971).

Jankowsky, H. D., Laudien, H., Precht, H.: Ist eine intrazelluläre Eisbildung bei Tieren tödlich? Versuche mit Polypen der Gattung Laomedea. Marine Biol. **3**, 73–77 (1969).

Kähler, H. H.: Über den Einfluß der Adaptationstemperatur und des Salzgehaltes auf die Hitze- und Gefrierresistenz von *Enchytraeus albidus* (Oligochaeta). Marine Biol. **5**, 315–324 (1970).

Kanwisher, J. W.: Freezing in intertidal animals. Biol. Bull. Marine Biol. Lab., Woods Hole **109**, 56–63 (1955).

Kanwisher, J. W.: Histology and metabolism of frozen intertidal animals. Biol. Bull. **116**, 258–264 (1959).

Kanwisher, J. W.: Freezing in intertidal animals. In: Cryobiology, p. 487–494. Ed. by H. T. Meryman. London: Academic Press 1966.

Kinne, O.: The effects of temperature and salinity on marine and brackish water animals. I. Temperature. Oceanogr. Marine Biol. Ann. Rev. **1**, 301–340 (1963).

Kinne, O.: Temperature: animals-intervertebrates. In: Kinne, O. (Ed.): Marine ecology. Vol. 1. Environmental factors, Part. 1, p. 407–514. London-New York: Wiley Interscience 1970.

Kohn, A.: Lysis of frozen and thawed cells of *Escherichia coli* by lysozyme and their conversion into spheroblasts. J. Bacteriol. **79**, 697–706 (1960).

Kühlmorgen-Hille, G.: The effects of the severe winter 1962/63 on the bottom fauna of the Kiel Bay. Ann. Biol. **20**, 98 (1963).

Laukner, G.: Zur Taxonomie, Ökologie und Physiologie von *Cardium edule* L. und *Cardium lamarcki* Reeve. Thesis, Math. Nat. Fac., University of Kiel (F.R.G.), 1972.

Lorenzen, S.: Die Meiofauna an der Grenze Watt/Salzwiese im Bereich der deutschen Nordseeküste. Thesis, Math. Nat. Fac., University of Kiel (F.R.G.), 1968.

Lovelock, J. E.: The mechanism of the protective action of glycerol against haemolysis by freezing and thawing. Biochem. Biophys. Acta **11**, 28–36 (1953).

Luyet, B. J., Gehenio, P. M.: The mechanism of injury and death by low temperature. Biodynamica **3**, 33–99 (1940).

Madsen, H.: A study of the littoral fauna of Northwest Greenland. Medd. Grønland **124**, 1–24 (1940).

Meryman, H. T. (Ed.): Cryobiology, p. 775. London: Academic Press 1966.

Meryman, H. T.: The exceeding of a minimum tolerable cell volume in hypertonic suspension as a cause of freezing injury. In: Wolstenholme, G. E. W, O'Connor, M. (Eds.): The frozen cell, p. 51–67. London: J. & A. Churchill 1970.

Meryman, H. T.: Osmotic stress as a mechanism of freezing injury. Cryobiology **8**, 489–500 (1971).

Nair, N. B., Leivestad, H.: Effect of low temperature on the vertical distribution of two wood-boring crustaceans. Nature **182**, 814–815 (1958).

Newell, R. C.: Biology of intertidal animals, p. 555. London: Logos Press Ltd. 1970.

Persidsky, M. D.: Lysosomes as primary targets in cryoinjury. Cryobiology **8**, 482–488 (1971).

Petersen, G. H.: The distribution of *Balanus balanoides* L. and *Littorina saxatilis* (Olivi) var. groenlandica Menke in northern west Greenland. Medd. Grønland **159** (No 9), 1–47 (1962a).

Petersen, G. H.: *Balanus balanoides* L. (Cirripedia) Life cycle and growth in Greenland. Medd. Grønland **159** (No. 12), 1–114 (1962b).

Precht, H., Christophersen, J., Hensel, H.: Temperatur und Leben. p. 415. Berlin-Göttingen-Heidelberg: Springer 1955.

Precht, H., Christophersen, J., Hensel, H.: Temperature and Life. (2. ed.). Berlin-Heidelberg-New York: Springer 1973 (in press).

Precht, H.: Über die Resistenzadaptation wechselwarmer Tiere an extremen Temperaturen und ihre Ursachen. Helgoländer Wiss. Meeresuntersuch. **9**, 392–411 (1964).

Precht, H., Basedow, T., Bereck, R., Lange, F., Thiede, W., Wilke, L.: Reaktionen und Adaptationen wechselwarmer Tiere nach einer Änderung der Anpassungstemperatur und der zeitliche Verlauf. Helgoländer Wiss. Meeresuntersuch. **13**, 369–401 (1966).

Prosser, C. L. (Ed.): Molecular mechanisms of temperature adaptation, p. 390. Washington, D. C. Am. Assoc. Advan. Sci., 1967.
Raschack, M.: Osmoregulation und Kälteresistenz mariner Knochenfische. Naturwiss. **54**, 97 (1967).
Raschack, M.: Untersuchungen über die Osmo- und Elektrolytregulation bei Knochenfischen aus der Ostsee. Int. Rev. Ges. Hydrobiol. **54**, 423–462 (1969).
Schlieper, C.: Physiology of brackish water. In: Remane, A., Schlieper, C. (Eds.): Biology of brackish water, p. 211–353. Stuttgart: Schweizerbart 1971.
Schlieper, C., Flügel, H., Theede, H.: Experimental investigations of the cellular resistance of marine temperate and tropical bivalves: Results of the Indian Ocean Expedition of the German Research Association. Physiol. Zool. **40**, 345–360 (1967).
Scholander, P. F., Flagg, W., Hock, R. J., Irving, L.: Studies on the physiology of frozen plants and animals in the arctic. J. Cellular Comp. Physiol. **42** (Suppl. 1), 1–56 (1953).
Scholander, P. F., van Dam, L., Kanwisher, J. W., Hammel, H. T., Gordon, M. S.: Supercooling and osmoregulation in arctic fish. J. Cellular Comp. Physiol. **49**, 5–24 (1957).
Smith, A. U.: The resistance of animals to cooling and freezing. Biol. Rev. **33**, 197–253. (1958).
Sömme, L.: Seasonal changes in the freezing-tolerance of some intertidal animals. Nytt. Mag. Zool. **13**, 52–55 (1966).
Southward, A. J.: Note on the temperature tolerances of some intertidal animals in relation to environmental temperatures and geographical distribution. J. Mar. Biol. Assoc. **37**, 305–346 (1958).
Tanno, K., Wada, J.: Frost resistance in intertidal animals and ecological observations of their habitats in overwintering period, a preliminary report. Low Temp. Sci., Ser. B **24**, 15–23 (1966). (Japanese with English summary).
Theede, H.: Vergleichende experimentelle Untersuchungen über die zelluläre Gefrierresistenz mariner Muscheln. Kieler Meeresforsch. **21**, 153–166 (1965).
Theede, H.: Probleme der Frostresistenz bei Meerestieren. Naturw. Rundschau **20**, 468–475 (1967).
Theede, H.: Experimentelle Untersuchungen über physiologische Unterschiede bei Evertebraten und Fischen aus Meer- und Brackwasser. Limnologica **7**, 119–128 (1969).
Theede, H.: Vergleichende ökologisch-physiologische Untersuchungen zur zellulären Kälteresistenz mariner Evertebraten. Marine Biol. **15**, 160–191 (1972).
Theede, H., Lassig, J.: Comparative studies on cellular resistance of bivalves from marine and brackish waters. Helgoländer Wiss. Meeresuntersuch. **16**, 119–129 (1967).
Tiedtke, B.: Über die ökologische Bedeutung eines extrem kalten Winters für die eulitorale Hartbodenfauna der Kieler Förde. Schriften Naturw. Ver. Schleswig-Holstein **35**, 33–60 (1964).
Tilbrook, P. J.: The terrestrial environment and invertebrate fauna of the maritime Antarctic. In: Holgate, M. W. (Ed.): Antarctic ecology. Vol. 2, p. 886–896. London and New York: Academic Press 1970.
Troshin, A. S. (Ed.): The cell and environmental temperature. (Transl. from. Russ.). Engl. ed. by Prosser, C. L., p. 549. Oxford: Pergamon Press 1967.
Umminger, B. L., Sub-zero temperatures and supercooling in *Fundulus heteroclitus*. Am. Zool. **7**, 731 (1967).
Umminger, B. L.: Physiological studies on supercooled killifish *(Fundulus heteroclitus)*. II. Serum organic constituentes and the problem of supercoolinT. J. Exp. Zoll. **172**, 409–424 (1969).
Umminger, B. L.: Low temperature resistance adaptations in the killifish, *Fundulus heteroclitus*. In: Vernberg, F. J. (Ed.): Physiological ecology of estuarine organisms. Columbia: University of South Carolina Press, in press.
Vernberg, F. J., Schlieper, C., Schneider, D. E.: The Influence of Temperature and Salinity on Ciliary Activity of Excised Gill Tissue of Molluscs from North Carolina. Comp. Biochem. Physiol. **8**, 271–285 (1963).
Vernberg, F. J., Vernberg, W. B.: Lethal limits and the zoogeography of the faunal assemblages of coastal Carolina waters. Marine Biol. **6**, 26–32 (1970).

Vogel, W.: Über die Hitze- und Kälteresistenz von Zoothamnium hiketes Precht (Ciliata, Peritr.). Z. Wiss. Zool. **173**, 344–378 (1966).

Williams, R. J.: Cryoprotective agents in intertidal mollusks. Abstract. Cryobiology **3**, 370 (1967).

Williams, R. J.: The mechanism of cryoprotection in the intertidal mollusk Mytilus Abstract. Cryobiology **4**, 250 (1968).

Williams, R. J.: Freezing tolerance in Mytilus edulis. Comp. Biochem. Physiol. **35**, 145–161 (1970).

Wolstenholme, G. E. W., O'Connor, M. (Eds.): The frozen cell, p. 316. London: J. & A. Churchill 1970.

Zhyubikas, J. J.: The resistance to cold in *Mytilus edulis* L., living on littoral and tidepools of east Murman. Vestn. Leningr. Univ. **15**, 27–31 (1968). (Russ., Engl. summary.)

Ziegelmeier, E.: Einwirkungen des kalten Winters 1962/63 auf das Makrobenthos im Ostteil der Deutschen Bucht. Helgoländer Wiss. Meeresuntersuch. **10**, 276–282 (1964).

Ecological Aspects of Cold Resistance in Spiders*
(A Comparative Study)

WALTER KIRCHNER

With 3 Figures

Introduction

The way in which European spiders spend the winter has, so far, been little investigated. For the majority of species the winter quarters are unknown, and exact information concerning the physiological adaptations which enable these animals to survive the low temperatures of winter is lacking.

Two investigations carried out several years ago on the orb-weaver *Araneus cornutus* Clerck (Kirchner, 1965; Kirchner and Kestler, 1969) helped to fill some gaps in our knowledge in this field. It was shown that this species can tolerate temperatures of —20 C and below, although high resistance to cold is confined to the winter. According to other authors (e.g. Engelhardt, 1964; Buche, 1966), some spiders are apparently killed by cooling to only a few degrees below zero. This was the starting point for the present author's comparative investigations into the cold resistance of a large number of spider species. It was hoped, among other things, to discover whether the degree of cold resistance is related to the particular habits of the species, especially to its choice of winter quarters. A preliminary answer was also sought to the question to what extent resistance to cold depends upon changes connected with annual rhythms or with age.

Material and Methods

Of the 800 species of spiders occurring in Germany, 50 species from 15 different families were investigated. It was unfortunately not always possible in winter to catch 5 or more individuals of the same species so that only single measurements are available for about half of the species. In view of the uncertainty attached to such values they will not be discussed in the following.

Most of the experimental animals were collected in the vicinity of Bonn (especially in the Siebengebirge and Eifel) and in Unterfranken. My thanks are due here to Prof. E. Kullmann (Kiel) and Dr. B. Krafft (Strasbourg) for providing me with spiders from other parts of Germany and from abroad.

Wherever possible, the cold resistance of the animals was measured within one to five days of their being caught. The spiders were kept, without food, in a refrigerator at +5 to +10 C until the commencement of the experiment.

* Supported by Deutsche Forschungsgemeinschaft.

The animals were cooled at a constant speed of 10 C per hour, and the temperature at which the body water of the spiders freezes was measured with a copper-constantan thermoelement. Details of the methods used can be found in Kirchner and Kestler (1969), Kirchner and Kullmann (1972).

Results

The following results are based upon more than 700 measurements of cold resistance:

In Central European Spiders, Death Brought about by Cold is Due to the Formation of Ice Crystals in the Body

In all indigenous spiders so far investigated, cold only leads to death at temperatures below 0 C, at the temperature at which the thermoelement indicates the formation of ice in the body of the experimental animal. No damage caused by cold can be detected prior to the freezing of the body water. Whether this is also the case if unfrozen experimental animals are exposed to temperatures below 0 C for longer than $2^1/_2$ days has not yet been tested experimentally.

The situation in tropical spiders is apparently different: although the body fluids of the funnel-web spider, *Agelena consociata* Denis, a native of Africa, do not freeze until —8 C, some individuals are unable to survive even a few hours at 0 C.

It was shown in *Araneus cornutus* (Kirchner and Kestler, 1969) that ice formation occurs in this species at temperatures much lower than those to be expected from the osmolarity of the haemolymph. This finding, together with the observation that indigenous spiders freeze at temperatures considerably lower than —2 C (Table 1), justifies the conclusion that cold resistance of Central European spiders is due to a more or less strongly developed ability to supercool.

The Degree of Cold Resistance is Species Specific

Individual species of spiders differ with respect to degree of cold resistance: frost-sensitive species freeze at —4 C, whereas species with a high resistance to cold tolerate temperatures down to —26 C (Table 1). As shown in Fig. 1 the supercooling point may vary considerably within a species: The values from species with low cold resistance (e.g. *Meta menardi*) exhibit, as a rule, less scatter than those from species with a moderate or high ability to supercool.

In some species (e.g. *Araneus cornutus* and *Theridion notatum*), it was possible to compare individuals from different Central European countries, but no noteworthy differences in the cold resistance of the various populations could be detected.

Similarly, no statistically significant differences between the frost resistance of males and females of identical age and from one and the same species have been detected.

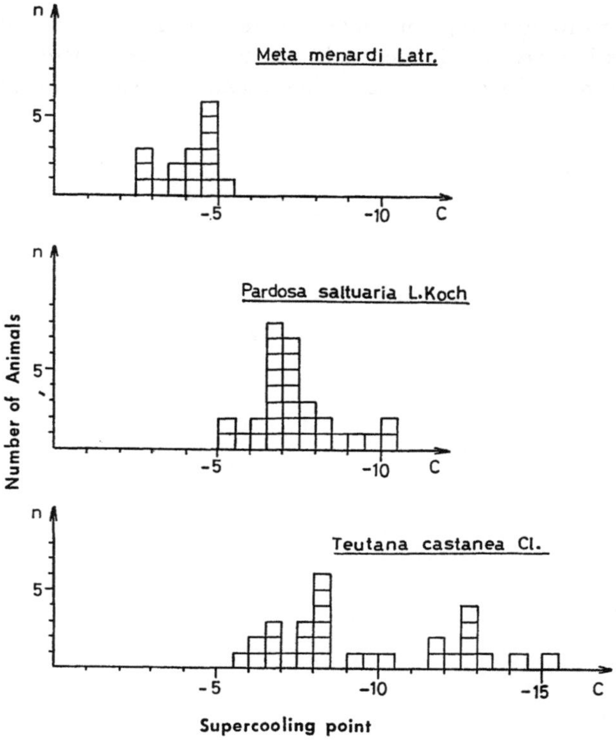

Fig. 1. Variability of supercooling point in three species of spiders. The number of individuals is given by the number of squares

The Degree of Cold Resistance Depends to Some Extent upon the Nature of the Winter Quarters

Various species of spiders may choose very different sites for passing the winter: in the vegetation (Lock, 1939; Duffey, 1962; Wiehle, 1927; Wiehle, 1931), in hollow plant stems (Kirchner, 1965; Tischler, 1966), in houses and caves, beneath bark, in tree stumps, in the soil or underneath stones (v. Kolosvary, 1939; Engelhardt, 1964; Buche, 1966; Kirchner and Kullmann, 1972, inter alia).

The cold stress for the animal may vary considerably according to the kind of site chosen: spiders spending the winter on plants have no form of protection from cold, and even the walls of hollow plant stems scarcely diminish the severity of the frost in the absence of snow, as the author's own measurements show (Fig. 2). On the other hand, spiders that spend the winter in the soil are much less exposed to cold (Kirchner and Kullmann, 1972) and even less exposed still are those spending the winter in houses or in caves.

To what extent are these conditions reflected in the animal's cold resistance? The supercooling points of various species of spider are given in Table 1, grouped according to type of winter quarters. It can be seen that spiders living in the vegetation have the highest cold resistance values whereas those living in soil or caves have the lowest. House spiders occupy a position somewhere in the middle.

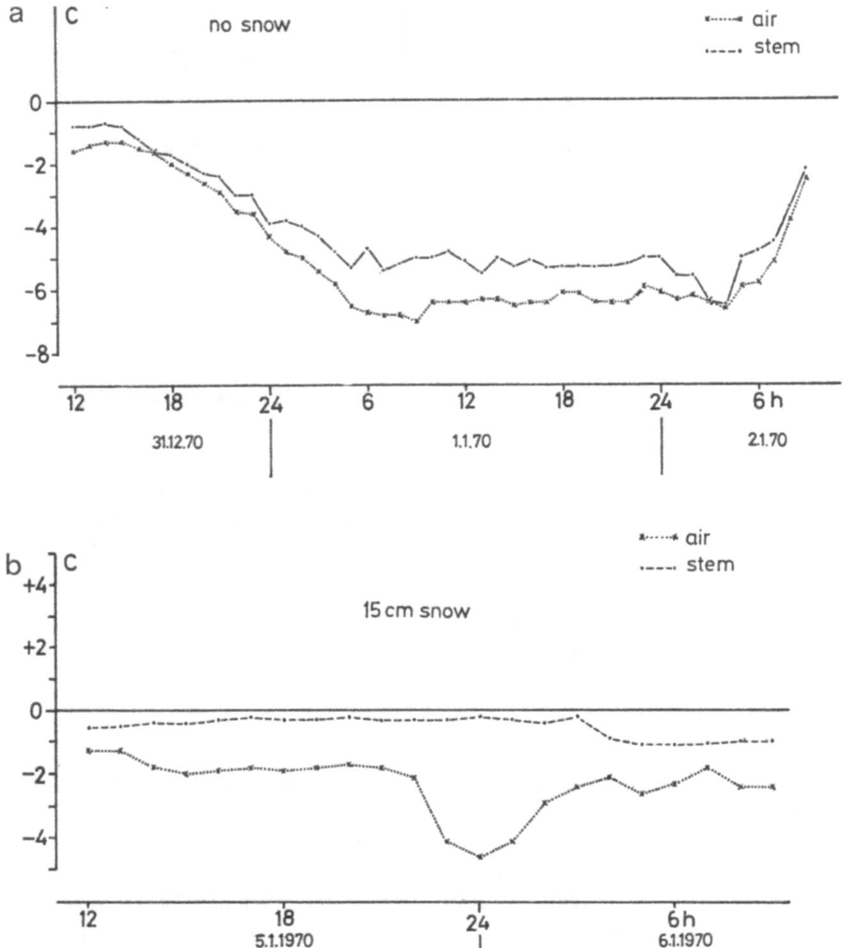

Fig. 2 a and b. Air temperatures measured 2 m above the ground and inside an umbelliferous stem lying on the ground, a without snow, b with a snow covering. Thermoelements were used for measuring

The Degree of Cold Resistance May Vary with Season and with Age

In a number of species the supercooling point was measured at other seasons besides winter, especially in summer. In some species the cold resistance underwent a change whilst in others it remained unaltered:

Table 1. The supercooling points (in degrees Centigrade) of different species of spider in winter

Winter quarters in the vegetation:	
Theridion notatum L. *(Theridiidae)*	—26.1 ± 1.0 (*n* = 19)
Philodromus spec.+) *(Philodromidae)*	—21.5 ± 5.1 (*n* = 4)
egg clutch of *Linyphia (Linyphiidae)*	—25.2 to —33.8 (*n* = 8)
Winter quarters in hollow plant stems:	
Araneus cornutus Clerck *(Araneidae)*	—22.8 ± 1.9 (*n* = 46)
Singa nitidula L. Koch *(Araneidae)*	—21.4 ± 3.3 (*n* = 4)
Clubiona phragmitis L. Koch *(Clubionidae)*	—16.1 ± 4.3 (*n* = 16)
Pachygnatha clercki Sundevall *(Tetragnathidae)*	—5.8 ± 0.6 (*n* = 14)
Winter quarters in caves:	
Meta menardi Latreille *(Araneidae)*	—4.0 ± 0.4 (*n* = 16)
Winter quarters underneath stones and in screes:	
Nesticus cellulanus Clerck *(Nesticidae)*	—4.7 ± 0.6 (*n* = 15)
Cicurina cicurea Fabricius *(Agelenidae)*	—6.7 ± 1.3 (*n* = 8)
Coelotes terrestris Wider *(Agelenidae)*	—6.2 ± 1.1 (*n* = 7)
Coelotes spec.+) *(Agelenidae)*	—6.0 ± 0.8 (*n* = 6)
Histopona torpida L. Koch *(Agelenidae)*	—6.5 ± 0.4 (*n* = 6)
Tegenaria spec.+) *(Agelenidae)*	—8.0 ± 2.6 (*n* = 4)
Pardosa lugubris Walckenaer *(Lycosidae)*	—6.8 ± 1.6 (*n* = 12)
Amaurobius fenestralis Stroem *(Amaurobiidae)*	—6.6 ± 0.4 (*n* = 5)
Winter quarters in tubes lined with silk in the soil:	
Eresus niger Petagna *(Eresidae)*	—16.6 ± 2.6 (*n* = 8)
Winter quarters under the bark of trees:	
Clubiona spec.+) *(Clubionidae)*	—12.9 ± 3.7 (*n* = 9)
Clubiona spec.+) *(Clubionidae)*	—15.9 ± 7.9 (*n* = 5)
Winter quarters inside houses:	
Teutana castanea Clerck *(Theridiidae)*	—9.5 ± 1.1 (*n* = 29)
Teutana triangulosa Walckenaer *(Theridiidae)*	—10.9 ± 1.7 (*n* = 8)
Theridion denticulatum Walckenaer *(Theridiidae)*	—11.4 ± 2.9 (*n* = 9)
Theridion tepidariorum L. Koch *(Theridiidae)*	—8.2 ± 2.1 (*n* = 11)

Confidence limit of the values given for $P = 0{,}05$.

Species marked with +) could not be identified accurately since only subadult animals were available.

Table 1 indicates the place in which the animals were collected in winter, but does not exclude the possibility that the species may spend the winter elsewhere.

The cold resistance of *A. cornutus* clearly follows an annual cycle. Whereas in winter the animal can withstand temperatures below —20 C the supercooling point in summer lies at about —8 C.

Typical of the annual cycle of *A. cornutus* is that the animals develop their high resistance to cold at the beginning of winter, and as soon as spring comes the values begin to sink again until the summer values are attained. During the winter the ability to supercool can be reduced experimentally to the summer values at any time, by keeping the animals at room temperature.

After hatching in August the theridiid, *Theridion notatum*, already possesses a high cold resistance, and not until the following summer is a reduction in the ability to supercool detectable in the adults. A reduction in cold resistance during the winter is not possible in this species.

Table 2. Supercooling points of several species of spider in winter and in summer

Species	Mean supercooling point (in degrees Centigrade) confidence limits for $P = 0{,}05$	
	winter	summer
Nesticus cellulanus	—4.7 ± 0.6 (n = 15)	—4.7 ± 0.6 (n = 12)
Coelotes terrestris	—6.2 ± 1.1 (n = 7)	—5.3 ± 0.4 (n = 14)
Histopona torpida	—6.5 ± 0.4 (n = 6)	—6.0 ± 0.5 (n = 9)
Pardosa lugubris	—6.8 ± 1.6 (n = 12)	—5.8 ± 0.5 (n = 14)

In the spiders listed in Table 2 no striking difference between summer and winter supercooling points could be observed. It is of interest that, according to existing observations, this group comprises exclusively spiders with a low cold resistance (i.e. chiefly soil spiders).

Discussion

The foregoing investigation has revealed that the cold resistance of Central European spiders is based upon their ability to supercool. Thus, the situation in the Araneae is similar to that in many insects (Salt, 1961; Asahina, 1966). The substances responsible for the ability to supercool have not so far been elucidated. Kirchner and Kestler (1969) showed that the glycerol content plays no important role in *Araneus cornutus* and the same is true of the protein content of the haemolymph (Kirchner, unpublished).

It was to be expected that the widely differing habits of the various species would in some way be reflected in their resistance to cold. In fact, closely related species with similar habits do usually resemble one another with respect to cold resistance. All *Araneus* species so far investigated (*A. cornutus*, *A. bituberculatus*, *A. ceropegius*, *A. cucurbitinus*) are apparently very resistant to cold. Members of the genus *Coelotes*, on the other hand, are, without exception, very sensitive to frost. The observation of Tretzel (1961), that *Coelotes terrestris* survives even if the temperature sinks two or three times to —20 C could not be confirmed in a single case on the present author's animals.

Not all related species exhibit a similar resistance to cold and marked differences can occur even within one and the same genus (as in the genus *Theridion*). Above all, extremes can be encountered within the same family of spider (e.g. the species *Araneus cornutus* and *Meta menardi* within the Araneidae (see Table 1).

Similarities in cold resistance between different species is in most cases due to a similar manner of spending the winter. The results of the present investigation suggest a connection between degree of cold resistance and the cold stress experienced in the winter quarters. On the basis of results so far obtained three groups of spider can be distinguished:

Group 1: Supercooling Point between —16 and —30 C. This group includes a number of species that spend the winter out of doors in the vegetation or in hollow plant stems. Since such quarters provide only a hiding place but no protection from cold a high cold resistance is essential if the animals are to survive the low temperatures. The egg cocoons of some spiders, which are attached to plants throughout the winter also belong to this group. Such an assumption is supported by the extremely high values found for the cold resistance of *Linyphia* eggs (Table 1), and the similar findings of Tischler (1967) for harvest men eggs.

Group 2: Supercooling Point between —3 and —7 C. Most cave- and soil-dwelling spiders belong to this group. Their low cold resistance is probably related to the fact that, apart from extremely cold winters with little snow, severe frost seldom reaches their winter quarters. Kirchner and Kullmann (1972) measured the microclimate in a scree and showed that at the most the air cooled down to —1 C despite an external temperature of —11 C. In regions with heavy snowfall soil-dwelling species are especially well protected from severe frost. Comparative measurements of the supercooling points on some diplopods, chilopods, isopods, insects and snails indicate that spending the winter in the soil renders a high cold resistance superfluous. In 12 of the 18 species investigated the supercooling point was between —3 C and —7 C and in three further species between —7 and —10 C. All species tested spend the winter in the same places as the soil-dwelling spiders.

Group 3: Supercooling Point between —7 and —15 C. Into this group fall some species that spend the winter beneath the bark of trees and, surprisingly, a number of house spiders, including even the greenhouse spider, *Theridion tepidariorum*. This suggests that they might also occur out of doors, and in fact, several species of house spider have been found in the open.

Not only in group 3 are species to be encountered in which no connection is apparent between the animals' cold resistance and the temperature to which they are subjected in their winter quarters; a few examples are present in groups 1 and 2. The tetragnathid, *Pachygnatha clercki*, for example, judging by its winter quarters, ough to be highly resistant to frost whereas it is one of the more frost-sensitive species. On the other hand, *Eresus niger*, which spends the winter in the soil in tubes lined with silk, is surprisingly resistant, in contrast to the majority of soil-dwelling spiders. These exceptions form such a small minority that they provide no strong argument against the postulated relationship between cold resistance and type of winter quarters. A final verdict, however, has to await further experiments.

An interesting result of the above investigation is the demonstration that the cold resistance of some spiders is subject to very marked temporal changes. So far, such variations have only been observed in group 1. In group 2, on the other hand, the supercooling point apart from statistically unverifiable variations remains constant throughout the entire year. High cold resistance during the winter thus seems to be a special physiological adaptation to the unfavourable temperatures of winter, of which not all spiders are capable. In this context the question arises as to why the animals of group 1 do not remain highly resistant to cold throughout the year. The findings in *Araneus corneus* (Fig. 3) that in this species high cold resistance is coupled with reduced metabolic intensity supply a possible preliminary answer to this question. A lowering of metabolic intensity would be disadvantageous, at least in the reproductive season and this might be the reason for the decrease in cold resistance and concomitant increase in metabolism observable prior to maturity.

This might also explain why Buche (1966) found only a minimal cold resistance in certain spiders that attain maturity in winter.

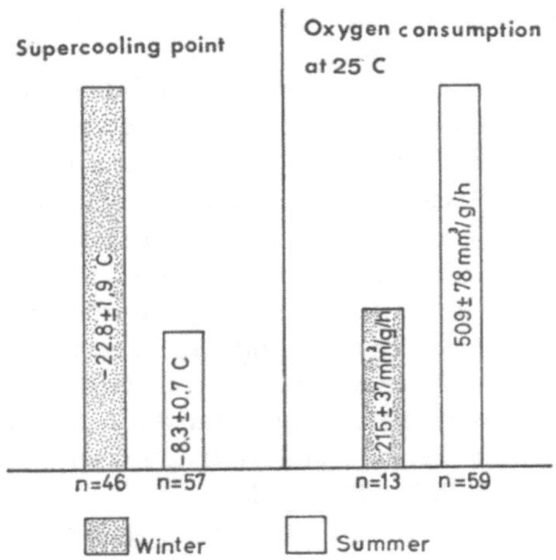

Fig. 3. Supercooling points and oxygen consumption of *Araneus cornutus* in winter and summer (confidence limit for $P = 0{,}001$). A Warburg apparatus was used for measuring oxygen consumption

The reduction of metabolism in winter, typical of *Araneus cornutus* and apparently found in other species (Dondale and Legendre, 1971), probably helps spiders to survive the cold season when food is scarce. It might also be necessary for the accumulation of substances in the spider's body which raise the supercooling point.

Further experiments are necessary to elucidate the chemical nature of these substances.

Summary

1. Death due to cold occurs in Central European spiders at the temperature at which ice forms in the body.

2. The cold resistance of these animals is based upon their ability to supercool.

3. The extent of cold resistance varies from species to species: closely related species often exhibit a similar degree of resistance to cold.

4. Spiders that spend the winter in the open or in hollow plant stems are usually very resistant to cold (supercooling point between —16 and —30 C). Soil- and cave-dwelling spiders, on the other hand, are more sensitive to frost (supercooling point between —3 and —7 C). Those spiders that choose to spend the winter beneath the bark of trees or inside houses usually occupy an intermediate position with regard to cold resistance.

5. The cold resistance of spiders that are highly resistant to low temperatures in winter may be subject to annual rhythms and may also vary with age. The supercooling point of species that tolerate only little cold remains more or less constant throughout the year.

References

Asahina, E.: Freezing and frost resistance in insects. In Cryobiology (Ed.: Meryman, H. T.), p. 451–486. London: Academic Press 1966.

Buche, W.: Beiträge zur Ökologie und Biologie winterreifer Kleinspinnen mit besonderer Berücksichtigung der Linyphiiden *Macrargus rufus rufus* (Wider), *Macrargus rufus carpenteri* (Cambridge) und *Centromerus silvaticus* (Blackwall). Z. Morphol. Ökol. Tiere **57**, 329–448 (1966).

Dondale, C. D., Legendre, R.: Winter diapause in a Mediterranean population of *Pisaura mirabilis* (Clerck). Bull. Brit. Arach. Soc. **2** (1), 6–10 (1971).

Duffey, E.: A population study of spiders in limestone grassland. The field layer fauna. Oikos **13**, 15–34 (1962).

Engelhardt, W.: Die mitteleuropäischen Arten der Gattung *Trochosa* C. L. Koch, 1848 (Araneae, Lycosidae). Morphologie, Chemotaxonomie, Biologie, Autökologie. Z. Morphol. Ökol. Tiere **54**, 219–392 (1964).

Kirchner, W.: Wie überwintert die Schilfradspinne *Araneus cornutus*? Frankfurt: Natur u. Mus. **95**, 163–170 (1965).

Kirchner, W., Kestler, P.: Untersuchungen zur Kälteresistenz der Schilfradspinne *Araneus cornutus* (Araneidae). J. Insect Physiol. **15**, 41–53 (1969).

Kirchner, W., Kullmann, E.: Ökologische Untersuchungen an einer Freilandpopulation von *Nesticus cellulanus* im Siebengebirge unter besonderer Berücksichtigung der Kälteresistenz. Decheniana **125**, 219–227 (1972).

Kolosváry, von G.: Neue Angaben über das Überwintern einiger Spinnenarten. Veröff. dts. Kolon.- u. Übersee-Mus. Bremen **2**, 268–271 (1939).

Lock, F.: Aus dem Leben der Spinnen. Ferd. Rau, Öhringen 1939.

Salt, R. W.: Principles of insect cold-hardiness. Ann. Rev. Entomol. **6**, 55–74 (1961).

Tischler, W.: Pflanzenstengel als Überwinterungsquartiere für Tiere der Agrarlandschaft. Faun.-Ökol. Mitt. **3**, 73–77 (1966).

Tischler, W.: Zur Biologie und Ökologie des Opilioniden *Mitopus morio* F. Biol. Zentr. **86**, 473–484 (1967).

Tretzel, E.: Biologie, Ökologie und Brutpflege von *Coelotes terrestris* (Wider) (Araneae, Agelenidae). Teil I: Biologie und Ökologie. Z. Morphol. Ökol. Tiere **49**, 658–745 (1961).

Wiehle, H.: Beiträge zur Kenntnis des Radnetzbaues der Epeiriden, Tetragnathiden und Uloboriden. Z. Morphol. Ökol. Tiere **8**, 468–537 (1927).

Wiehle, H.: Araneidae. In: Dahl: Die Tierwelt Deutschlands, Teil 23. Jena 1931.

Cold Resistance and Freezing in *Arianta arbustorum* L. (Pulmonata)

HELMA STÖVER

With 6 Figures

Introduction

The literature on cold resistance of molluscs deals almost exclusively with marine species (Kanwisher, 1955; Williams, 1970; Theede, 1972). Several marine lamellibranchs and gastropods tolerate the formation of ice in the tissues up to about 67% of their total water content. In freezing resistant intertidal molluscs this value is reached only at very low temperatures and after long periods of frost.

The present investigation was stimulated by the lack of knowledge concerning the cold resistance of terrestrial gastropods (see, e.g. Blažka, 1955). Experiments with *A. arbustorum* were carried out in order to answer the following questions:

1. To what extent does *A. arbustorum* survive sub-zero temperatures?
2. Is *A. arbustorum* frost resistant or does it tolerate subzero temperatures by supercooling?
3. To what extent are the snail's tissues damaged by ice formation?
4. Are there differences in resistance between summer and winter animals, and between populations from low and high altitudes?

Material and Methods

Collection, Maintenance, Survival: *A. arbustorum* was collected near Innsbruck, in the Inn valley (low-altitude populations), or at the Seegrube (high-altitude population at about 1900 m). The animals were used immediately or kept, in summer, in the laboratory for several days in flower pots with soil, but in winter in a cold room at 2–4 C with short day photoperiod (8L/16D). Some of the animals were kept moist at room temperature and were fed daily with lettuce. Food remains and faeces were removed every morning. Over the weekend the animals were kept dry and without food.

In the survival experiments the snails were observed for at least five days since it sometimes took that long before it became clear whether a specimen had survived or not. In doubtful cases an animal was considered alive if it showed withdrawal movements when pinched with a pair of pliers.

Temperature was measured with thermistors that had been inserted into the shell 1.5 turns from the opening, and fastened with dental wax. The thermistors were connected to a *Knauer* temperature-measuring instrument and the changes in current were recorded on paper. The snails were suspended on the thermistor leads in a tightly closed vial that was lowered into a cooling bath at —10 ± 0.02 C (*Lauda* Ultra Kryostat UK 20L). It took about two minutes for the temperature inside the vial to equilibrate with that of the cooling bath.

Dry Weight and Water Content were determined by separating soft parts and shell and drying at 105 C for about 17 hours. All the dry weights given are those of the soft parts only.

Haemolymph: The shell was pierced at about 1.25 turns from the opening, and the haemolymph collected by tearing the vein leading to the heart. Blood samples from individual snails served for the determination of sodium and potassium (with an *EEL* flame photometer) and osmolarity (with a *Knauer* osmometer).

Inulin Space: A crawling snail was gripped with pliers so that it could not withdraw into its shell. 0.1 ml of saline (50 mM NaCl, 5 mM KCL, 12 mM $CaCl_2$) with 0.05% inulin were injected into the foot sinus. After one hour the inulin concentration in the haemolymph was determined according to Führ, Kaczmarczyk und Krüttgen (1955) and the inulin space was calculated from the dilution of the sample.

The *amount of ice* formed in the tissues after freezing was determined according to Kanwisher (1955).

Results

Field Observations: In winter *A. arbustorum* was found on meadows around Innsbruck, mostly under boards and fruit boxes, preferring slanting sites which allow better drainage of melt water. *A. arbustorum* apparently does not burrow into the soil. The shell aperture which is directed upward is sealed by several layers of mucuous membranes. At sub-zero temperatures the shells of living animals are frozen to the ground, but upon opening them the soft parts of the snail are found to be unfrozen.

Temperature Recordings in the —10 C Bath; Ice Formation in the Tissues: The supercooling points recorded are probably higher than those tolerated under natural conditions since in the laboratory the animals were disturbed by handling, implantation of the thermistors and by vibrations. It was found that snails not only survived supercooling but also brief ice formation in certain tissues.

Two types of temperature curve may be distinguished (Fig. 1a, b).

Type I: Ice formation begins at the relatively high temperature of —1 to —2.5 C, the release of the heat of crystallization leading at first to a few small spikes and then to a plateau at a slightly higher temperature. The plateau is maintained – depending on the size of the snail – for 40–50 minutes, after which time the curve drops to the temperature of the cooling bath (see, e.g. Asahina 1966). After removal of the snail from the cooling bath the foot is inevitably found to be frozen and protruding from the shell. If the animals are removed as long as the temperature plateau is maintained they always survive. Snails whose foot had just been frozen often started to crawl about within 10 minutes.

Type II: Under the experimental conditions the animals supercool to a minimum of —9 C. After removal of the snail from the cooling bath the foot is found to be

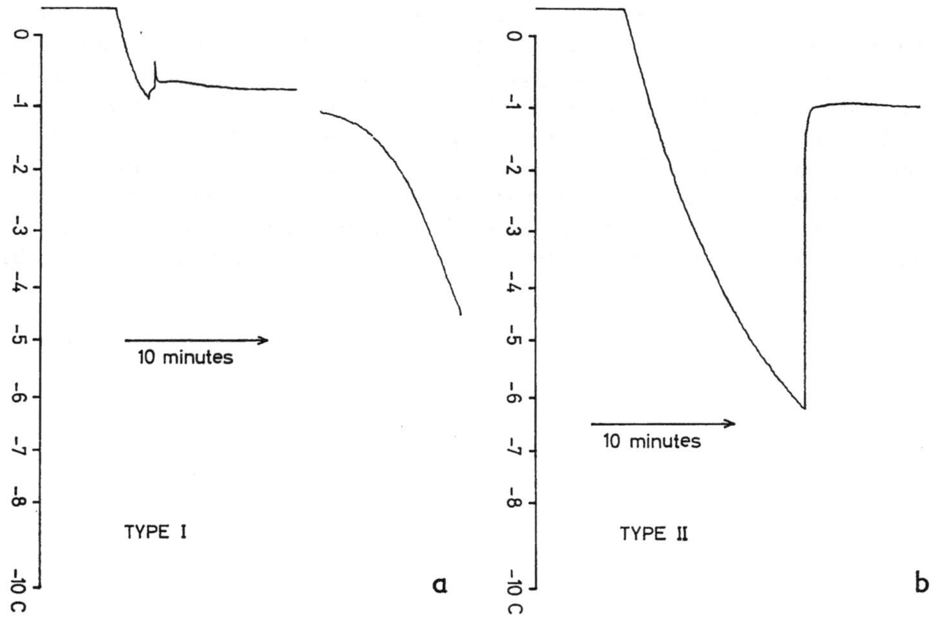

Fig. 1. a Temperature record of a snail freezing according to type I. b Temperature record of a snail freezing according to type II (see text). The vial in which the snail was suspended on the thermistor lead was transferred from room temperature to the cooling bath of —10 C

withdrawn and soft even some time after the beginning of ice formation. It can be concluded that ice formation begins in another tissue of the body. The lower the supercooling point the more ice is formed immediately after the onset of freezing and the greater is the damage to the snail. The damage is irreversible even a short time after the temperature plateau has been attained.

Table 1. Supercooling temperature and water content in *Arianta arbustorum*. The data are based on active and inactive animals measured at different seasons

Type of freezing	Supercooling temperature (C)	g water/g dry weight (without shell) $\bar{x} \pm$ S.D.	*n*
I	to —2	6.31 ± 0.97	11
	—2.1 to —3	6.29 ± 1.32	14
II	—4.1 to —5	6.46 ± 1.21	11
	—5.1 to —6	5.37 ± 1.00	17
	—6.1 to —7	4.56 ± 0.56	12
	—7.1 to —8.8	4.75 ± 0.57	14

Table 1 shows that moist animals freeze predominantly according to type I, more desiccated animals according to type II. However, the course of freezing is probably determined by the state of activity of the animals and not by their water content. Active snails, before starting to crawl, will squeeze a large amount of haemolymph into the foot sinus, and as a consequence will freeze according to type I. That water content and degree of activity are correlated in snails has already been established by Howes and Wells (1934; see also Wieser and Fritz, 1971).

The relationship between supercooling point, time spent at —10 C after the onset of freezing and survival is summarized in Fig. 2. The curve separating surviving from non-surviving animals approximates the survival limit. The survival limits of various populations are compared in Fig. 3, from which it may be concluded that freezing resistance increases from summer to winter, and that in summer the

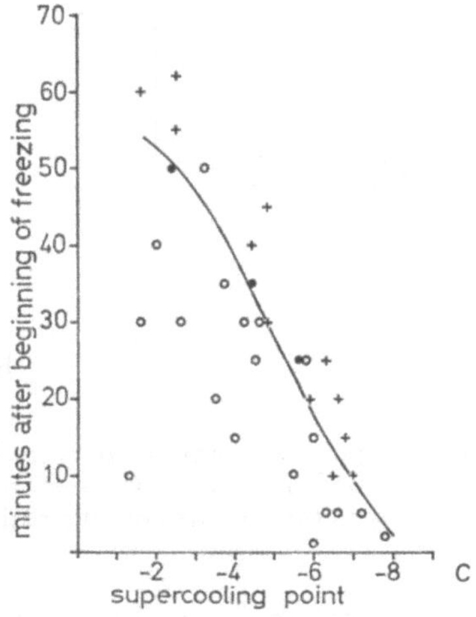

Fig. 2. Survival of *Arianta arbustorum* supercooling to different temperatures. The snails were removed from the cooling bath (—10 C) at different intervals after the beginning of ice formation. Open circles: surviving; crosses: dead; closed circles: dead within next days. The curve approximately separates surviving from non-surviving animals. November 1970

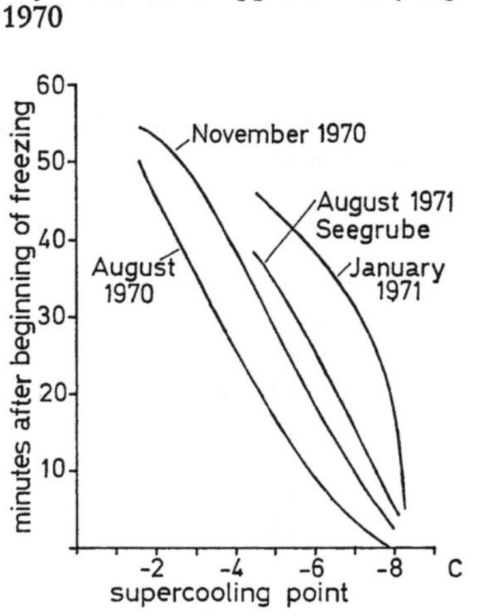

Fig. 3. Survival limits of various experimental groups; as in Fig. 2 but individual data not shown. Low altitude populations: August 1970, $n = 53$; November 1970, $n = 34$; January 1971, $n = 28$. High altitude population ("Seegrube"): August 1971, $n = 33$

high altitude population ("Seegrube") is more resistant than the low altitude population.

The Amount of Ice Formed and Total Body Water: In frost resistant intertidal molluscs Kanwisher (1955) found the amount of ice formed in the tissues to be a function of the temperature with which the animal had been equilibrated. This is not the case in *A. arbustorum* which only survives ice formation for a brief time. Once ice formation has started, about 80 % of the total body water freezes and at the end of the process the animal is dead. The relative amount of ice formed is independent of the water content of the snail. The data on which these conclusions are based are summarized in Table 2.

Table 2. Amount of ice and total body water in *Arianta arbustorum*; $\bar{x} \pm$ S.D

Month	State of animals	g water/ g dry weight	Measuring temperature (C)	Ice in % of total body water	*n*
February	activated	5.13 ± 0.58	— 5	82 ± 4	4
	hibernating	3.95 ± 0.43	—10	79 ± 8	5
	hibernating	4.08 ± 0.53	—15	80 ± 1	3
April	active	5.69 ± 0.50	—15	79 ± 8	10
	desiccated	5.02 ± 0.74	—15	80 ± 5	10

Potassium Concentration in the Haemolymph after Freezing: The mean concentration of potassium in the haemolymph of terrestrial pulmonates varies less than that of other cations (Burton, 1968). After ice formation the concentration of K^+ in the haemolymph of *A. arbustorum* increases and the size of this increase is correlated with the animal's survival limit.

Table 3. Potassium concentration in the haemolymph of controls and of animals frozen according to type I

Month		Potassium in mval/l	*n*
June 1971	controls	3.2 ± 0.9	10
	after 10′ freezing	3.9 ± 1.0	30
October 1971	controls	4.7 ± 1.4	25
	after 15′ freezing	4.6 ± 1.2	22

All animals survive freezing according to type I for 10 to 15 minutes, and no change in the concentration of potassium can be detected in the haemolymph of these animals (Table 3). However, if freezing occurs according to type II, and if the animals are removed from the cooling bath 5–10 minutes after onset of freezing, the potassium concentration in the haemolymph is found to be the higher the lower the supercooling temperature (Fig. 4).

Moreover, in summer animals, potassium concentration rises more steeply than in autumn animals although in the experiments summarized in Fig. 4 the interval between onset of freezing and removal from the cooling bath was 10 minutes in the latter and only 5 minutes in the former. This seasonal difference is in accord with

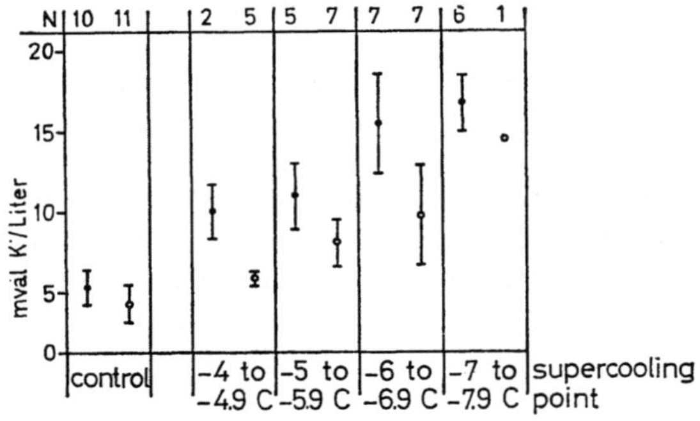

Fig. 4. Potassium concentration ($\bar{x} \pm$ S.D.) in the haemolymph of control (left column) and of experimental animals that had frozen according to type II. Closed circles: July 1970, removed from cooling bath 5 minutes after beginning of ice formation; open circles: October 1970, removed 10 minutes after beginning of ice formation. Number of animals indicated in upper row

the survival experiments (Fig. 3). Whether there is a critical potassium threshold in the haemolymph above which death always occurs, is a question difficult to answer since the sampling of the haemolymph represents an additional stress to the experimental animals. In any case, concentrations up to 11 mval K^+/l were tolerated by all animals examined whereas no specimen survived conditions which led to potassium concentrations of 17 mval/l or more. The surplus potassium in the haemolymph must have come from the cells, which leads to the question of whether a concomitant fall in haemolymph sodium can be detected.

Sodium Concentration and Osmolarity of the Haemolymph after Freezing: Despite its variability the concentration of sodium in the haemolymph of *A. arbustorum* is positively correlated with osmolarity from spring to autumn (Stöver, 1973). In winter, however, these two concentrations are not correlated with each other, but the numbers of sodium and of osmotically active particles in the inulin space are. These relationships may be used to detect some effects of frost damage.

Freezing According to Type I: The water content of the snails was increased by allowing them to crawl in a dish with tap water. This treatment caused them to freeze according to type I. In June the following groups were put into the cooling bath and removed 10 minutes after onset of freezing:

		n
K_1: Moist controls (unfrozen)		10
V_1: Blood sampling immediately	after removal	10
V_2: Blood sampling 10 minutes	from	10
V_3: Blood sampling 20 minutes	cooling bath	10
K_2: Dry controls (unfrozen)		4

The K_2 controls had also been immersed in tap water but were transferred to a dry dish 1–1,5 hours prior to blood sampling. For the experimental animals the whole treatment of cooling, freezing, thawing and blood sampling never exceeded one hour. Table 4 demonstrates that in the course of freezing the osmolarity of the haemolymph, but not the sodium concentration, increases somewhat. This slight shift in proportionality might be due to a loss of water together with some sodium during thawing.

In October, 21 experimental animals survived ice formation according to type I for 15 minutes. No significant shift in the ratio of sodium concentration and osmolarity in the haemolymph could be detected.

Table 4. Sodium concentration and osmolarity in the haemolymph of *Arianta arbustorum*. Experimental animals after 10 minutes freezing according to type I. Abbreviations explained in the text. $\bar{x} \pm$ S.D

	mosmol/l	mval Na/l	*n*
K_1	126 ± 10	40 ± 6	10
V_1	134 ± 8	38 ± 4	10
V_{10}	145 ± 10	40 ± 4	10
V_{20}	153 ± 13	41 ± 3	10
K_2	140 ± 5	45 ± 3	4

Freezing According to Type II: In July, 20 animals that had supercooled to —4.4 to —7.1 C were left to freeze for 5 minutes. After removal from the cooling bath the foot of each snail was found to be unfrozen. 8 specimens survived. As shown in Fig. 5 (left) the sodium concentration at a given osmolarity was lower in the experimental animals than in the controls. If, however, the sum of [Na^+] plus [K^+] (in mval/l) is plotted against osmolarity this difference between controls and experimental animals disappears (Fig. 5, right). The loss of sodium from the haemo-

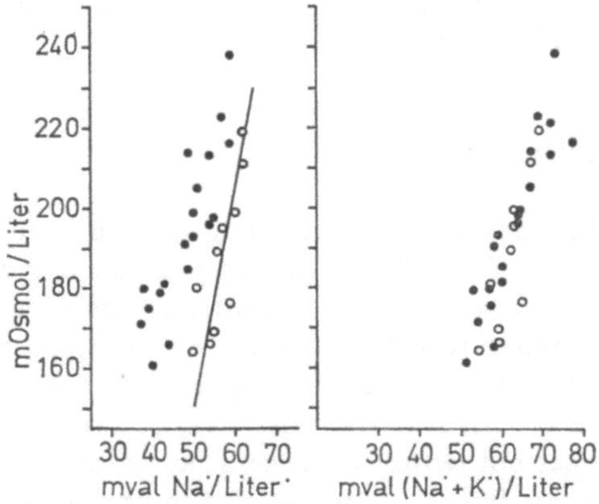

Fig. 5. Relationship between concentration of Na^+ (left) or of Na^+ plus K^+ (right) and osmolarity of the haemolymph. Open circles: controls; closed circles: experimental animals that had frozen according to type II. July 1970

lymph appears to be compensated by the gain in potassium. In December the amount of sodium and of osmotically active particles was calculated by determining sodium concentration, osmolarity and inulin space for each snail. In the unfrozen controls the relationship between sodium and osmotically active particles followed the equation

$$\mu\text{osmol} = 21 + 2.5\,\mu\text{val Na}^+ \quad (n = 14;\ r = 0{,}97).$$

In Fig. 6 this function is compared with data from animals that had frozen for 10 minutes according to type II ($n = 14$). At a given osmolarity the experimental snails clearly contain less sodium ions in their inulin space than the controls. Of the 14 experimental animals 8 survived the cold treatment.

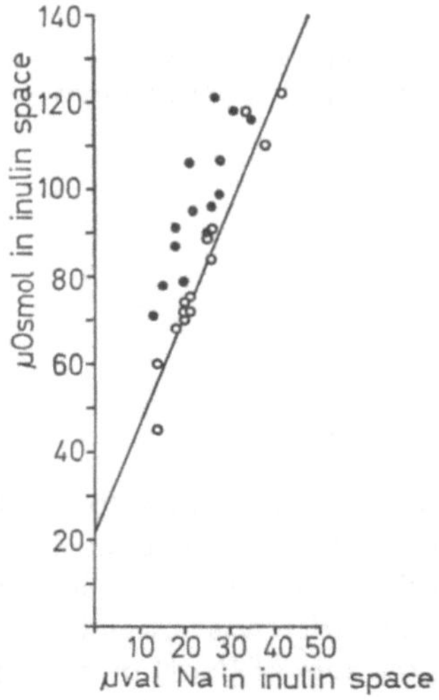

Fig. 6. Relationship between number of sodium and of osmotically active particles in inulin space of control (open circles) and of experimental animals that had frozen according to type II (closed circles) in December 1970. Regression line of control animals: μosmol $= 21 + 2.5\,\mu$val Na^+

Discussion

In snails freezing according to type I, ice formation in the foot is almost certainly extracellular. From the temperature curves it is inferred that within 1–4 minutes after onset of freezing a small amount of ice is formed which leads to slow dehydration of the cells. Up to a critical level this process of dehydration is tolerated by the cells of the foot tissues. During thawing the snail loses some water, perhaps in the form of mucus. It is also possible that a small portion of the extracellularly frozen water does not return to the cells but is extruded through the skin. If a larger amount of water freezes, as is the case in animals freezing according to type II, dehydration is more rapid and damage to the snails is correspondingly greater. In these instances the origin of ice formation in the body is unknown, but it definitely is not the foot since this was found to be unfrozen in all animals that had been removed from the cooling bath a few minutes after onset of freezing.

The changes in the concentration of sodium and potassium in the haemolymph indicate that membranes were damaged as a result of ice formation. Since the duration of cold treatment was chosen so as to be not quite or just lethal one may even surmise that changes in the distribution of cations are amongst the first effects of frost damage. Similar observations have been reported by Heber and Santarius

(1971), Mazur (1966), Theede (1972) and others. Up to a point the frost damage is reversible, but the process of complete recovery may take several days.

The movements of potassium and sodium as a result of ice formation suggest that a coupled Na^+/K^+ carrier mechanism has been damaged. Grainger (1969), however, observed that exposure of *A. arbustorum* to heat leads to drastic changes of sodium but not of potassium levels. After one hour at 40 C the sodium concentration in the haemolymph had dropped to one half, whereas changes in potassium concentration remained insignificant. None of these animals survived. This raises the interesting possibility that freezing and heat treatment affect different components of membrane-dependent processes.

In nature the ability of *A. arbustorum* to survive short periods of freezing of certain tissues would not appear to have any adaptive value. As far as can be ascertained, snails, like many arthropods (see, e.g. Salt, 1964; Kirchner, this symposium) survive sub-zero temperatures only by supercooling. Winter and summer animals do not differ in their supercooling points as determined in the laboratory. This would seem to confirm the suggestion by Remmert and Wisniewski (1970) that one of the differences between arctic and alpine ectothermic animals is that the latter have to be prepared for sub-zero temperatures throughout the year, the former only in winter. However, although there are no seasonal differences in cold resistance, in winter the ability of *A. arbustorum* to withstand sub-zero temperatures is favoured by the following circumstances:

1. The snails are inactive and contain less water than summer animals (see, e.g. Wieser and Fritz, 1971);
2. The diaphragm closing the aperture of the shell prevents the seeding effect of ice crystals from the environment.

Summary

1. *A. arbustorum* survive sub-zero temperatures both in winter and in summer by supercooling. Under the laboratory conditions chosen (rapid cooling to —10 C) ice formation in tissues is tolerated for a maximum of 55 minutes.
2. Two types of freezing can be distinguished: *Type I* is found in active, moist snails; supercooling points are high (—1 to —2.5 C) and ice formation begins in the protruding foot. The animals die at the end of the temperature plateau, 40–55 minutes after onset of freezing. *Type II* is found in inactive, desiccated snails; supercooling points are lower (—3.5 to —9 C) and ice formation begins elsewhere than in the foot. The animals die during the temperature plateau: The lower the supercooling point the earlier they die.
3. The slight freezing resistance of this species is greater in winter than in summer, and greater in high altitude (1900 m) than in low altitude populations. However, seasonal differences appear to be due to the inactivity and lower water content of the winter animals.
4. One of the first effects of frost damage is a rise in the level of potassium in the haemolymph. Changes in steady state leading to increases in concentration from the normal value of approximately 4 mval/l up to approximately 10 mval/l are

tolerated by the snails. No animal that survived had more than 17 mval K^+/l in its haemolymph.

5. The increases in potassium levels in the haemolymph are compensated by decreases in sodium levels. This suggests interference of the freezing process with membranes and, more specifically, with a Na^+/K^+ carrier mechanism.

References

Asahina, E.: Freezing and Frost Resistance in Insects. In: Meryman, H. T. (Ed.): Cryobiology, p. 451–486. London and New York: Academic Press 1966.

Blažka, P.: Temperaturadaptation des Gesamtmetabolismus bei der Weinbergschnecke *Helix pomatia*. Zool. Jahrb., Abt. Allgem. Zool. Physiol. **65**, 430–438 (1955).

Burton, R. F.: Ionic Balance in the Blood of Pulmonata. Comp. Biochem. Physiol. **25**, 509–516 (1968).

Führ, J., Kaczmarczyk, J., Krüttgen, C. D.: Eine einfache kolorimetrische Methode zur Inulinbestimmung für Nieren-Clearance-Untersuchung bei Stoffwechselgesunden und Diabetikern. Klin. Wochschr. **33**, 729 (1955).

Grainger, J. N. R.: Heat Death in *Arianta arbustorum*. Comp. Biochem. Physiol. **29**, 665–670 (1969).

Heber, U., Santarius, K. A.: Empfindlichkeit und Resistenz der Zelle gegen Frost. Umschau Wiss. Tech. **71**, 930 (1971).

Howes, N. H., Wells, G. P.: The Water Relations of Snails and Slugs. J. Exptl. Biol. **11**, 327–351 (1934).

Kanwisher, J. W.: Freezing in Intertidal Animals. Biol. Bull. (Woods Hole) **109**, 56–63 (1955).

Mazur, P.: Physical and Chemical Basis of Injury in Singlecelled Micro-organisms Subjected to Freezing and Thawing. In: Meryman, H. T. (Ed.): Cryobiology, p. 214–315. London and New York: Academic Press 1966.

Remmert, H., Wisniewski, W.: Low Resistance to Cold of Polar Animals in Summer. Oecologia **4**, 111–112 (1970).

Salt, R. W.: Terrestrial Animals in Cold: Arthropods. In: Handbook of Physiology, Section 4, p. 349–355, Washington: American Physiological Society 1964.

Stöver, H.: Über den Wasser- und Elektrolythaushalt von *Arianta arbustorum* (L.). J. Comp. Physiol. **83**, 51–61 (1973).

Theede, H.: Vergleichende ökologisch-physiologische Untersuchungen zur zellulären Kälteresistenz mariner Evertebraten. Marine Biol. **15**, 160–191 (1972).

Wieser, W., Fritz, H.: Seasonal changes of metabolism in *Arianta arbustorum* (Gastropoda; Pulmonata): The Cholinesterase of the Blood. Comp. Biochem. Physiol. **39 A**, 63–73 (1971).

Williams, R. J.: Freezing Tolerance in *Mytilus edulis*. Comp. Biochem. Physiol. **35**, 145–161 (1970).

Subject Index

MIX
Papier aus verantwortungsvollen Quellen
Paper from responsible sources
FSC® C105338

If you have any concerns about our products,
you can contact us on
ProductSafety@springernature.com

In case Publisher is established outside the EU,
the EU authorized representative is:
Springer Nature Customer Service Center GmbH
Europaplatz 3, 69115 Heidelberg, Germany

Printed by Libri Plureos GmbH
in Hamburg, Germany